Technical Writing

Principles, Strategies, and Readings

SIXTH EDITION

Diana C. Reep
The University of Akron

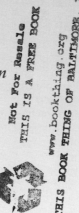

PEARSON
Longman

New York   San Francisco   Boston
London   Toronto   Sydney   Tokyo   Singapore   Madrid
Mexico City   Munich   Paris   Cape Town   Hong Kong   Montreal

For Canyon and Presley

Publisher: Joseph Opiela
Senior Supplements Editor: Donna Campion
Media Supplements Editor: Jenna Egan
Senior Marketing Manager: Melanie Craig
Production Manager: Bob Ginsberg
Project Coordination, Text Design, and Electronic Page Makeup:
  GGS Book Services, Atlantic Highlands
Senior Cover Design Manager/Cover Designer: Nancy Danahy
Cover Photo: Copyright © Getty Images, Inc.
Manufacturing Manager: Mary Fischer
Printer and Binder: Courier Corporation/Stoughton
Cover Printer: Courier Corporation/Stoughton

For permission to use copyrighted material, grateful acknowledgment is made to the
copyright holders on pp. 575–578, which are hereby made part of this copyright page.

Library of Congress Cataloging-in-Publication Data
Reep, Diana C.
  Technical writing: principles, strategies, and readings / Diana C. Reep.—6th ed.
    p. cm.
  Includes bibliographical references and index.
  ISBN 0-321-33350-0
    1. English language—Technical English. 2. Technical writing—Problems, exercises, etc.
  3. English language—Rhetoric. 4. College readers. I. Title.

PE1475.R44 2006
808'.0666—dc22
                                                              2005045915

Please visit our website at http://www.ablongman.com

ISBN 0-321-33350-0

3 4 5 6 7 8 9 10—CRS—08 07

Contents

PART 1   Technical Writing: Ways of Writing   1

1   Technical Writing on the Job   3

2   Collaboration and Ethics   29

11   Formal Report Elements   273

12   Short and Long Reports   307

15   Career Communication   434

16   Oral Presentations   456

PART 2 Technical Writing: Advice from the Workplace   473

Appendix A   Guidelines for Grammar, Punctuation, and Mechanics   549

Appendix B   Frequently Confused Words   566

Appendix C   Internet Resources for Technical Communication   570

Thematic Contents
for Part 2 Readings

Etiquette

Graphics: Oral Presentations

Graphics: Written Communication

International Communication

Job Search

Legal Liability in Technical Writing

Listening

Meetings

Oral Presentations

Proofreading

Style

Web Design and Usability

Preface

The sixth edition of *Technical Writing: Principles, Strategies, and Readings* preserves its unique combination of instructional chapters covering the standard technical communication topics, models illustrating a variety of documents and purposes, exercises providing challenging in-class and out-of-class writing assignments, discussion topics, and articles from professional journals and the Internet offering advice on communication topics. This new edition also retains the flexible organization that instructors said they prefer, and it can easily be used in conjunction with an instructor's personal teaching materials. *Technical Writing: Principles, Strategies, and Readings* is designed for students who study technical writing as part of their career preparation in science, business, engineering, social services, and technical fields. Because of its practical features and guidelines, many students keep the book as a reference for their later on-the-job writing.

CHAPTERS

The 16 chapters in this sixth edition cover the standard topics of technical communication: the writing process, ethics and collaboration, audience analysis, organization, revision, graphics, writing for the Web, definition, description, instructions, and process explanations, formal report elements, short and long reports, traditional types of reports, general correspondence, job search, and oral presentations.

Chapters include (1) checklists and guidelines for planning, drafting, and organizing documents; (2) sample outlines for specific document types; (3) models of printed documents and Web pages; (4) assignments for in-class and out-of-class writing; (5) chapter summaries that highlight major points from the chapter; and (6) lists of relevant readings in Part 2 to supplement the chapters and provide further discussion topics.

STRATEGY BOXES

Strategy Boxes appear in each chapter and provide tips on a variety of topics that are not covered specifically in the chapter discussions. The Strategy Boxes include tips for handling common technology (e.g., cell phones, voice mail), business etiquette, and style. These Strategy Boxes provide further topics for class discussion.

MODELS

This edition includes **models** in every chapter, representing a variety of technical documents and suggested outlines for specific types of documents.

Several models feature successive drafts of a document to show the changes a writer may make during the writing and design process. The end-of-chapter models feature commentary that explains the purpose of the document and the rhetorical strategies used by the writer. Discussion questions for each end-of-chapter model ask students to analyze document features. The writing exercises are specific to the model or to the type of document under discussion. Because no writing is ever "perfect," these models provide the basis for discussion of possible style, format, and design strategies.

WRITING ASSIGNMENTS

New writing assignments appear at the end of every chapter. All end-of-chapter exercises are organized in three categories: **Individual Exercises, Collaborative Exercises,** and **Internet Exercises.** These convenient categories allow instructors to choose exercises more readily depending on the purpose of the assignment. Many assignments include choices based on the preferences of the instructor.

The writing assignments vary in difficulty and may require students (1) to revise a document; (2) to analyze and critique a document; (3) to develop a full, original document; (4) to read an article in Part 2 and incorporate that information into class discussion or a writing task; (5) to collaborate with other students in drafting a document; (6) to use the Internet to find information for a print report; or (7) to analyze the design of Web pages of documents on the Internet.

READINGS

Part 2 features short articles written by technical communication professionals that have appeared both in print and on the Internet. The "Thematic Contents for Part 2 Readings" provides an easy way for instructors and students to find an article on a specific topic. Also, each chapter includes a list of the articles in Part 2 that complement the chapter topic.

APPENDIXES

This edition features three useful appendixes. **Appendix A: Guidelines for Grammar, Punctuation, and Mechanics** covers basic guidelines for correctness.

Appendix B: Frequently Confused Words explains differences in meaning and usage of similar words. **Appendix C: Internet Resources for Technical Communication** lists Web sites that provide information for effective oral presentations, writing style and design elements for Web pages, job search tactics, and professional associations for technical fields.

NEW TO THIS EDITION

I have updated material and added new elements in every edition of *Technical Writing: Principles, Strategies, and Readings* to reflect the changing atmosphere of technical writing and to provide instructors with new pedagogical material. This sixth edition includes the following:

- **Ethics**—Chapter 2, "Collaboration and Ethics," has been expanded to include discussion of plagiarism, copyright infringement, and Web logs.

- **Collaboration**—Because more companies have international operations, Chapter 2, "Collaboration and Ethics," now includes information about writing teams who collaborate over long distances.

- **Email Interviews**—Tips for interviewing sources by email are included in Chapter 3, "Audience."

- **Models**—All the models of Web pages are the latest available at the time this textbook went to press. New models of reports, emails, and graphic aids are included in this edition.

- **Exercises**—New exercises appear in every chapter.

- **Writing for the Web**—A new Chapter 7, "Writing for the Web," includes discussions of planning a Web site, organizing Web pages, effective language and style, and designing Web page elements for users with special needs.

- **Documentation**—Chapter 11, "Formal Report Elements," includes new samples of correct APA documentation style.

- **Email**—Discussion of email advantages and disadvantages in Chapter 14, "Letters, Memos, and Email," has been expanded. Also added is a discussion of the danger of fraud from email schemes.

- **Oral Presentations**—Chapter 16, "Oral Presentations," now includes information about making effective PowerPoint presentations.

- **Readings**—New readings cover organizing email and office communication, developing effective home pages, PowerPoint presentations, an ethics code, and writing product information for company Web sites.

- **Internet Sources**—Appendix C, "Internet Resources for Technical Communication," has been updated with new Web sites for the job search, professional organizations, oral presentations, and technical writing for the Internet.

- **Strategy Boxes**—New Strategy Boxes appear throughout the textbook on such topics as cell phone use, telephone job interviews, instant messaging, telephone manners, protecting computers, supporting coworkers, and questions for a job interview.

INSTRUCTOR'S RESOURCES

The instructor's manual for this textbook offers suggestions to those instructors who wish to try new approaches with familiar topics and to new instructors who are teaching technical communication for the first time. The manual includes suggestions for course policies, sample syllabi, a directory of document types that appear in the chapters, chapter-by-chapter teaching suggestions, reading quizzes for each chapter, and a list of academic journals that cover research in technical communication. Included are additional exercises for class use, including tested favorites from earlier editions.

STUDENT'S RESOURCES

The following supplements are available to package at a discount with student copies of the sixth edition of *Technical Writing*.

PRINT RESOURCES FOR TECHNICAL COMMUNICATION STUDENTS

- *Resources for Technical Communication* (ISBN 0-321-27870-4). This handy resource contains over 50 sample documents and 10 case studies. Organized by genre for easy reference, introductory materials and activities are also included.

- *Visual Communication,* Second Edition (ISBN 0-321-09981-8). Susan Hilligoss' popular text introduces document-design principles and features practical discussions of space, type, organization, pattern, graphic elements, and visuals. The second edition includes more on electronic documents and on-screen presentations.

- *Workplace Literacy,* Second Edition (ISBN 0-321-12737-4). *Workplace Literacy* provides detailed guidelines for writing in workplace communities. The booklet offers strategies for understanding the issues, audiences, and rhetorical conventions of this community, as well as giving concrete advice and models for writing in the workplace.

- *Research Navigator Guide for English,* by H. Eric Branscomb and Doug Gotthoffer (ISBN 0-321-20277-5). Designed to teach students how to conduct high-quality online research and to document it properly. *Research Navigator* guides provide discipline-specific academic resources as well as helpful tips on the writing process, online research, and finding and citing valid sources. Free when packaged with any Longman text, *Research Navigator* guides include an access code to Research Navigator™-providing access to thousands of academic journals and periodicals, the *New York Times* Search by Subject Archive, Link Library, Library Guides, and more.

MULTIMEDIA RESOURCES FOR TECHNICAL COMMUNICATION STUDENTS

- **MyTechCommLab** (www.mytechcommlab.com). MyTechCommLab offers the best multimedia resources for technical writing in one, easy-to-use place. Students will find guidelines, tutorials, and exercises for Writing, Grammar, and Research, as well as Exchange, Longman's new online peer and instructor review program. MyTechCommLab is appropriate for any technical writing course where instructors want to give their students additional resources in technical writing, grammar, and research and/or want to do online peer and instructor review of papers. MyTechCommLab is available in website, CourseCompass, Blackboard, and WebCT platforms.

TO STUDENTS

Use this book as a resource. Your instructor will assign specific chapters, readings, and exercises, but you also can explore new topics in the chapters and in the readings.

- The models show typical on-the-job documents and Web pages. Remember that no document is ever "perfect." Writers do many drafts before the final version. As you analyze a model for writing, design, and strategy, you may have a new idea for the document. Discussing possible revisions helps you develop your own perceptions about the kinds of choices writers have to make in preparing technical documents.

- Use the chapter checklists as guides when planning your writing. Checklists are useful tools for getting started and for reviewing the final draft.

- Use the readings in Part 2 as an extra resource beyond your technical communication class. If you have a job interview, read the relevant articles in Part 2 to help you prepare.

- The appendixes are for your convenience in checking your written work and for finding Internet resources.

Consider this book as a tool you can use for communication tasks both in school and on the job.

ACKNOWLEDGMENTS

I am grateful to the many people who have contributed to this book through six editions. My thanks to those who provided technical writing models and information about on-the-job writing practices: Maggie R. Kohls, AMEC Construction Management, Inc., Chicago, Illinois; Trisha L. Pergande, General Mills, Minneapolis, Minnesota; Sidney Dambrot, Ford Motor Company, Brook Park, Ohio; Charles Groves, Canton, Ohio; Sue Hum, The University of Texas at San Antonio; Jacquelyn Biel, Kompas/Biel Associates, Milwaukee, Wisconsin; Debra Canale, Roadway Express, Akron, Ohio; Michele A. Oziomek, Lewis Research Center, Cleveland, Ohio; Mary H. Bailey, Construction Specifications Institute, Alexandria, Virginia; Ellen Heib, Reinhart, Boerner, Van Dueren and Norris, Attorneys at Law, Milwaukee, Wisconsin; Steven C. Stultz, Babcock & Wilcox Co., Barberton, Ohio; Michelle Merritt, Ciba Corning Diagnostics, Oberlin, Ohio; Brian S. Fedor, Cleveland, Ohio; and Michael J. Novachek, Akron, Ohio. My thanks to Gerald Alred, University of Wisconsin–Milwaukee and Thomas Dukes, The University of Akron, for their long support for this textbook. My special gratitude remains for the late Faye Dambrot's help and encouragement.

Thanks to the reviewers of this edition for their suggestions: Dr. Cynthia Hallett, Bennett College for Women; Janet Kirchner, Southeast Community College; Thomas H. Miles, West Virginia University; Arlene Rodriguez, Springfield Technical Community College; Terry L. Schifferns, Central Community College; Elisa Stone, Salt Lake Community College; and Kris Walker, Tennessee Technological University.

First at Allyn and Bacon and now at Longman, Joseph Opiela, Senior Vice President and Publisher, has guided this project from the initial idea through this sixth edition. I am deeply indebted to his leadership. My appreciation also to Eden Kram, Editorial Assistant, for her work on this edition. Finally, my intense gratitude to Sonia Dial, who cheerfully typed, retyped, sorted, and copied multiple drafts of every page of every edition.

DIANA C. REEP
dreep@uakron.edu

PART 1

Technical Writing: Ways of Writing

CHAPTER 1

Technical Writing on the Job

WRITING IN ORGANIZATIONS

No matter what your job is, writing will be important to your work because you will have to communicate your technical knowledge to others, both inside and outside the organization. Consider these situations: An engineer writes an article for a professional engineering journal, describing a project to restructure a city's sewer pipelines. A police officer writes a report for every arrest and incident that occurs on a shift. An artist writes a grant application asking for funds to create a large, postmodern, steel sculpture. A dietitian writes a brochure about choosing foods low in cholesterol for distribution to participants in a weight-control seminar. Anyone who writes about job-related information prepares technical documents that supply information to readers who need it for a specific purpose.

Surveys indicate that writing responsibilities represent a significant portion of the workday. A 2004 survey of 120 American companies, employing nearly 8 million people, reported that two-thirds or more of the salaried employees at 70% of the companies have some responsibility for writing in their job descriptions.[1] The same survey found that between one-fifth and one-third of hourly employees also had some writing responsibilities. A survey of 60 front-line manufacturing supervisors—none with college degrees—found that 70% of the supervisors spent up to 14 hours a week writing.[2] Engineers who had been working three to five years reported that they spent 64% of their time in some form of communication, with 32% of their time dedicated to written communication.[3]

Writing skills are important in getting a job in the first place. More than half the companies in the 2004 survey reported that writing skills were almost always taken into account when hiring employees. More than half the companies also considered writing skills when making decisions about promotion.[4] A survey of employers hiring entry-level management graduates indicated that, for prospective employees, communication skills were second in importance only to problem-solving skills.[5] All these studies show that companies value writing skills in employees, and such skills are important for successful careers.

Reader/Purpose/Situation

Three elements to consider in writing any technical documents are reader, purpose, and writing situation. The reader of a technical document seeks information for a specific purpose, and the writer's goal is to design a document that will serve the reader's needs and help the reader understand and use the information efficiently. The writing situation consists of both reader and purpose, as well as such factors as the sponsoring organization's size, budget, ethics, deadlines, policies, competition, and priorities. Consider this example of a writing situation.

Lori Vereen, an occupational therapist, must write a short article about the new Toddler Therapy Program, which she directs, at Children's Hospital. Her article will be in the hospital's monthly newsletter, which is sent to people who have donated funds in the past ten years. Newsletter readers are interested in learning about hospital programs and new medical technology, and many of the readers will donate funds for specific programs. Reading the articles in the newsletter helps them decide how to allocate their donations.

Lori understands her readers' purpose in using the newsletter. She also knows that the hospital's management wants to encourage readers to donate to specific programs (writer's purpose). Although she could include in her article much specialized information about new therapeutic techniques for children with cerebral palsy, she decides that her readers will be more interested in learning how the children progress through the program and develop the ability to catch a ball or draw a circle. Scientific details about techniques to enhance motor skills will not interest these readers as much, or inspire as many donations, as will stories about children needing help. Lori's writing situation is restricted further by hospital policy (she cannot use patients' names), space (she is limited to 700 words), and time (she has one day to write her article).

Like Lori, you should consider reader, purpose, and situation for every on-the-job writing task you have. These three elements will influence all your decisions about the document's content, organization, and format.

Diversity in Technical Writing

Science and engineering were once thought to be the only subjects of technical writing, but this limitation no longer applies. All professional fields require technical writing, the communication of specialized information to those who need to use it. The following sentences from technical documents illustrate the diversity of technical writing:

- "Text-based navigation works better than image-based navigation because it enables users to understand the link destinations." (This sentence is from an online report about effective Web design.)

- "Retired employees who permanently change their state of residence must file Form 176-B with the Human Resources Office within 90 days of moving to transfer medical coverage to the regional plant serving their area." (This sentence is from a manufacturing company's employee handbook. Readers are employees who need directions for keeping their company benefit records up to date.)

- "Be sure your coffeemaker is turned OFF and unplugged from the electrical outlet." (This sentence is from a set of instructions packed with a new product. Readers are consumers who want to know how to use and care for their new coffeemaker properly.)

- "The diver must perform a reverse three-and-a-half somersault tuck with a 3.4 degree of difficulty from a 33-ft platform." (This sentence is from the official regulations for a college diving competition. The readers are coaches who need to train their divers to perform the specific dives required for the meet.)

- "For recover installations, pressure-relieving metal vents should be installed at the rate of 1 for every 900 square feet of roof area." (This sentence is from a catalog for roofing products and systems. The readers are architects and engineers who may use these products in a construction project. If they do, they will include the manufacturer's specifications for the products in their own construction design documents.)

Although these sentences involve very different topics, they all represent technical writing because they provide specific information to clearly identified readers who will use the information for a specific purpose.

WRITING AS A PROCESS

The process of writing a technical document includes three general stages—planning, multiple drafting, and revising—but remember, this process is not strictly linear. If you are like most writers, you will revise your decisions about the document many times as you write. You may get an entirely new idea about format or organization while you are drafting, or you may change your mind about appropriate content while you are revising. You probably also will develop a personal writing process that suits your working conditions and preferences. Some writers compose a full draft before tailoring the document to their readers' specific needs; others analyze their readers carefully before gathering any information for the document. No one writing system is appropriate for all writers, and even experienced writers continue to develop new habits and ways of thinking about writing.

One Writer's Process

Margo Keaton is a mechanical engineer who works for a large construction firm in Chicago. Presently, her major project is constructing a retirement community in a Chicago suburb. As the project manager, she meets frequently with the general contractor, subcontractors, and construction foremen. At one meeting with the general contractor and six subcontractors, confusion arises concerning each subcontractor's responsibility for specific construction jobs, including wiring on control motors and fire-safing the wall and floor openings. Because no one appears willing to take the responsibility for the jobs or to accept statements at the meeting as binding, Margo realizes that, as project manager, she will have to determine each subcontractor's precise responsibility.

When she returns to the office she shares with five other engineers, she considers the problem. Responsibility became confused because some subcontractors were having trouble keeping their costs within their initial bids for the construction project. Eliminating some work would ease their financial burden. Then too, no one, including the general contractor, is certain about who has definite responsibilities for the wiring system and for fire-safing the openings. Margo's task is to sort out the information and write a report that will delegate responsibility to the appropriate subcontractors. As Margo thinks about the problem, she realizes that her audience for this report consists of several readers who will each use the report differently. The general contractor needs the information to understand the chain of responsibility, and the subcontractors need the report as an assignment of duties. All the readers will have to accept Margo's report as the final word on the subject.

While analyzing the situation, Margo makes notes on the problem, the report objectives, and the information she must include. She decides that she needs to check two sets of documents:

- The original specifications and original bids
- Correspondence with the subcontractors

She calls the design engineer, who offers to send her the original specifications via email. She next lists everyone she regards as her readers at this point in her writing process.

The next morning, Margo checks her email and finds the specifications sent by the design engineer. She then calls up her computer files for the project and locates the correspondence with the subcontractors. She finds several progress reports and the minutes of past meetings, but they do not mention the specific responsibilities under dispute. After she checks all her material, she decides to enlarge her list of readers to include the design engineer, whose specifications she intends to quote, and the owner's representative, who attended all the construction meetings and heard the arguments over responsibilities. Although the design engineer and the owner's representative are not directly involved in the argument over construction responsibilities, Margo knows they want to keep up to date on all project matters.

Margo needs to compile and organize relevant information, so she opens two new files—FIRESAFE and WIRING. As she reads through the specifications and her own correspondence files, she selects relevant paragraphs and data and saves them to her clipboard. She then sorts and enters these items in either the FIRESAFE file or the WIRING file.

Margo makes her first decision about organization because she knows that some of her readers will resent any conclusions in her report that assign them more responsibility. The opening summary she usually uses in short reports will not suit this situation. Some of her readers might become angry and stop reading before they reach the explanation based on the specifications.

Instead, Margo decides to structure the report so that the documentation and explanation come first, followed by the assignment of responsibilities. She plans to quote extensively from the specifications.

In her opening she will review the misunderstanding and remind her readers why this report is needed. Because the general contractor ultimately has the task of enforcing assignment of responsibilities, she decides to address the report to him and send copies of it to all other parties.

The information in the FIRESAFE and WIRING files is not in any particular order, so Margo prints a hard copy of each file and makes notes about organization in the margins. She groups her information according to specific task, such as wiring the control motors, and puts a code in the margin. All the information about wiring the control motors is coded "B." She also notes in the margins the name of the subcontractor responsible for each task.

Because the original design specification indicated which subcontractor was to do what and yet everyone seems confused, she decides that she should not only quote from the documents but also paraphrase the quotations to ensure that her readers understand. By the end of the report, she will have assigned responsibility for each task in the project.

When she is fairly confident about her organizational decisions, she reorders the material in the FIRESAFE and WIRING files, adds her marginal notes, such as "put definition here," and prints another hard copy of both files. After making a few more notes in the margins, Margo opens a new file on her computer and begins a full draft of her report, using the information she has gathered but rewriting completely for her readers and purpose. As she writes her draft, she also chooses words carefully because her readers will resent any tone that implies they have been trying to avoid their responsibilities or that they are not intelligent enough to figure out the chain of responsibility for themselves. She writes somewhat quickly because she has already made her major organizational decisions. The report turns out to be longer than she expected because of the need to quote and paraphrase so extensively, but she is convinced that such explanations are crucial to her purpose and her readers' understanding. Looking at her first draft again, she changes her mind about the order of the sections and decides to discuss the least controversial task first to keep her readers as calm as possible for as long as possible.

Her second draft reflects the organization she thinks her readers will find most helpful. Margo's revision at this stage focuses on three major questions:

- Is all the quoted material adequately explained?

- Does she have enough quotations to cover every issue and clearly establish responsibilities for each task?

- Is her language neutral, so no one will think she is biased or dictatorial?

She decides that she does have enough quoted material to document each task and that her explanations are clear. Finally, she runs her spell-check and grammar-check programs. The spell-check stops at every proper name, and she reviews her records to ensure she has the correct spelling. The grammar-check stops at the use of passive voice, but Margo wants passive voice in some places to avoid an accusatory tone when she talks about tasks that have not been completed.

Because the situation is controversial, she takes the report to her supervisor and asks him whether he thinks it will settle the issue of job responsibilities. When Margo writes a bid for a construction project, her supervisor always reads the document carefully to be sure that special conditions and costs applying to the project are covered thoroughly. For this report, he assumes that Margo has included all the pertinent data. He suggests adding a chart showing division of responsibilities.

Margo agrees and creates a chart to attach to her report. Knowing her spell-check and grammar-check cannot find all errors, she then edits for *surface accuracy*—clear sentence structure and correct grammar and punctuation. She reads the report aloud, so she can hear any problems in sentence structure or word choice. Her edit reveals a dangling modifier and three places where she inadvertently wrote "assign" instead of "assigned." She makes all the corrections, prints a final draft, and gives her report to Kimberly, the secretary, who will make copies and distribute them.

STRATEGIES—Telephone Interview

Many companies and recruiters are conducting preliminary job interviews by telephone before inviting a candidate for an in-person interview. You must succeed on the telephone before you can advance. Remember these guidelines:

- Arrange for a quiet place where you sit comfortably to take the call.
- Keep your résumé and application materials in front of you for quick reference.
- Make notes on key points while you talk. Be sure and get the name and title of the person calling you.
- If the caller does not mention a face-to-face interview, ask for one at the close of the conversation. Indicate you are interested in the position.

Stages of Writing

Margo Keaton's personal writing process enables her to control her writing tasks even while working at a busy construction company where interruptions occur every few minutes. She relies on notes and lists to keep track of both information and her decisions about a document because she often has to leave the office in the middle of a writing task. Her notes and lists enable her to pick up where she left off when she returns.

Aside from using personal devices for handling the writing process, most writers go through the same three general stages as they develop a document. Remember that all writers do not go through these stages in exactly the same order, and writers often repeat stages as they make new decisions about the content and format of a document.

Planning

In the planning stage, a writer analyzes the reader, purpose, and writing situation; gathers information; and tentatively organizes the document. All these activities may recur many times during the writing.

Analyzing Readers.   No two readers are exactly alike. They differ in knowledge, needs, abilities, attitudes, relation to the situation, and their purpose in using the document. All readers are alike, however, in that they need documents that provide information they can understand and use. As a writer, your task is to create a document that will fit the precise needs of your readers. Margo Keaton's readers were all technical people, so she was free to use technical terms without defining them. Because she knew some readers would be upset by her report, she made her first organizational decision— she would not use an opening summary. Detailed strategies for thinking about readers are discussed in Chapter 3. In general, however, consider these questions whenever you analyze your readers and their needs:

- Who are my specific readers?

- Why do they need this document?

- How will they use it?

- Do they have a hostile, friendly, or neutral attitude toward the subject?

- What is the level of their technical knowledge about the subject?

- How much do they already know about the subject?

- Do they have preferences for some elements, such as tables, headings, or summaries?

Analyzing Purpose.   A document should accomplish something. Remember that a document actually has two purposes: (1) what the writer wants the reader to know or do and (2) what the reader wants to know or do. The writer of instructions, for example, wants to explain a procedure so that readers can perform it. Readers of instructions want to follow the steps to achieve a specific result. The two purposes obviously are closely related, but they are not identical. Margo Keaton wrote her report to assign responsibilities (writer's purpose), as well as to enable her readers to understand the situation and plan their actions accordingly (reader's purpose). A writer should consider both in planning a document. Furthermore, different readers may have different purposes in using the same document, as Margo Keaton's readers did. In analyzing your document's purpose, you may find it helpful to think first of these general purposes:

To Instruct.   The writer tells the reader how to do a task and why it should be done. Documents that primarily instruct include training and operator manuals, policy and procedure statements, and consumer instructions. Such documents deal with

- The purpose of the procedure
- Steps in performing the procedure
- Special conditions that affect the procedure

To Record.   The writer sets down the details of an action, decision, plan, or agreement. The primary purpose of minutes, file reports, and laboratory reports is to record events both for those currently interested and for others who may be interested in the future. Such documents deal with

- Tests or research performed and results
- Decisions made and responsibilities assigned
- Actions and their consequences

To Inform (for Decision Making).   The writer supplies information and analyzes data to enable the reader to make a decision. For decision making, a reader may use progress reports, performance evaluations, or investigative reports. Such documents deal with

- Specific facts that materially affect the situation
- The influence the facts have on the organization and its goals
- Significant parts of the overall situation

To Inform (without Decision Making).   The writer provides information to readers who need to understand data but do not intend to take any action or make a decision. Technical writing that informs without expectation of action by the reader includes information bulletins, literature reviews, product descriptions, and process explanations. Such documents deal with

- The specific who, what, where, when, why, and how of the subject
- A sequence of events showing cause and effect
- The relationship of the information to the company's interests

To Recommend.   The writer presents information and suggests a specific action. Documents with recommendation as their purpose include simple proposals, feasibility studies, and recommendation reports. Such documents deal with

- Reasons for the recommendation
- Expected benefits
- Why the recommendation is preferable to an alternative

To Persuade.   The writer urges the reader to take a specific action or reach a specific conclusion about an issue. To persuade, the writer must convince the reader that the situation requires action and that the information in the report is relevant and adequate for effective decision making. A report recommending purchase of a specific piece of equipment, for instance, may present a simple cost comparison between two models. However, a report that argues the need to close a plant in one state and open a new one in another state must persuade readers about the practicality of such a move. The writer will have to (1) explain why the facts are relevant to the problem, (2) describe how they were obtained, and (3) answer potential objections to the plan. Documents with a strong persuasive purpose include construction bids, grant applications, technical advertisements, technical news releases, and reports dealing with sensitive topics, such as production changes to reduce acid rain. Such documents emphasize

- The importance or urgency of the situation
- The consequences to the reader or others if a specific action is not taken or a specific position is not supported
- The benefits to the reader and others if a specific action is taken or a specific position is supported

To Interest.   The writer describes information to satisfy a reader's intellectual curiosity. Although all technical writing should satisfy readers' curiosity, writing that has interest as its main purpose includes science articles in popular magazines, brochures, and pamphlets. Such documents deal with

- How the subject affects daily life

- Amusing, startling, or significant events connected to the subject

- Complex information in simplified form for general readers

The general purpose of Margo Keaton's report was to inform her readers about construction responsibilities. Her specific purpose was to delegate responsibilities so that the subcontractors could work efficiently. The readers' specific purpose was to understand their duties. Remember that technical documents have both a general purpose and a specific purpose relative to the writing situation. Remember also that documents generally have multiple purposes because of the specific needs of the readers. For instance, a report that recommends purchasing a particular computer model also must provide enough information about capability and costs so that the reader can make a decision. Such a report also may include information about the equipment's design to interest the reader and may act as a record of costs and capability as of a specific date.

Strategies for analyzing readers' purpose are included in Chapter 3. Consider these questions in determining purpose:

- What action (or decision) do I want my reader to take (or make)?

- How does the reader intend to use this document?

- What effect will this document have on the reader's work?

- Is the reader's primary use of this document to be decision making, performing a task, or understanding information?

- If there are multiple readers, do they all have the same goals? Will they all use the document in the same way?

- Do my purpose and my reader's purpose conflict in any way?

Analyzing the Writing Situation.   No writer on the job works completely alone or with complete freedom. The organization's environment may help or hinder your writing, and it certainly will influence both your document and your writing process. The organizational environment in which you write includes (1) the roles and authority both you and your readers have in the organization and in the writing situation; (2) the communication atmosphere— that is, whether information is readily available to employees or only to a few top-level managers; (3) preferences for specific documents, formats, or types of information; (4) the organization's relationship with the community, customers, competitors, unions, and government agencies; (5) government regulations controlling both actions and communication about those actions; and (6) trade or professional associations with standards or ethical codes the organization follows. You can fully understand an organization's environment only by working in it because each is a unique combination of individuals, systems, relationships, goals, and values.

When Margo Keaton analyzed her writing situation, she realized that (1) her readers were hostile to the information she was providing, (2) work on the retirement community could not continue until the subcontractors understood and accepted her information about construction responsibilities, and (3) the delay caused by the dispute among subcontractors jeopardized her own position because she was responsible for finishing the project on time and within budget. Margo's writing situation, therefore, included pressures from readers' attitudes and time constraints. In analyzing your writing situation, consider these questions:

- Is this subject controversial within the organization?

- What authority do my readers have relative to this subject?

- What events created the need for this document?

- What continuing events depend on this document?

- Given the deadline for this document, how much information can be included?

- What influence will this document have on company operations or goals?

- Is this subject under the control of a government agency or specific regulations?

- What external groups are involved in this subject, and why?

- Does custom indicate a specific document for this subject or a particular organization and format for this kind of document?

Gathering Information.   Generally, you will have some information when you begin a writing project. Some writers prefer to analyze reader, purpose, and situation and then gather information; others prefer to gather as much information as possible early in the writing process and then decide which items are needed for the readers and purpose. Information in documents should be (1) accurate, (2) relevant to the readers and purpose, and (3) up-to-date or timely. Margo Keaton's information gathering was focused on existing internal documents because she had to verify past decisions. Many sources of information are available beyond your own knowledge and company documents. Chapter 12 discusses how to find information from outside sources.

Organizing the Information.   As you gather information, you will probably think about how best to organize your document so that your readers can use the information efficiently. As Margo Keaton gathered information for her report, she separated it into major topics in two computer files. She then grouped the information according to specific tasks and made notes about additions. She printed a hard copy of her reordered notes for easy reference before starting her first full draft. General strategies for organizing documents are covered in Chapter 4. In organizing information, a writer begins with two major considerations: (1) how to group the information into specific topics and (2) how to arrange the information within each topic.

Grouping Information into Topics.   Arranging information into groups requires looking at the subject as a whole and recognizing its parts. Sometimes you know the main topics from the outset because the subject of the document is usually organized in a specific way. In a report of a research experiment, for example, the information would probably group easily into topics, such as research purpose, procedure, specific results, and conclusions. Always consider what groupings will help the reader use the information most effectively. In a report comparing two pieces of equipment, you might group information by such topics as price, capability, and repair costs to help your reader decide which model to purchase. If you were writing the report for technicians who will maintain the equipment, you might group information by such topics as safety factors, downtime, typical repairs, and maintenance schedules. Consider these questions when grouping information:

- Does the subject matter have obvious segments? For example, a process explanation usually describes a series of distinct stages.

- Do some pieces of information share one major focus? For example, data about equipment purchase price, installation fees, and repair costs might be grouped under the major topic "cost," with subtopics covering initial cost, installation costs, and maintenance costs.

- Does the reader prefer that the same topics appear in a specific type of document? For example, some readers may want "benefits" as a separate section in any report involving recommendations.

Arranging Information within a Topic.   After grouping the information into major topics and subtopics, organize it effectively within each group. Consider these questions:

- Which order will enable the reader to understand the material easily? For example, product descriptions often describe a product from top to bottom or from bottom to top so that readers can visualize the connecting parts.

- Which order will enable the reader to use this document? For example, instructions should present the steps in the order in which the reader will perform them. Many managers involved in decision making want information in descending order of importance so that they can concentrate on major issues first.

- Which order will help the reader accept this document? For example, when Margo Keaton organized her sections, she held back information her readers were not eager to have until the end of the report.

By making a master list of your topics in order or by reorganizing your notes in computer files as Margo Keaton did, you will be able to visualize your document's structure. The master list of facts in order is an *outline*. Although few writers on the job take the time to make a formal outline with Roman numerals or decimal numbering, they usually make some kind of informal outline because the information involved in most documents is too lengthy or complicated to be organized coherently without using a guide. Some writers use lists of the major topics; others prefer more detailed outlines and list every major and minor item. Outlines do not represent final organization decisions. You may reorganize as you write a draft or as you revise, but an outline will help you control a writing task in the midst of on-the-job interruptions. After a break in the writing process, you can continue writing more easily if you have an outline.

Multiple Drafting

Once you have tentatively planned your document, a rough draft is the next step. At this stage, focus on thoroughly developing the information you have gathered, and do not worry about grammar, punctuation, spelling, and fine points of style. Thinking about such matters during drafting will interfere with your decisions about content and organization. Follow your initial plan of organization, and write quickly. When you have a completed draft, then think about revision strategies. A long, complicated document may require many drafts, and most documents except the simplest usually require several drafts. Margo Keaton wrote her first rough draft from beginning to end using the notes she had organized; other writers may compose sections of the document out of order and then put them in order for a full draft. Keep your reader and purpose firmly in mind while you write because ideas for new information to include or new ways to organize often occur during drafting.

Revising and Editing

Revision takes place throughout the writing process, but particularly after you have begun drafting. Read your draft and rethink these elements:

- *Content*—Do you need more facts? Are your facts relevant for the readers and purpose?

- *Organization*—Have you grouped the information into topics appropriate for your readers? Have you put the details in an order that your readers will find easy to understand and use? Can your readers find the data easily?

- *Headings*—Have you written descriptive headings that will guide your readers to specific information?

- *Openings and closings*—Does your opening establish the document's purpose and introduce the readers to the main topic? Does your closing provide a summary, offer recommendations, or suggest actions appropriate to your readers and purpose?

- *Graphic aids*—Do you have enough graphic aids to help your readers understand the data? Are the graphic aids appropriate for the technical knowledge of your readers?

- *Language*—Have you used language appropriate for your readers? Do you have too much technical jargon? Have you defined terms your readers may not know?

- *Reader usability*—Can your readers understand and use the information effectively? Does the document format help your readers find specific information?

After you are satisfied that you have revised sufficiently for your readers and purpose, edit the document for correct grammar, punctuation, spelling, and company editorial style. For Margo Keaton, revision centered on checking the report's content to be sure that every detail relative to assigning construction responsibilities was included. She also asked her supervisor to read the report to check that the information was clear and that the tone was appropriate for the sensitive issue. Her final editing focused on grammar and punctuation. Remember that no one writes a perfect first draft. Revising and editing are essential for producing effective technical documents.

■ *CHAPTER SUMMARY*

This chapter discusses the importance of writing on the job and the writing process. Remember:

- Writing is important in most jobs and often takes a significant portion of job-related time.

- Technical writing covers many subjects in diverse fields, and every writing task involves analyzing reader, purpose, and writing situation.

- Each writer develops a personal writing process that includes the general stages of planning, multiple drafting, and revising and editing.

- The planning stage of the writing process includes analyzing the reader, purpose, and writing situation; gathering information; and organizing the document.

- The multiple drafting stage of the writing process involves developing a full document and redrafting as writers rethink their original planning.

- The revising and editing stage of the writing process includes revising the document's content and organization and editing for grammar, punctuation, spelling, and company style.

SUPPLEMENTAL READINGS IN PART 2

Garhan, A. "ePublishing," *Writer's Digest*, p. 495.

Nielsen, J. "Ten Steps for Cleaning Up Information Pollution," *Alertbox*, p. 520.

ENDNOTES

1. National Commission on Writing, "Writing: A Ticket to Work . . . Or a Ticket Out." Retrieved September 27, 2004, from http://www.writingcommission.org/pr/writing_for_employ.htm.

2. Mark Mabrito, "Writing on the Front Line: A Study of Workplace Writing," *Business Communication Quarterly* 60.3 (September 1997): 58–70.

3. Pneena Sageev and Carol J. Romanowski, "A Message from Recent Engineering Graduates in the Workplace: Results of a Survey on Technical Communication Skills," *Journal of Engineering Education* 90.4 (October 2001): 685–692.

4. National Commission on Writing, "Writing: A Ticket to Work . . . Or a Ticket Out."

5. Callum J. Floyd and Mary Ellen Gordon, "What Skills Are Most Important? A Comparison of Employee, Student, and Staff Perceptions," *Journal of Marketing Education* 20.4 (August 1998): 103–109.

MODEL 1-1   Commentary

This model shows a writer's first two outlines for an administrative bulletin at a manufacturing company. Because the company has defense contracts, many written and visual materials include classified information that must be coded according to U.S. Department of Defense regulations. The company administrative bulletin will describe the correct procedures for marking classified materials. Readers are the six clerks who work in the Security Department and are responsible for marking the materials correctly. The clerks are accustomed to following company guidelines in administrative bulletins.

The writer begins planning the document by reading through the U.S. Department of Defense regulations and noting those that apply to his company. He jots down a list of the items as he reads, creating his first rough outline of the content that must appear in the document.

Next, he interviews several Security Department clerks about the kinds of classified materials they handle and their preferences for written guidelines. In his office, the writer drafts a second and longer outline on his computer.

In his second list, he adds specific detail to each item. Based on his interviews with the clerks, he now lists the most common classified materials first and the least common last. He also records two special items that he must explain in the bulletin: (1) materials that cannot be marked in the usual manner and (2) documents that include paragraphs classified at different levels.

Discussion

1. Assume you are the writer in this situation. Discuss what kinds of information you would want to have about your readers and about how they intend to use the bulletin.

2. In his second list, the writer includes more detail under each item. Identify the types of information he added to his list. Discuss why he probably chose to draft a second list at this stage rather than write his first full draft of the bulletin.

3. After talking to the clerks, the writer rearranged items in his second list according to how frequently they appear. Discuss what other principle he appears to be using in grouping his items.

MARKING CLASSIFIED INFORMATION
- BOOKS, PAMPHLETS, BOUND DOCS
- CORRESPONDENCE + NONBOUND DOCS
- ARTWORK
- PHOTOS, FILM, MICROFILM
- LETTERS OF TRANSMITTAL
- SOUND TAPES
- MESSAGES
- UNMARKABLE MATERIAL
- CHARTS, TRACINGS, DRAWINGS
AUTHORITY FOR CLASSIFYING
- CONTRACTING AGENCY
CLASSIFIED — ALL INFO CONNECTED TO PROJECT
OTHER INFO — COMPANY MAY CLASSIFY —
REGIONAL OFFICE REVIEWS
MARKING PARAGRAPHS FOR DIFFERENT CLAS-
SIFICATIONS
- PREFERRED
- OR STATEMENT ON FRONT
- OR ATTACH CLASSIFICATION GUIDE FOR
CONTENT

NOTE: MATERIAL IN PRODUCTION
- EMPLOYEES NOTIFIED OF CLASSIFI-
CATION

MODEL 1-1   First Outline

Military Security Bulletin

Authority to classify—contracting agency
Classified info—according to Form 264
Info not in contract—company can classify
If needed—regional office should review

Marking classified info—no typing—use date—classification—
name and address of facility

1. bound docs—books, pamphlets: top and bottom, covers,
   title, first and last pages
2. correspondence and nonbound: top and bottom; if parts
   differ, highest classification prevails
3. letters of transmittal: first page
4. charts, tracings, and drawings: under legend, title block or
   scale, top and bottom
5. artwork: top and bottom of board and page
6. photos, films, microfilms: outside of container beginnings
   and ends of rolls, title block
7. sound tapes: on containers and beginning and end of
   recording
8. messages: top and bottom, first and last words of
   transmitted oral message
9. unmarkable material: tagged, production employees
   notified
10. paragraphs: if varying classifications in force, each
    marked for degree
    —or statement on front or in text
    —or attach guide for each part

**Mark each paragraph if possible.

Second Outline

MODEL 1-2   Commentary

This model shows the first two drafts of the administrative bulletin. The writer's first draft is based on his expanded outline, and he follows company style guidelines by numbering items and giving the bulletin a title and number. He also capitalizes "company" wherever it appears. He develops each point from his second outline into full, detailed sentences, including instructions for each item and special considerations for specific items.

Before writing his second draft, the writer asks the senior clerk to read his first version and suggest changes. The writer then revises based on her comments.

In the second draft, the writer uses a headline that clarifies the purpose of the bulletin and includes headings in the document to separate groups of information. In addition to numbering the classified materials, the writer uses boldface for key words that identify each item.

Item 10 in the first draft becomes a separate section in the second draft because the senior clerk noticed that the information about marking paragraphs applies to all written material.

Discussion

1. Discuss the changes in format and headings the writer made from his first to his second draft. Why will these changes be helpful to the readers?

2. In his discussion with the clerks, the writer learned that they prefer to divide their work so that each handles only specific types of classified materials. Discuss further changes in format and organization that might help the clerks use the bulletin efficiently. For instance, what changes would help a clerk who worked only with items 1, 2, and 3?

3. Compare the first draft and second draft as if you were a clerk who must mark a 20-page booklet in which the sections have different classifications. Discuss how the organization changes will help you find the information you need.

Military Security Guide—Bulletin 62A

The contracting federal agency shall have the authority for classifying any information generated by the Company. All information developed or generated by the Company while performing a classified contract will be classified in accordance with the specifications on the "Contract Security Classification Specification," Form 264. Information generated by the Company shall not be classified unless it is related to work on classified contracts; however, the facility management can classify any information if it is believed necessary to safeguard that information in the national interest. Moreover, the information classified by Company management should be immediately reviewed by the regional security office.

MARKING CLASSIFIED INFORMATION AND MATERIAL

All classified information and material must be marked (not typed) with the proper classification, date of origin, and the name and address of the facility responsible for its preparation.

1. Bound Documents, Books, and Pamphlets shall be marked with the assigned classification at the top and bottom on the front and back covers, the title page, and first and last pages.

2. Correspondence and Documents Not Bound shall be marked on the top and bottom of each page. When the separate components of a document, such as sections, etc., have different classifications, the overall classification is the highest one for any section. Mark sections individually also.

3. Letters of Transmittal shall be marked on the first page according to the highest classification of any component. A notation may be made that, upon removal of classified material, the letter of transmittal may be downgraded or declassified.

4. Charts, Drawings, and Tracings shall be marked under the legend, title block, or scale and at the top and bottom of each page.

5. Artwork shall be marked on the top and bottom margins of the mounting board and on all overlays and cover sheets.

6. Photographs, Films, and Microfilms shall be marked on the outside of the container. In addition, motion picture films shall be marked at the beginning and end of each roll and in the title block.

MODEL 1-2   First Draft

2.

7. Sound tapes shall be marked on their containers and an announcement made at the beginning and end of the recording.

8. Messages, sent electronically, such as email, shall have the classification marking at the top and bottom of each page. In addition, the first and last word of the message shall be the classification.

9. Classified Material that cannot be marked shall be tagged with the classification and other markings. Material still in production that cannot be tagged requires that all employees be notified of the proper classification.

10. Paragraphs in documents, bound or nonbound, which are of different classifications, shall be marked to show the degree of classification, if any, of the information contained therein. Or a statement on the front of the document or in the text shall identify the parts of the document that are classified and to what degree. Or an appropriate classification guide shall be attached to cover the classified contents of the document. Marking paragraphs individually is the preferred method. Neither of the other two alternatives may be used until it is determined that paragraph marking is not possible.

Military Security Guide—Bulletin 62A

MARKING CLASSIFIED INFORMATION AND MATERIAL

Authority

The contracting federal agency shall have the authority for classifying any information generated by the Company. All information developed or generated by the Company while performing a classified contract will be classified in accordance with the specifications on the "Contract Security Classification Specification," Form 264. Information generated by the Company shall not be classified unless it is related to work on classified contracts; however, management can classify any information that may affect the national interest. The information classified by Company management should be immediately reviewed by the regional security office.

Required Marking

All classified information and material must be marked (not typed) with (1) the proper classification, (2) date of origin, and (3) the name and address of the facility responsible for its preparation.

Types of Information and Material

1. **Bound Documents, Books, and Pamphlets** shall be marked with the assigned classification at the top and bottom on the front and back covers, the title page, and first and last pages.

2. **Correspondence and Documents Not Bound** shall be marked on the top and bottom of each page. If the components of a document, such as sections or chapters, have different classifications, the overall classification is the highest one for any section. Individual sections shall be marked also.

3. **Letters of Transmittal** shall be marked on the first page according to the highest classification of any component. A notation may be made that, upon removal of classified material, the letter of transmittal may be downgraded or declassified.

4. **Charts, Drawings, and Tracings** shall be marked under the legend, title block, or scale and at the top and bottom of each page.

Second Draft

Military Security Guide—Bulletin 62A   2.

5. **Artwork** shall be marked on the top and bottom margins of the mounting board and on all overlays and cover sheets.

6. **Photographs, Films, Microfilms, and Disks** shall be marked on the outside of the container. Disks shall be marked on labels. In addition, motion picture films shall be marked at the beginning and end of each roll and in the title block.

7. **Sound Tapes** shall be marked on their containers and an announcement made at the beginning and end of the recording.

8. **Electronic Messages** shall have the classification marking at the top and bottom of each hard-copy page. In addition, the first and last word of the message shall be the classification.

9. **Classified Material that cannot be marked** shall be tagged with the classification and other markings. All employees shall be notified of the proper classification for material still in production that cannot be tagged.

Individual Paragraphs

Paragraphs of documents, bound or nonbound, which are of different classifications, shall be marked to show the degree of classification. Material shall be marked in one of three ways:

- Individual paragraphs shall be marked separately.

- A statement on the front of the document or in the text shall identify the parts of the document that are classified and to what degree.

- An appropriate classification guide shall be attached to provide the classifications of each part of the document.

Marking paragraphs individually is the preferred method. Neither of the other two alternatives may be used until facility management determines that paragraph marking is not possible.

Chapter 1   Exercises

1. Find a consumer brochure or set of instructions for a common household product. Evaluate how well the document suits the intended reader (a consumer) and purpose (writer's purpose and readers' purpose). Write a short analysis of the features you believe the writer used for the intended readers and purposes. Prepare to explain your analysis to your classmates.

2. Find a local newspaper and a national newspaper, such as *The Wall Street Journal* or *USA Today*. Analyze the content of the business news sections in the two publications, and identify the intended readers. Explain the differences in content and style that reveal the intended readers for each publication. Prepare to give a brief, oral presentation of your findings. Oral presentations are covered in Chapter 16.

3. Interview someone you know who has a job that requires some writing. Ask the person about his or her personal writing process. What kinds of documents and for whom does the person write? How do company requirements affect the writing process? Is team writing or management review usually involved? What activity in the writing process does the person find most difficult? How much rewriting does the person do for a typical writing task? What part of the person's writing process would he or she like to change or strengthen? Write a memo to your instructor describing how this person handles a typical writing task. Use the memo format shown in Chapter 14.

4. Find a professional article written for people in your field. Identify the intended reader and purpose. Write a memo to your instructor describing the elements in the article that helped you identify the reader and purpose. Use the memo format shown in Chapter 14.

5. In groups, assume you are an intern in the office of the Director of Development at your school. Your office works to get financial support from people in the community for special projects on campus. The director wants to write a letter to alumni of your school, asking them to donate money toward one of the following new building projects: a library, a student center building, a basketball arena, and scholarships for students who are in the top 10% of their high school classes. Discuss the reader, purpose, and writing situation you would have to consider in order to write these letters. Compare the results of your analyses with those of the other groups.

6. In groups, make notes on your individual writing processes. Decide which steps fit the outline of the writing process in this chapter. Identify any individual elements, such as needing a peanut butter sandwich to get started. Discuss how these individual preferences may contribute to the person's writing process.

7. In groups, assume you are an intern in the Office of Admissions at your school. The director has decided to prepare two new brochures covering special features of your school. One brochure is for high school juniors, and the other brochure is for parents. Discuss the intended readers for these brochures, and decide what differences in content would be necessary for these readers. Compare your results with those of the other groups.

INTERNET EXERCISES

8. Find the Web sites of four companies that employ people in the field you are studying. Check the Web site for a job openings page, and note any communication skills requested in the job descriptions. Make a list of these communication skills for class discussion.

9. Find the Web site of a professional association related to your major. Identify the purpose of the Web site by considering the special features (e.g., job openings, convention information) included. Make a list of the special features for class discussion.

CHAPTER 2

Collaboration and Ethics

WRITING WITH OTHERS

Writing on the job usually means collaborating with others. A survey of members of six professional associations, such as the American Institute of Chemists and the American Consulting Engineers Council, found that 87% of those responding said they sometimes wrote in collaboration with others, and 98% said that effective writing was "very important" or "important" to successfully doing their work.[1] A recent survey of engineers found that they spent 32% of their time working in teams.[2]

You may informally ask others for advice when you write, or you may write a document with a partner or a team of writers. When writers work together to produce a document, problems often arise beyond the usual questions about content, organization, style, and clarity. Overly harsh criticism of another's work can result in a damaged ego; moreover, frustration can develop among writers when their writing styles and paces clash. In addition to other writers, you also may collaborate with a manager who must approve the final document, or you may work with a technical editor who must prepare the document for publication.

Writing with a Partner

Writing a document with a partner often occurs when two people report research or laboratory tests they have conducted jointly. In other situations, two people may be assigned to a feasibility study or field investigation, and both must write the final report. Sometimes one partner will write the document and the other will read, approve, and sign it. More often, however, the partners will collaborate on the written document, contributing separate sections or writing the sections together. Here is a checklist for writing with a partner:

- Plan the document together so that you both clearly understand the document's reader and purpose.

- Divide the task of gathering information so that neither partner feels overworked.

- Decide who will draft which sections of the document. One partner may be strong in certain topics and may prefer to write those sections.

- Consult informally to clarify points or to change organization and content during the drafting stage.

- Draft the full document together so that you both have the opportunity to suggest major organizational and content changes.

- Edit individually for correctness of grammar, punctuation, usage, and spelling, and then combine your results for the final draft. Or, if one partner is particularly strong in these matters, that person may handle the final editing alone.

Writing on a Team

In addition to writing alone or with a partner, writers frequently work on technical documents with a group. An advantage to writing with a team is that several people, who may be specialists in different areas, are working on the same problem, and the final document will reflect their combined knowledge and creativity—several heads are better than one. A disadvantage of writing with a group is that conflicts arise among ideas, styles, and working methods—too many cooks in the kitchen. An effective writing team must become a cohesive, problem-solving group in which conflict is used productively.

Team Planning

The initial meetings of the writing team are important for creating commitment to the project as well as mutual support. All team writers should feel equally involved in planning the document. Here is a checklist for team planning:

- Select someone to act as a coordinator of the team, call meetings, and organize the agenda. The coordinator also can act as a discussion leader, seeking consensus and making sure that all questions are covered. The team coordinator can set up a spreadsheet to track the activities and progress of individual members. The coordinator also can use a software calendar or email to remind members to enter their progress reports on the spreadsheet.

- At meetings, select someone to take notes on decisions and assignments. These notes should be distributed to team members as soon as possible.

- Clarify the writing problem so that each team member shares the same understanding of reader and purpose.

- Generate ideas about strategy, format, and content without evaluating them until all possible ideas seem to be on the table. This brainstorming should be freewheeling and noncritical so that members feel confident about offering ideas.

- Discuss and evaluate the ideas about content and graphics, and then narrow the suggestions to those that seem most suitable.

- Arrive at a group agreement on overall strategy, format, and content for the document.

- Organize a tentative outline for the major sections of the document so that individual writers understand the shape and boundaries of the sections they will write.

- Divide the tasks of collecting information and drafting sections among individual writers.

- Establish procedures for exchanging drafts by email or fax.
- Schedule meetings for checking progress, discussing content, and evaluating rough drafts.
- Schedule deadlines for completing rough drafts.

Working with Designers

For many writing projects, writers work closely with designers or graphic artists to produce the finished product. The team coordinator and one or two other team members should meet with the designer early in the planning process. Designers and writers approach projects from different perspectives. Be sure the designer understands the project. Here is a checklist for collaborating with a designer:

- Explain the budget for the project. Designers usually want to use the highest production values. If you have a specific budget for printing and design, tell the designer.
- Describe your audience and purpose for the document.
- Describe any graphic ideas the team thought would be appropriate for the document.
- Describe the key idea or theme for the project.
- Clarify terms. Are you producing a policy handbook, an operating manual, or a sales brochure? Designers need to know.
- Explain how and where you will distribute the document.
- Establish a schedule for viewing rough designs.

After the planning meeting with the designer, the coordinator should confirm the team understanding of what was agreed on with a memo or email. That correspondence will be useful in later meetings with the designer.

Team Drafting

Meetings to work on the full document generally focus on the drafts from individual writers. Team members can critique the drafts and offer revision suggestions. At these meetings, the group also may decide to revise the outline, add or omit content, and further refine the format. Here is a checklist for team drafting:

- Be open to suggestions for changes in any area. No previous decision is carved in stone; the drafting meetings should be sessions that accommodate revision.

- Encourage each team member to offer revision suggestions at these meetings. Major changes in organization and content are easier to incorporate in the drafting stage than in the final editing stage.

- Ask questions to clarify any points that seem murky.

- Assign the coordinator to assemble the full draft, and if needed, send it for review to people not on the team, such as technicians, marketing planners, lawyers, and upper-level management.

- Schedule meetings for revising and editing the full draft.

Team Revising and Editing

The meetings for revising and editing should focus on whether the document achieves its purpose. This stage includes final changes in organization or content as well as editing for accuracy. Here is a checklist for revising and editing:

- Evaluate comments from outside reviewers, and incorporate their suggestions or demands.

- Assign any needed revisions to individual writers.

- Assign one person to evaluate how well the format, organization, and content fit the document's reader and purpose.

- Check that the technical level of content and language is appropriate for the audience.

- Decide how to handle checking grammar, punctuation, and spelling in the final draft. One writer who is strong in these areas might volunteer to take on this editing.

Working at a Distance

Because more companies today have international operations and new communication technology allows fast and secure contact, many writing teams are working with members in different cities or even different countries. Members of such a "remote team" often feel isolated because of the physical distances and lack of face-to-face contact. Having coleaders in different locations may help keep the communication on track and maintain focus on the project. The basic principles of team writing are still relevant, but members of remote teams also need to consider the following:

- *Time zones* increase the difficulty of exchanging information within a specific time frame. When sending information or asking for feedback, remote team members must allow for the difference in hours between locations. In Los Angeles, California, for example, 2:30 P.M. on Tuesday is 1:30 A.M. on Wednesday in Athens, Greece, and 8:30 A.M. on Wednesday in Sydney, Australia.

- *National holidays* will affect work schedules. For example, Toronto is in the same time zone as New York, but the Toronto office will be closed on July 1 for Canada Day. The New York office will be open, however, on July 1 and closed on July 4, when the Canadian office will be open.

- *Teleconferencing and videoconferencing* can create a greater sense of personal contact between team members than email or other written communication. Team leaders should schedule these conferences on a regular basis if possible.

Although each team member works alone much of the time, appropriate use of communication technology and attention to differences in time and culture will keep the team together and assure the project's success.

Handling Conflict among Writers

Even when writers work well together, some conflict during a project is inevitable; however, conflict can stimulate effort and creativity. Remember these strategies for turning conflict into productive problem solving:

- Discuss all disagreements in person if possible. If the team is scattered in distant locations, a teleconference or videoconference might be in order. Avoid using third parties to discuss the conflict.

- Handle the conflict immediately. If the disagreement is between the team and one member, the team members should discuss the problem as a group. If the conflict is between two members, those two members should meet and discuss their differences in a professional manner.

- Define the issue clearly so that you are discussing relevant points.

- Assume that others are acting in good faith and are as interested as you in the success of the project.

- Avoid feeling that all conflict is a personal challenge to your skills and judgment.

- Listen carefully to what others say; do not interrupt them before they make their points.

- Paraphrase the comments of other people to make sure you understand their meaning.

- Acknowledge points on which you can agree.

- Discuss issues as if a solution or compromise is possible instead of trying to prove other people wrong.

By keeping the communication channels open, you can eliminate conflict and develop a document that satisfies the goals of the writing team.

STRATEGIES—Videoconferences

As virtual conferences become more frequent, the following tips will help keep meetings running smoothly:

- Introduce all the participants at the beginning of the conference just as you would if all were in the same room.
- Because the entire conference room may not be visible to viewers, announce when someone is leaving or arriving.
- Use large, standing name cards in front of each speaker with names in bold, dark ink.
- Place microphones at convenient locations, and avoid distracting noises such as pen tapping or paper tearing.

Writing for Management Review

The procedure called *management review*—in which a writer composes a document that must be approved and perhaps signed by a manager—holds special problems. First, management review almost inevitably involves criticism from a writer's superior. Second, the difference in company rank between manager and writer makes it difficult for the two to think of themselves as partners. Finally, the writer often feels responsible for the real work of producing the document, while the manager is free to cast slings and arrows at it without having struggled through the writing process. Management review works best when writers feel that they share in making the decisions instead of merely taking orders as they draft and redraft a document. Here are guidelines for both writers and managers who want a productive management review:

- Hold meetings in a quiet workplace where the manager will not be interrupted with administrative problems.

- Discuss the writing project from the beginning so that you both agree on purpose, reader, and expected content.

- Review the outline together before the first draft. Changes made at this point are less frustrating than changes made after sections are in draft.

- As the project progresses, meet periodically to review draft sections and discuss needed changes.

- Be certain that meetings do not end before you both agree on the goals of the next revision and the strategies for changes.

The manager involved in a management review of a document is ultimately responsible for guiding the project. In some cases, the manager is the person who decides what to write and to whom. In other situations, the manager's primary aim is to direct the writer in developing the final document's content, organization, and format. Here are guidelines for a manager who wants a successful management review:

- Criticize the problem, not the person.

- Avoid criticism without specifics. To say, "This report isn't any good," without pinpointing specific problems will not help the writer revise.

- Criticize the project in private meetings with the writer; do not discuss the review process publicly.

- Analyze the situation with the writer rather than dictate solutions or do all the critiquing yourself.

- Project an objective, calm frame of mind for review meetings.

- Concentrate on problem solving rather than on the writer's "failure."

- Have clear-cut reasons for requesting changes, and explain them to the writer. Avoid changes made for the sake of change.

- If the standard company format does not fit the specific document very well, encourage the writer to explore new formats.

- Seek outside reviewers for technical content or document design before a final draft.

- Listen to the writer's ideas, and be flexible in discussing writing problems.

- Begin the review meetings by going over the successful parts of the document.

Most managers dislike having to give criticism, but subordinates dislike having to accept it even more. Dwelling on the criticism itself will interfere with the writing process. Here are guidelines for a writer involved in a management review:

- Concentrate on the issues that need to be solved, not on your personal attachment to the document.

- Accept the manager's criticism as guidance toward creating an improved document rather than as an attack on you.

- Remember that every document can be improved; keep your ego out of the review process.

- Ask questions to clarify your understanding of the criticism. Until you know exactly what is needed, you cannot adequately revise.

Writing for Multiple Reviewers

Many projects require multiple reviewers. Engineers may have to check descriptions and process explanations. Marketing specialists may review the document for sales features and appeal to customers. Public Relations may be concerned with publicity and community response. Trainers may need to check the material for usability. The legal department may need to review the entire document to be sure it does not violate libel laws or leave the company vulnerable to lawsuits. Multiple reviews sometimes put the writer in the difficult position of getting conflicting advice for revision. If you have multiple reviewers, remember these guidelines:

- Contact all necessary reviewers early in the planning stage, and determine how they want to handle the review process. Some reviewers might prefer to review specific sections before you finish a first full draft.

- Set deadlines with each reviewer for responses.

- Create a form or checklist for reviewers. This form will help you compare responses and spot inconsistencies.

- Handle conflicting suggestions for revision promptly. If the conflict is over a technical point, ask the technical people to resolve it. If the disagreement comes over a question of format or relevance, you will have to consider the suggestions and, perhaps, seek other opinions or hold a reviewer meeting before making your decision.

Writing with a Technical Editor

A writer of technical documents also may work with a technical editor during the review. Documents with a large and diverse audience, such as policy and procedures manuals or consumer booklets, usually require a technical editor. The editor may work with the writer in the development stage to (1) help plan the content and organization, (2) clarify the purpose and audience, and (3) anticipate difficulties in format and production or spot potential legal problems.

In some cases, the editor may not see the document until it has gone through several drafts and technical reviews. If the schedule permits, the editor should begin work before the final technical review to avoid last-minute conflicts between necessary changes and pending deadlines.

The writer should present a draft that is accurate and free of errors in grammar, punctuation, and spelling. The editor will review several areas. First, the editor will review the document's structure, ensure that graphic aids are placed appropriately, and check for consistent terminology throughout the sections. Second, the editor will mark the draft for clarity; revise sentences that seem overly long, garbled, ambiguous, or awkward; and review word choice. Third, the editor will check punctuation, grammar, mechanics, and spelling. Last, the editor will mark for standard company style. Both the editor and writer need to think of themselves as a team rather than as adversaries. Here are some guidelines for writers and editors working together:

- Prepare for editorial conferences. The writer should receive the editor's marked copy before a scheduled conference so the writer can review changes and comments and prepare for the meeting.

- Discuss how editorial changes clarify the document's meaning or serve the reader's purpose rather than focusing on personal preferences in style or format.

- Do not try to cover all editorial matters for a long document in one editorial conference. Schedule separate meetings for format, content, grammar and mechanics, organization, and consistency.

- Check to ensure that editorial changes do not alter the document's meaning or appropriate tone.

- Be open-minded and flexible. Listen to each other's ideas, and ask each other questions about purpose, audience, strategy, and content.

After each editorial conference, the writer incorporates the agreed-upon changes. The editor will review the production checklist to be sure the final draft is complete, correct, and properly formatted. The editor is responsible for overseeing the production process. Remember, the editor and writer have a common goal—to produce a document that serves its readers efficiently.

WRITING ETHICALLY

Ethics is a broad term that refers to a set of moral principles. These principles vary according to someone's culture, religion, and personal values. Many companies now have a written corporate code of ethics that includes statements outlining the company's goals and its responsibilities to customers, employees, dealers, stockholders, and others.

In producing technical documents, writers who want to behave ethically should consider how their communication choices affect readers, the company, and other writers on the project. Review the following when writing a document:

- Are your facts accurate and up-to-date?

- Have you chosen angles or close-ups that do not obscure the overall subject of the photo or drawing?

- Have you used software, color, type, and icons to enhance but not distort information?

- Have you included examples or metaphors to aid the reader's understanding?

- Have you organized material in a way to help readers use it?

- Have you included enough information to avoid harming the reader or damaging equipment?

- Have you warned readers of all possible hazards in specific terms?

- Have you presented comparable facts (e.g., sales figures) similarly to avoid manipulating the reader's understanding?

- Have you differentiated clearly between fact and opinion?

- Have you checked for potentially misleading word choice, such as "minor leak" referring to a flow that could cause serious damage if left unchecked?

- Are your research sources reliable and appropriate for an unbiased presentation of data?

- Is your information presented in a context that helps readers understand the implications of the facts?

- Have you tested the document for usability, or if no interaction with the reader is possible, have you anticipated the reader's needs?

- Have you included any potentially libelous statements or misrepresentations that could cause the company to lose its reputation or result in lawsuits?

- Have you used confidential or classified company information inappropriately in the document?

- Have you been receptive to other writers' suggestions for the document and worked with them in a cooperative spirit?

A recent survey asked technical communicators the question, "What would you advise a new technical communicator to do if he or she were asked to do something that he or she thought was unethical?" The top three answers were (1) talk to the supervisor, (2) talk to the person who requested the action, and (3) talk to colleagues.[3]

As a writer, consider the ethical aspects of each document from the perspectives of those who provide information and those who use it, and seek advice if you have questions.

Plagiarism

Plagiarism is the use of information discovered and reported by others or the use of their exact words without acknowledging the sources. Plagiarism is a serious ethical violation because the person committing plagiarism is taking credit for the work produced by others. To avoid plagiarism, you must give proper credit to the sources of the information you are using. Give credit for the following:

- A quotation using the exact words of the original
- A paraphrase rewriting the words of the original
- A theory or opinion from a specific source
- Charts, tables, and other graphics that you did not produce yourself
- A fact that is not generally known

The last item is sometimes difficult to assess. For example, the fact that the planet Saturn has rings is general information. The theory that the rings resulted from a comet striking one of Saturn's moons and blowing it into bits, which then formed the rings, however, is not general information and needs a reference.[4]

The following is a paragraph from a book about American architecture from 1879 to 1901:

> In the vestibule off Fifth Avenue, the Caen stone walls were carved to look like hung fabric with fringed edges. The grand stairway to the second floor was a tour de force of trophies, fruit, masks, and cherubs holding aloft the letter V, all exquisitely carved in Caen stone.[5]

When a student used this source for a report, she wrote as follows:

> In the entrance area off Fifth Avenue, the Caen stone walls were carved so they looked like drapes with fringed edges. The grand stairway to the second floor was decorated with trophies, fruit, masks, and cherubs holding aloft a V for Vanderbilt. These all were carved exquisitely in Caen Stone too.

The writer here has plagiarized because she has failed to give a source for her information and has copied most of the words from the original. Changing a few words in the original material does not eliminate the plagiarism.

The same rules apply to using material off the Internet. Even though Internet material is available to millions of readers, you must cite your sources if you want to use information from it in your own writing. Chapter 11 covers the appropriate style for references.

Copyright

Copyright is the legal protection for the creators of original works, including written materials, music, art and graphics, films, and television programs. Creators of material have control over the following:

- Distribution of copies, including sales and rentals

- Performances of the works, such as plays or music

- Displaying the works, such as films and television programs

- Using the original for other works, such as a sequel or a film based on a novel

It is illegal to violate the creator's copyright by using these works for your own benefit without getting permission and paying the appropriate fees. Copyright violators have been sued for damages and made to pay large penalties. Copyright law also covers works on the Internet, and the same rules apply.

Works in the public domain can be reproduced and distributed without penalty. The term *public domain* refers to materials produced by the U.S. government or materials on which the copyright protection has expired. When using works in the public domain, you must cite your sources so that the reader knows where the information originated. For full information about copyright, check the Internet at http://www.copyright.gov.

Web Logs

The Internet has created new ways to engage in unethical behavior. A *Web log,* or a "blog," is a Web site structured like an on-line diary, with one or more people posting messages. Companies often maintain blogs so that employees can comment on projects or workplace issues. Such blogs can be quite helpful for team projects as employees exchange information. Many individuals also create blogs so they can comment on any topics they wish.

Blogs can create ethical problems. People who use them to criticize companies or other people can face lawsuits for Internet slander. Putting intimate information about other people on the Web violates their privacy, and posting intimate information about yourself violates good taste. Companies can search the Web for key words and find blogs that mention the company, their products, and their personnel. Some employees have lost their jobs because of their comments on a blog. In May 2004, a staff member in a U.S. Senate office was fired for posting offensive material on a blog.[6]

Ethical dilemmas can appear in many forms. We should all strive to respect the rights of others and to behave ethically in all activities.

CHAPTER SUMMARY

This chapter discusses guidelines for effective collaborative writing on the job and the ethical questions every writer should consider. Remember:

- Writing on the job often requires collaborating with others by writing with a partner, writing on a team, writing for management review, or writing with a technical editor.

- Conflict among writers working together can be productive if writers exchange ideas freely.

- A writer must consider the ethical implications in the content and design of every technical document.

- Plagiarism and copyright violations are also unethical practices.

SUPPLEMENTAL READINGS IN PART 2

Allen, L., and Voss, D. "Ethics in Technical Communication," *Ethics in Technical Communication*, p. 475.

Caher, J. M. "Technical Documentation and Legal Liability," *Journal of Technical Writing and Communication*, p. 486.

International Association of Business Communicators (*IABC*), "Code of Ethics for Professional Communicators," *IABC Website*, p. 507.

Kostelnick, C., and Roberts, D. D. "Ethos," *Designing Visual Language*, p. 509.

Nielsen, J. "Ten Steps for Cleaning Up Information Pollution," *Alertbox*, p. 520.

Wicclair, M. R., and Farkas, D. K. "Ethical Reasoning in Technical Communication: A Practical Framework," *Technical Communication*, p. 541.

ENDNOTES

1. Andrea Lunsford and Lisa Ede, "Why Write . . . Together: A Research Update," *Rhetoric Review* 5.1 (Fall 1986): 71–81.

2. Pneena Sageev and Carol J. Romanowski, "A Message from Recent Engineering Graduates in the Workplace: Results of a Survey on Technical Communication Skills," *Journal of Engineering Education* 90.4 (October 2001): 686–692.

3. Sam Dragga, "A Question of Ethics: Lessons from Technical Communicators on the Job," *Technical Communication Quarterly* 6.2 (Spring 1997): 161–178.

4. Fred Guterl, "Saturn Spectacular," *Discover* 25.8 (August 2004): 36–43.

5. John Foreman and Robbe Pierce Stimson, *The Vanderbilts and the Gilded Age: Architectural Aspirations 1879–1901* (New York: St. Martin's Press, 1991), p. 38.

6. Harry Wessel, "Businesses Go Bullish on Blogs," *Akron Beacon Journal* (July 26, 2004): D1.

MODEL 2-1   Commentary

This model shows two drafts of a safety bulletin for workers who are renovating buildings that contain lead-based paint. In his first assignment at the company, the writer was told to prepare readable bulletins for the work site. In a planning meeting, the project manager and the writer agreed that they needed separate bulletins for the workers, the engineers, and the contractors, who all had different on-the-job requirements. After writing his first draft of the bulletin for the workers, the writer submitted it to the project manager for review. The project manager sent the draft back with the following comments.

This draft looks like a good start. I have some questions:

1. You have the items separated into before, during, and after work, but they don't stand out very well. Maybe underlining or all capitals? Headings?
2. Rearrange the first section. If a worker fails the physical, there is no need for the safety course. The last section looks out of order too.
3. The suit information is hard to read—reformat?

Let's get together when you have a second draft.

The writer considered the project manager's comments and then wrote the second draft.

Discussion

1. Discuss the changes the writer made in his second draft in response to the project manager's comments. How do the changes make the bulletin more useful to workers on the job site?

2. Discuss the project manager's comments. How helpful are they in directing the writer? Does the project manager sound objective? Flexible? Supportive?

3. Assume you are the project manager. In groups, discuss what suggestions you would make about the second draft. Draft a third version based on those suggestions. Discuss your results with the class.

BULLETIN 128—SAFETY GUIDELINES FOR WORKERS RENOVATING BUILDINGS CONTAINING LEAD-BASED PAINT

In accordance with OSHA Regulations CFR 1910.1200 and 1910.1025, all workers must complete medical testing and safety training before work begins. Testing and training must be finished no later than one week before renovation begins. The following is included:

- A six-hour Lead Paint Abatement Health and Safety Training course conducted by site engineers

- a thorough physical examination

- A baseline whole blood test

- a Respiratory Protection Program and test administered **one day prior to the beginning work**

During daily work, OSHA regulations require workers to wear the following equipment (approved by OSHA):

- full-body disposable TYVEK suits with hoods, gloves, booties, and goggles with protective side shields (suits must be worn prior to entering the work area and remain on until the area passes final clearance inspection). Disposable suits may be worn **once** and then discarded in designated bins. Nylon clothing may be worn under TYVEK suits but must be laundered separately.

- Half-faced respirators approved by OSHA and NIOSH must be worn before entering the work area and remain on until the washing area.

Workers may not eat, drink, smoke, or apply cosmetics in the work areas. These activities must be at least 100 feet from any work area.

After work, the following clean-up procedures must be followed:

- Properly dispose of all protective suits, safety equipment, and respirators.

- Shower immediately.

- Wash and dry hands and face thoroughly before entering shower.

MODEL 2-1   First Draft

BULLETIN 128—SAFETY GUIDELINES FOR WORKERS RENOVATING BUILDINGS CONTAINING LEAD-BASED PAINT

Before Work Begins

OSHA Regulations CFR 1910.1200 and 1910.1025 require the following medical testing and safety training to be completed **no later than one week before renovation begins:**

1. A baseline whole blood test

2. A thorough physical examination

3. A six-hour course, Lead Paint Abatement Health and Safety Training, conducted by site engineers

4. A Respiratory Protection Program and test administered **one day prior to beginning work**

During Daily Work

Workers must wear the following equipment approved by OSHA and NIOSH **before entering the work area and until reaching the washing area:**

- Full-body disposable TYVEK suits with hoods, gloves, booties, and goggles with protective side shields
  Note: (1) Disposable suits may be worn **once** and discarded in designated bins; (2) Nylon clothing may be worn under the TYVEK suit, but it must be laundered separately.

- Half-faced respirators

Workers may **not** eat, drink, smoke, or apply cosmetics within 100 feet of any work areas.

After Work

Workers must follow these clean-up procedures:

1. Properly dispose of all protective suits, safety equipment, and respirators.

2. Wash and dry hands and face thoroughly before entering the shower.

3. Shower immediately.

Second Draft

Chapter 2 Exercises

1. Identify the ethical problems in the following situations. Discuss with the class how you might handle the situation. *Or,* if your instructor prefers, write a memo to your instructor that explains how you would handle these problems. Memo format is covered in Chapter 14.

   a. Marcy McBain is writing a brochure for Hills and Lakes Center, an expensive summer camp for children ages 10 to 16. The director tells Marcy to be sure to describe the camp counselors as "highly experienced." Marcy knows that only the head counselor is highly experienced. The other counselors usually are new to the job each year.

   b. Brad Stalsky, the Director of Communications at the Cedars Wellness Center, needs information about good eating habits for the clients at the Center. He goes to the library and finds three articles that fit his needs. After copying the articles at the library, Brad tells his assistant to make multiple copies and put them in the packets of nutrition information the Center sells for $2.00 each.

   c. Jake Farrell needs to prepare a report on landfill management for his boss. He is behind in his work and needs the report tomorrow. He searches the Internet and finds an up-to-date report on the Web site of the Association of Landfill Management. He also finds four relevant maps on the Web site of the U.S. Department of the Interior. Jake downloads the report and the maps and then pastes them into the company report format under his name without citing any sources.

   d. Brittany Curtiz needs a research paper on the fishing industry in Peru. She searches the Internet and finds a four-page report on the Web site of the Center for South American Studies at the University of Santa Clara. Brittany downloads the report and writes a new opening paragraph. She adds one citation for information on page 2 of the report.

2. Read "Ethics in Technical Communication" by Allen and Voss in Part 2, and then find an advertisement for a technical product. Analyze the advertisement, and decide whether the writer has made ethical choices in language, content, and graphics. Bring the advertisement to class for discussion.

3. Read "Ethical Reasoning in Technical Communication: A Practical Framework" by Wicclair and Farkas in Part 2. In groups, discuss Case 3 described in the article. Assume your group represents the technical writer's friend at another government agency, and draft a memo advising the writers how to handle the ethical dilemma. Compare your group's draft with the drafts of the

other groups. Next, discuss your group's collaborative process. Did someone take the role of coordinator? How were decisions made? How was the drafting process handled? Compare your group's collaborative process with that of the other groups.

4. Find a sales flyer or sales letter. In groups, act as the manager reviewing the materials. Draft some suggestions for changes. Discuss your collaborative process. Did someone take the lead? How did the group come to agreement on the suggested changes? Were there disagreements, and if so, how were they resolved?

INTERNET EXERCISES

5. Read the IABC "Code of Ethics for Professional Communicators" in Part 2. Find the codes of ethics for two more companies or institutions on the Internet. Study these codes for similarities and differences. Are the differences related to the types of companies or institutions? What similarities in the codes do you see? Discuss your evaluations with the class.

6. Search the Internet for Web sites with information on international business ethics. Make notes for class discussion about the types of information available, and discuss your findings with the class.

7. Find the Web site for copyright law (http://www.copyright.gov), and decide whether the following are under copyright protection:

    a. A book manuscript written a year ago but never published

    b. A book written in 1980 by an author now deceased

    c. A chart on the Web site of the U.S. Department of State

    d. A recording of a popular song in 1992 by a group that disbanded in 1996

    e. A book published in 2002 after the death of the author

    f. An architect's designs for a shopping mall posted on his Web site

    g. The slogan ("Start Today and Never Stop") used by a fitness center

Discuss your conclusions with the class. *Or*, if your instructor prefers, write a memo to your instructor stating your conclusions about each item. Identify the section on the Web site that led you to your answers. Memo format is covered in Chapter 14.

CHAPTER 3

Audience

ANALYZING READERS

Each reader represents a unique combination of characteristics and purpose that will affect your decisions about document content and format. To prepare an effective technical document, therefore, analyze your readers during the planning stage of the writing process. Consider your readers in terms of these questions:

- How much technical knowledge about the subject do they already have?
- What positions do they have in the organization?
- What are their attitudes about the subject or the writing situation?
- How will they read the document?
- What purpose do they have in using the document?

If you have multiple readers for a document, you also need to consider the differences among your readers.

Subject Knowledge

Consider how much information your readers already have about the main subject and subtopics in the document. In general, you can think of readers as having one of these levels of knowledge about any subject:

- *Expert level*—Readers with expert knowledge of a subject understand the theory and practical applications as well as most of the specialized terms related to the subject. Expert knowledge implies years of experience and/or advanced training in the subject. A scientist involved in research to find a cure for emphysema will read another scientist's report on that subject as an expert, understanding the testing procedures and the discussion of results. A marketing manager may be an expert reader for a report explaining possible strategies for selling a home appliance in selected regions of the country.

  Expert readers generally need fewer explanations of principles and fewer definitions than other readers, but the amount of appropriate detail in a document for an expert reader depends on purpose. The expert reader who wants to duplicate a new genetic test, for instance, will want precise information about every step in the test. The expert reader who is interested primarily in the results obtained will need only a summary of the test procedure.

- *Semiexpert level*—Readers with semiexpert knowledge of a subject may vary a great deal in how much they know and why they want information. A manager may understand some engineering principles in a report but probably is more interested in information about how the project affects company planning and budgets, subjects in which the manager is an expert. An equipment operator may know a little about the scientific basis of a piece of machinery but is more interested in information about handling the equipment properly. Other readers with semiexpert knowledge may be in similar fields with overlapping knowledge. A financial analyst may specialize in utility stocks but also have semiexpert knowledge of other financial areas. Semiexpert readers, then, may be expert in some topics covered in a document and semiexpert in other topics. To effectively use all the information, the semiexpert reader needs more definitions and explanations of general principles than the expert reader does.

- *Nonexpert level*—Nonexpert readers have no specialized training or experience in a subject. These people read because (1) they want to use new technology or perform new tasks or (2) they are interested in learning about a new subject. Nonexperts using technology for the first time or beginning a new activity are such readers as the person using a stationary exercise bike, learning how to play golf, or installing a heat lamp in the bathroom. These readers need information that will help them use equipment or perform an action. They are less interested in the theory of the subject than in its practical application. Nonexperts who read to learn more about a subject, however, often are interested in some theory. For instance, someone who reads an article in a general science magazine about the disappearance of the dinosaurs from Earth will probably want information about scientific theories on the cause of the dinosaurs' disappearance. If the reader becomes highly interested in the topic and reads widely, he or she then becomes a semiexpert in the subject, familiar with technical terms, theories, and the physical qualities of dinosaurs. For nonexpert readers, include glossaries of technical terms, checklists of important points, simple graphics, and summaries.

All readers, whatever their knowledge level, have a specific reason for using a technical document, and they need information tailored to their level of knowledge. Remember that one person may have different knowledge levels and objectives for different documents. A physician may read (1) a report on heart surgery as an expert seeking more information about research, (2) a report recommending new equipment for a clinic as a semiexpert who must decide whether to purchase the equipment, (3) an article about space travel as a nonexpert who enjoys learning about space, and (4) an owner's manual for a new camera as a consumer who needs instructions for operating the camera.

Position in the Organization

Your reader's hierarchical position in the company and relationship with you also are important characteristics to consider. Readers are either external (outside the company) or internal (inside the company). Those outside the company include customers, vendors, stockholders, employees of government agencies or industry associations, competitors, and the general public. The interests of all these groups center on how a document relates to their own activities. Within all companies, the hierarchy of authority creates three groups of readers:

- *Superiors*—Readers who rank higher in authority than the writer are superiors. They may be executives who make decisions based on information in a document. Superiors may be experts in some aspects of a subject, such as how cost projections will affect company operations, and they may be semiexperts in the production systems. If you are writing a report to superiors about a new company computer system, your readers would be interested in overall costs, the effect of the system on company operations, expected benefits company wide, and projections of future computer uses and needs.

- *Subordinates*—Readers who rank lower in authority than the writer are subordinates. They may be interested primarily in how a document affects their own jobs, but they also may be involved in some decision making, especially for their own units. If your report on the new computer system is for subordinates, you will probably emphasize information about specific models and programs, locations for the new computers, how these computers support specific tasks and systems, and how the readers will use the computers in their jobs.

- *Peers*—Those readers on the same authority level as the writer are peers, although they may not be in the same technical field as the writer. Their interests could involve decision making, coordinating related projects, following procedures, or keeping current with company activities. Your report on the new computer system, if written for peers, might focus on how the system will link departments and functions, change current procedures, and support company or department goals.

Personal Attitudes

Readers' personal responses to a document or a writing situation often influence the document's design. As you analyze your writing situation, assess these considerations for your readers:

- *Emotions*—Readers can have positive, negative, or neutral feelings about the subject, the purpose of the document, or the writer. Even when readers try to read objectively, these emotions can interfere. If your readers have a negative attitude about a subject, organizing the information to move from generally accepted data to less accepted data or starting with shared goals may help them accept the information.

- *Motivation*—Readers may be eager for information and eager to act. On the other hand, they may be reluctant to act. Make it easy for your readers to use the information by including items that will help them. Use lists, tables, headings, indexes, and other design features to make the text more useful. Most important, tell readers why they should act in this situation.

- *Preferences*—Readers sometimes have strong personal preferences about the documents they must use. Readers may demand features such as lists or charts, or they may refuse to read a document that exceeds a specified length or does not follow a set format. If you discover such preferences in your readers, adjust your documents to suit them.

Reading Style

Technical documents usually are not read, nor are they meant to be read, from beginning to end, like a mystery novel. Readers read documents in various ways, determined partly by their need for specific information and partly by personal habit. These reading styles reflect specific readers' needs:

- *Readers use only the summary or abstract.* Some readers want only general information about the subject, and the abstract or opening summary will serve this purpose. An executive who is not directly involved in a production change may read only the abstract of a report. This executive needs up-to-date information about the change, but only in general form. A psychologist who is looking for research studies about abused children may read the abstract of an article to decide whether the article contains the kind of information needed. Relying on the abstract for an overview or using the abstract to decide whether to read the full document saves time for busy readers.

- *Readers check for specific sections of information.* A reader may be interested only in some topics covered in a lengthy document. A machine operator may read a technical manual to find the correct operating procedures for a piece of equipment. A design engineer may look for a description of the machine. A service technician will turn to the section on maintenance. Long reports and manuals often have multiple readers who are interested only in the information relevant to their jobs or departments. Use descriptive headlines to direct these readers to the information they need.

- *Readers scan the document, pausing at key words and phrases.* Sometimes readers quickly read a document for a survey of the subject, but they concentrate only on information directly pertaining to topics that affect them. The manager of an insurance company's annuity department may scan a forecasting report to learn about company planning and expected insurance trends over the next ten years. The same manager, however, will read carefully all information that in any way affects the annuity department. Such information may be scattered throughout the report, so the manager will look for key words. To help such a reader, use consistent terms throughout.

- *Readers study the document from beginning to end.*   Readers who need all the information to make a decision are likely to read carefully from beginning to end. A reader who is trying to decide what automobile to buy may read the manufacturers' brochures from beginning to end, looking for information to aid in the decision. Someone who needs to change an automobile tire will read the instructions carefully from beginning to end in order to perform the steps correctly. For these readers, highlight particularly important information in lists or boxes, direct their attention to information through headings, and summarize main points to refresh their memories.

- *Readers evaluate the document critically.*   Someone who opposes a project or the writer's participation in it may read a document looking for information that can be used as negative evidence. The multiple readers of a report recommending a company merger, for instance, may include those opposed to such a plan. Such readers will focus on specific facts they believe are inadequate to support the recommendation. If you know that some of your readers are opposed to the general purpose of your report, anticipate criticisms of the plan or of your data, and include information that will respond to these criticisms.

Multiple Readers

Technical documents usually have multiple readers. Sometimes the person who receives a report will pass it on informally to others. Manuals may be used by dozens of employees in all departments of an organization. Readers inside and outside a company study annual reports. Consumer instruction manuals are read by millions. This diversity among readers presents extra problems in document design. In addition to analyzing readers based on technical knowledge, positions in the organization, attitudes, and reading styles, consider whether readers are primary or secondary.

Primary Readers

Primary readers will take action or make decisions based on the document. An executive who decides whether to accept a recommendation to change suppliers and purchase equipment is a primary reader. A consumer who buys a blender and follows the enclosed instructions for making drinks is a primary reader. The technical knowledge, relationship to the organization and to the writer, and personal preferences of these two primary readers are very different, but each person will use the document directly.

If your document has multiple primary readers, plan for those who need the most help in understanding the subject and the technical language. For instance, when preparing a bulletin for dozens of regional sales representatives, some of whom have been with the company many years and some of whom are newly hired, include the amount of detail and the language most appropriate for the least experienced in the group. New sales representatives need full explanations, but experienced people can skim through the bulletin and read the sections they have a specific interest in if you include descriptive headings.

Secondary Readers

Secondary readers do not make decisions or take direct action because of the document, but they are affected or influenced by it. A technician may be asked to read a report about changing suppliers and to offer an opinion. If the change takes place, the technician will have to set up maintenance procedures for the new equipment. A report suggesting new promotion policies will have one or more primary readers who are authorized to make decisions. All the employees who have access to the report, however, will be secondary readers with a keen interest in the effect of any changes on their own chances for promotion. Remember that secondary readers are not necessarily secondary in their interest in a subject, only in their power or authority to act on the information.

In some situations, the differences among primary readers or between primary and secondary readers may be so great that you decide to write separate documents. For example, one report of a medical research program cannot serve both the general public and experts. The two groups of readers have entirely different interests and abilities for dealing with the specialized medical information. Even within a company, if the amount of technical information needed by one group will shut out other readers, separate documents may be the easiest way to serve the whole audience.

International Readers

Growing international trade, company branch offices in foreign markets, and booming immigration into the United States have created new challenges for writers. Many documents you write may be used by people for whom English is a second language or who must rely on translations. If you have international readers, you must research their specific communication customs and expectations to write an effective document. Even if your international readers use English to conduct business, you need to find out about cultural customs so you can avoid offending or confusing people who rely on the documents you write. Editing written materials for international readers requires attention to details. Here are only a few areas for you to consider when developing documents for use by international readers:

- *Graphics*—In the United States, we expect readers to scan pictures or drawings from left to right, but some cultures scan graphics in a circular direction or up and down. "Universal" symbols are not necessarily universal. One British software company used the wise old owl as an icon for the help file. In India, however, when people refer to a person as an owl, they mean the person is mentally disturbed.[1]

- *Colors*—Cultures attach different meanings to colors. In the United States, a blue ribbon is first prize, but in England, first prize is a red ribbon. Red, however, represents witchcraft and death in some African countries. In Japan, white flowers symbolize death, but in Mexico, purple flowers symbolize death.[2]

- *Shapes*—The meaning of shapes and gestures differs among countries. The OK sign we make in the United States by putting thumb and forefinger together is a symbol for money in Japan and is considered vulgar in Brazil.[3]

- *Numbers*—Some countries do not use an alphabet with the telephone system. Therefore, telephone numbers or fax numbers should be given only in figures, even if the company uses a word in its telephone number for advertising in the United States.[4]

- *Format*—Acceptable document format varies by country. Dates are often given in a different style from that in the United States. The day May 15, 1998, for instance, is written as 15.Mai 1998 in Germany. The time 10:32 P.M. is 22.32 in France.[5] Correspondence salutations and closings differ from country to country. You also must adopt the formality or informality expected by your international readers.

Writing for the Web means your writing can travel worldwide in an instant. If you are writing material likely to attract international readers, remember these general principles:

- Avoid all slang or business jargon. Web site visitors quickly move on if they do not understand the terms. "Part-time workers" has a specific meaning in the United States, but readers from the Middle East may not understand it.[6]

- Keep paragraphs short and to the point.

- Do not try to eliminate technical terms. Use specific language.

- Spell out even well-known American initialisms (e.g., FBI, IRS) to be sure the readers understand them.[7]

- If your Web site is going to be translated into another language, your text will probably take more space. The text expands because English contains a lot of short words and often requires fewer words to convey an idea than other languages do. The differences in language also can affect your page design for such features as menu buttons or drop-down menus. Longer text can force readers to do more scrolling as well.[8]

The world's cultures contain thousands of variations in communication style. Taking the time to research the customs of your international readers shows respect, but it also is the only way to ensure that your writing fulfills its purpose.

STRATEGIES—International Etiquette

If you visit another country for business purposes, do some research on specific cultural differences. Consider the following general tips for international business dealings:

- Business cards are very important, especially in Asian cultures. Accept and offer business cards in a serious manner. Take a moment to study the card you receive. If you visit one place frequently, have your cards printed in English on one side and in the other language on the other side.

- Conservative business dress is always the best choice.

- Use last names and courtesy titles (e.g., Herr Krueger, Dr. O'Reilly) until you are told to do something else.

- In most countries, mealtimes are not appropriate for business discussions. These times are for getting to know each other.

FINDING OUT ABOUT READERS

Sometimes you will write for readers whom you know well, but how do you find out about readers whom you do not know well or whom you will never meet? Gathering information about your readers and how they plan to use a document can be time-consuming, but it is essential at the planning stage of the writing process. Use both informal inquiries and formal interviews to identify readers' characteristics and purposes.

Informal Checking

When you are writing for people inside your organization, you often can find out about them and their purposes by checking informally in these ways:

- Talk in person or by telephone with the readers themselves or with those who know them, such as the project director, the person who assigned you the writing task, and people who have written similar documents. If you are writing to a high-level executive, you may not feel comfortable calling the executive, but you can talk with others who are familiar with the project, the purpose of the document, and the reader's characteristics.

- Check the readers' reactions to your drafts. You can find out how your readers will use the information in your document by sharing drafts during the writing process. If you are writing a procedures manual, sharing drafts with the employees who will have to follow the procedures will help you clarify how they intend to use the document and what their preferences are.

- Analyze your organization's chain of responsibility relative to the document you are writing. For example, the marketing director may supervise 7 regional managers who, in turn, supervise 21 district managers who, in turn, supervise the sales representatives. Perhaps all are interested in some aspects of your document. By analyzing such organizational networks, you can often identify secondary readers and adjust your document accordingly.

- Brainstorm to identify readers' characteristics for groups. If you are writing a document for a group of readers, such as all registered nurses at St. Luke's Hospital, list the characteristics that you know the readers have in common. You may know, for example, that the average nurse at St. Luke's has been with the hospital for 4.5 years, has a B.S.N. degree, grew up in the area, and attends an average of two professional training seminars a year. Knowing these facts about your readers will help you tailor your document to their needs.

- Construct a user profile that focuses on readers' attitudes. Are the readers being forced to use your information? Do they have any resistance to this particular technology or purpose? What do users need to accomplish with your document? Will they be distracted or interrupted while they use your document? How will the work environment affect their ability to use your document?[9]

Interviewing

Writers on the job use interviews as part of the writing process in two ways: (1) to gather information about the subject from experts and technicians and (2) to find out readers' purposes and intended uses of a document. Whether you are interviewing experts for information about the subject or potential readers to decide how to design the document, a few guiding principles will help you control the interview and use the time effectively.

Preparing for the Interview

Interviewers who are thoroughly prepared get the most useful information while using the least amount of time.

- *Make an appointment for the interview.* Making an appointment shows consideration for the other person's schedule and indicates that the interview is a business task, not a casual chat. Be on time, and keep the interview within the estimated length.

- *Do your background research before the interview.* Interview time should be as productive as possible. Do not try to discuss the document before having at least a general sense of the kinds of information and overall structure required. You are then in a position to ask more specific questions about content or ask for and use opinions about document design.

- *Prepare your questions ahead of time.* Write out specific questions you intend to ask. Ask the expert for details that you need to include in the document. Project yourself into your readers' minds, and imagine how the document might serve their purposes. Then, frame your questions to pinpoint exactly how your readers will respond to specific sections, headings, graphic aids, and other document features.

- *Draft questions that require more than yes-or-no answers.* Interview time will be more productive if each question requires a detailed answer. Notice the following ineffective questions and their more effective revisions:

  Ineffective: Do you expect to use the manual daily for repair procedures?

  Effective: For which repair and maintenance procedures do you need to consult the manual?

This revision will get more specific answers than the first version.

  Ineffective: What do you think operators need in a manual like this one?

  Effective: Which of the following features are essential in an operator's manual for this equipment? (List all that you can think of and then ask if there are others.)

The first version may get a specific answer, but it will not necessarily be comprehensive. Check all possibilities during an interview to avoid repeating questions later.

Conducting the Interview

Think of the interview as a meeting that you are leading. You should guide the flow of questioning, keep the discussion on the subjects, and cover all necessary topics.

- *Explain the purpose of the interview.*   Tell the person why you want to ask questions and how you think the interview will affect the final document.

- *Break the ice if necessary.*   If you have never met the person before, spend a few minutes discussing the subject in general terms so that the two of you can feel comfortable with each other.

- *Listen attentively.*   Avoid concentrating on your questions so much that you miss what the other person is saying. Sometimes in the rush to cover all the topics, an interviewer is more preoccupied with checking off the questions than with listening carefully to the answers.

- *Take notes.*   Take written notes so that you can remember what the reader told you about special preferences or needs. If you want to tape the interview, ask permission before arriving. Remember, however, that some people are intimidated by recording devices.

- *Group your questions by topic.*   Do not ask questions in random order. Thoroughly cover each topic, such as financial information, before moving on to other subjects.

- *Ask follow-up questions.*   Do not stick to your list of questions so rigidly that you miss asking obvious follow-up questions. Be flexible, and follow a new angle if it arises.

- *Ask for clarification.*   If you do not understand an answer, ask the person for more detail immediately. It is easier to clarify meaning on the spot than days later.

- *Maintain a sense of teamwork.*   Although you need to control the interview, think of the meeting as a partnership to solve specific questions of content and format.

- *Keep your opinions out of the interview.*   What you think about the project or the purpose of the document is not as relevant during an interview as what the other person thinks about these topics. Concentrate on gathering information rather than on debating issues.

- *Ask permission to follow up.* End the interview by requesting a chance to call with follow-up questions if needed.

- *Thank the interviewee.* Express your appreciation for the person's help and cooperation.

Immediately after the interview, write up your notes in a rough draft so that you can remember the details.

In some cases, you may need to use email for interviewing. Email can be helpful, but it cannot entirely replace an in-person interview. You might find email helpful to reach very busy or important people, and email can effectively reach several people at the same time if you need to ask them the same question.[10]

If you use email for interviews, here are some guidelines:

- Begin your email by identifying yourself and your project.

- State clearly that you are asking for an answer by email.

- Suggest that the email answer will save the interviewee time.

- Ask for only a small, factual piece of information.

- Use email for brief follow-up questions after an in-person interview.

- If you do not receive an answer within a week, send a reminder email repeating the original question.

Handling Challenging Situations

Busy engineers, scientists, and other experts do not always fully cooperate with technical communicators at the first request. They might refuse to schedule a meeting or refuse to give you enough time for a complete interview because they are busy and do not perceive writing as important work. Although they may make time for an interview with a journalist, they often do not regard an interview with a technical communicator as equally important. Maintain a professional attitude if you have difficulty getting cooperation from subject experts. Here are some guidelines:

- Enhance your interview request by explaining how the interview fits the overall goals of the company for the activity or product you are working on.

- Explain how the material will also benefit the expert's own work.

- If you arrive for an interview and see that the expert is dealing with an unexpected crisis, offer to reschedule. Do, however, set a definite time before you leave.

- Adjust to the expert's style of communicating. Someone may answer a question and then digress into related issues that you planned to cover later. Do not insist on following your question outline, but do cover all your questions.

To be successful in interviewing, adapt to the circumstances, maintain a professional attitude, and be sensitive to the other person's situation.

TESTING READER-ORIENTED DOCUMENTS

Along with analyzing and interviewing readers, writers sometimes evaluate audience needs through user tests of a document while it is still in draft. Manuals and instructions are particularly good candidates for user tests because their primary purpose is to guide readers in performing certain functions and because they are meant for groups of readers. When potential readers test a document's usefulness before the final draft stage, a writer can make needed changes in design and content based on actual experience with the document. When you test usability of documents or Web sites, select participants who match your intended audience. Videotape the testing session if possible, both as a record of the test and as a reference for future tests. There are four main types of usability tests.

Readers Working as a Group

Some usability studies feature focus groups. A focus group usually consists of 6 to 12 people who work together to discuss and evaluate the material being tested. A moderator usually questions the participants and guides the discussion. If the focus group is composed of subject experts, they may submit written opinions to the moderator before meeting, so all participants can review the opinions ahead of time.[11] A company focus group usually has participants from all units of the company. These focus groups usually have 20 to 25 participants. The discussions can help participants understand the needs and purposes of all the organization's units.[12] Focus groups have several advantages:

- More people can participate at the same time.

- Group analysis often is part of a company's decision-making process, so a group reaction is more desirable.

- Participants can discuss an issue rather than just react to it.

Focus groups do have several limitations:

- Some people automatically take the lead in a group; others tend to say very little in a group.

- Participants may agree with the first opinion just to avoid a disagreement.

- The moderator or researcher must interpret the diverse responses in a meaningful way.

Readers Answering Questions

Asking readers to answer questions after reading a document helps a writer decide whether the content will be easily understood by the audience. A utility company may want to distribute a brochure that explains to consumers how electricity use is computed and how they can reduce their energy consumption. Through a user test, the writer can determine if the brochure answers consumers' questions. Several versions of the same brochure also can be tested to determine which version readers find most useful. The test involves these steps:

1. A group of typical consumers reads the draft or drafts of the brochure.

2. The consumers then answer a series of questions on the content and, perhaps, try to compute an energy-consumption problem based on the information in the brochure.

3. The consumers also complete a demographic questionnaire, which asks for age, sex, job title, education level, length of time it took to complete the questions, and their opinions as to the readability of the brochure.

4. The writer then analyzes the results to see how correctly and how fast the consumers answered the questions and what the demographic information reveals about ease of use for different consumer groups.

5. Finally, the writer revises any portions of the brochure that proved difficult for readers and then retests the document.

Readers Performing a Task

In another situation, a writer may be most concerned about whether readers will be able to follow instructions and perform a task correctly. A manual may be tested in draft by operators and technicians who will use the final version. This user test follows the same general pattern as the preceding one, but the emphasis is on performing a task:

1. Selected operators follow the written instructions to perform the steps of a procedure.

2. The length of time to complete the task is recorded, as well as how smoothly the operators proceeded through the steps of the procedure.

3. The operators next complete a demographic questionnaire, comment on how effective they found the instructions, and point out areas that were unclear.

4. Analysis of the results focuses on correctness in completing the task, the length of time required, and the portions of the document that received most criticism from the readers.

5. The writer revises the sections that were not clear to the readers and retests the document.

Readers Thinking Aloud

Another type of user test requires readers to think aloud as they read through a document. This method (called *protocol analysis*) allows the writer or design-review team to record readers' responses everywhere in the document, analyze readers' comments, and revise content, structure, and style accordingly. Readers may be puzzled by the terminology, the sequence of items, too much or too little detail, the graphics, or sentence clarity. The writer can then revise these areas before the final draft.

CHAPTER SUMMARY

This chapter discusses analyzing readers and testing reader use of documents. Remember:

*   Writers analyze readers according to the level of their readers' technical knowledge, positions in the organization, attitudes, reading styles, and positions as primary or secondary readers.

*   Readers generally have one of three levels of technical knowledge—expert, semiexpert, and nonexpert.

*   A reader's position in the organization may be as the writer's superior, subordinate, or peer.

*   A reader's response to a document may be influenced by emotions, motivation, and preferences.

*   Readers may differ in their reading styles. Some read only the abstract or introduction, and some look for specific sections or topics. Others read from beginning to end, and some may read to criticize the information or project.

- When a document has multiple readers, the writer must develop content, organization, and style to serve both primary and secondary readers.

- When writing for international readers, a writer must research cultural differences in communication style.

- Writers find out about their readers through informal checking and formal interviewing.

- Writers test document effectiveness by asking readers to (1) work as a group, (2) read the document and answer questions about content, (3) read the document and then perform the task explained in the document, and (4) talk about their reactions to the document.

SUPPLEMENTAL READINGS IN PART 2

Frazee, V. "Establishing Relations in Germany," *Global Workforce,* p. 492.

Garhan, A. "ePublishing," *Writer's Digest,* p. 495.

Nielson, J. "Ten Steps for Cleaning Up Information Pollution," *Alertbox,* p. 520.

Weiss, E. H. "Taking Your Presentation Abroad," *Intercom,* p. 535.

ENDNOTES

1. Greg Bathon, "Eat the Way Your Mama Taught You," *Intercom* 46.5 (May 1999): 22–24.

2. "Recognizing and Heeding Cultural Differences Can Be Key to International Business Success," *Business America* 115.10 (October 1994): 8–11.

3. Ibid.

4. Ernest Plock, "Understanding Cultural Traditions Is Critical When Doing Business with the Newly Independent States," *Business America* 115.10 (October 1994): 14.

5. William Horton, "The Almost Universal Language: Graphics for International Documents," *Technical Communication* 40.4 (1993): 682–692.

6. Larae D. Lundgren, "The Technical Communicator's Role in Bridging the Gap between Arab and American Business Environments," *Journal of Technical Writing and Communication* 28.4 (1998): 335–343.

7. Steve Outing, "Think Locally, Write Globally," *Writer's Digest* (July 2001): 52–53.

8. Kirk R. St. Amant, "Designing Web Sites," *Intercom* 50.5 (May 2003): 15–18.

9. Heather Lazzaro, "How to Know Your Audience," *Intercom* 48.9 (November 2001): 18–20.

10. Michael Bugeja, "To E or Not to E," *Writer's Digest* 83.4 (April 2003): 32–34, 59.

11. Rien Elling, "Revising Safety Instructions with Focus Groups," *Journal of Business and Technical Communication* 11.4 (October 1997): 451–468.

12. Elizabeth G. Filippo, "Teaming Up to Define Your Users," *Intercom* 49.9 (November 2002): 11–14.

MODEL 3-1   Commentary

This Web page is from the U.S. Environmental Protection Agency (U.S. EPA) Web site (http://www.epa.gov). This page is one section of a unit on reducing solid waste.

Discussion

1. In groups, discuss the features of the page and what they tell you about the intended reader. Discuss whether the intended reader is an expert or non-expert reader. Discuss the links at the top of the page. How helpful are they for the intended reader? Why is a "print version" link included?

2. Do an Internet search for recycling information on Web sites other than the U.S. EPA. Compare two sites for (a) intended readers, (b) graphics, and (c) language. What specific elements on the Web sites help you analyze the differences in readers? How useful are the graphics?

3. Draft the text for a Web page that answers the question "Why study?" for your major. Sketch a design for the Web page. Compare your results with those of others in your class.

Too Much Trash
A Basic Solution
Making it Work
The 4 Principles
The 12 Tips
Conclusion

The Four Basic Principles

Individual consumers can substantially reduce solid waste by following these basic principles:

 REDUCE the amount of trash discarded.

 REUSE containers and products.

 RECYCLE - use recycled materials, and compost.

 RESPOND to the solid waste dilemma by reconsidering waste-producing activities and by expressing preferences for less waste.

EPA Home | Privacy and Security Notice | Contact Us

Last updated on Friday, March 29th, 2002
URL: http://www.epa.gov/epaoswer/non-hw/reduce/catbook/the4.htm

---

MODEL 3-1   EPA Web Page

MODEL 3-2   Commentary

This brochure is designed to help patients with arthritis understand their disease. Other readers might be family members or friends of the patients. The brochure begins by defining the disease and the types of arthritis that typically occur. The drawing shows readers the differences between a normal hip and an arthritic one. The brochure concludes with an explanation of the kinds of treatment available and offers the assurance that the disease is treatable if not curable.

Discussion

1. Discuss the section headings. Why are they appropriate for the readers of the brochure?

2. Discuss the purpose of the photo in the brochure. Is it worth the space it uses, or should more information about the disease be included?

3. This brochure is written for multiple readers—all nonexperts in the subject. In groups, decide what information you would include in a 500-word summary to be read by the spouses of patients with arthritis. What information would you omit? Draft the 500-word summary, and compare your draft with those written by other groups.

4. Individually, or in groups if your instructor prefers, develop a reader test as discussed in this chapter under "Testing Reader-Oriented Documents." Bring to class a short brochure intended for general readers and distributed by a utility company, a charitable organization, or a consumer-products company. Develop five questions designed to test a reader's understanding of the information in the brochure. Follow the testing procedure described under "Readers Answering Questions." Analyze the results, and identify any sections of the brochure that could be improved. Rewrite one section of the brochure based on your reader test. Submit both the original brochure and your revision of one section to your instructor.

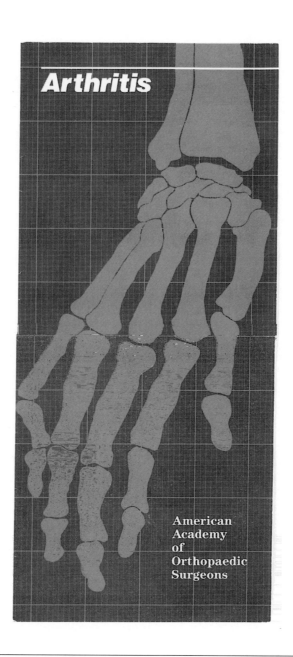

MODEL 3-2

What is arthritis?

The word "arthritis" means joint inflammation (*arthr* = joint; *itis* = inflammation). As many as 36 million people in the United States have some form of arthritis. It is a major cause of lost work time and causes serious disability in many persons. Although arthritis is mainly a disease of adults, children may also have it.

What is a joint?

A joint is a special structure in the body where the ends of two or more bones meet. For example, a bone of the lower leg called the tibia and the thigh bone, which is called the femur, meet to form the knee joint. The hip is a simple ball and socket joint. The upper end of the femur is the ball. It fits into the socket, a part of the pelvis called the acetabulum.

The bone ends of a joint are covered with a smooth, glistening material called hyaline cartilage. This material cushions the underlying bone from excessive force or pressure and allows the joint to move easily without pain. The joint is enclosed in a capsule with a smooth lining called the synovium. The synovium produces a lubricant, synovial fluid, which helps to reduce friction and wear in a joint. Connecting the bones are ligaments, which keep the joint stable. Crossing the joint are muscles and tendons, which also help to keep the joint stable and enable it to move.

What is inflammation?

Inflammation is one of the body's normal reactions to injury or disease. When a part of the body is injured, infected, or diseased, the body's natural defenses work to repair the problem. In an injured or diseased joint, this results in swelling, pain, and stiffness. Inflammation is usually temporary, but in arthritic joints, it may cause long-lasting or permanent disability.

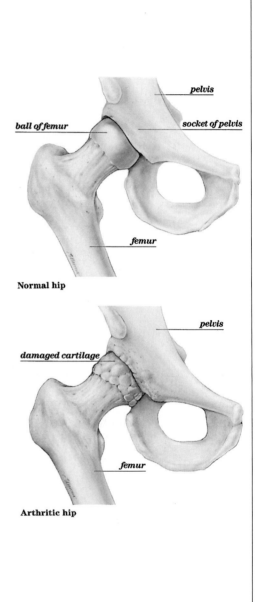

Normal hip

Arthritic hip

Types of arthritis

What is osteoarthritis?

There are more than 100 different types of arthritis. One major type is osteoarthritis, which is sometimes called degenerative joint disease. It occurs to some extent in most people as they age. Osteoarthritis can begin earlier as a result of a joint injury or overuse. Often, weightbearing joints such as the knee, hip, and spine, are involved. The wrists, elbows, and shoulders are usually not affected by osteoarthritis. These joints may be affected if they are used extensively in either work or recreational activities, or if they have been damaged from fractures or other injuries.

In osteoarthritis, the cartilage surface covering the bone ends becomes rough and eventually wears away. In some cases, an abnormal bone growth called a "spur" can develop. Pain and swelling result from joint inflammation. Continued use of the joint produces more pain. It may be relieved somewhat by rest.

What is rheumatoid arthritis?

Rheumatoid arthritis is a chronic, or long-lasting disease that can affect many parts of the body, including the joints. In rheumatoid arthritis, the joint fluid contains chemical substances that attack the joint surface and damage it. Inflammation occurs in response to the disease. The joints most commonly involved are those in the hands, wrists, feet, and ankles, but large joints such as hips, knees, and elbows may also be involved. Swelling, pain, and stiffness are usually present even when the joint is not used. These are the main signs of rheumatoid arthritis and its related diseases. Rheumatoid arthritis can affect persons of all ages, even children, although more than 70% of persons with this disease are over thirty. Many joints of the body may be involved at the same time.

How is arthritis diagnosed?

Diagnosing arthritis includes noting the patient's symptoms, performing a physical examination, and taking x-rays, which are important in showing the extent of damage to the joint. Blood tests and other laboratory tests, such as examination of the joint fluid, may help to determine the type of arthritis the patient has. Once the diagnosis has been made, treatment can begin.

In most cases, people with arthritis can continue to perform the activities of daily living.

How is arthritis treated?

The goals of treatment are to provide pain relief and also to maintain or restore function to the arthritic joint. There are several kinds of treatment:

Medication–Many medications, including aspirin and other anti-inflammatory drugs, may be used to control pain and inflammation in arthritis. The physician chooses a particular medication by taking into account the type of arthritis, its severity, and the patient's general physical health. At times, injections of a substance called cortisone directly into the joint may help to relieve pain and swelling. It is important to know, however, that repeated frequent injections into the same joint can have undesirable side effects.

Joint protection–Canes, crutches, walkers, or splints may help relieve the stress and strain on arthritic joints. Protection of the joint also includes learning methods of performing daily activities that are the least stressful to painful joints. In addition, certain exercises and other forms of physical therapy, such as heat treatments, are used to treat arthritis. They can help to relieve stiffness and to strengthen the weakened muscles around the joint. The type of physical therapy depends on the type and severity of the arthritis and the particular joints involved.

Surgery–In general, an orthopaedist will perform surgery for arthritis when other methods of non-surgical treatment have failed to give the patient sufficient relief. The physician and patient will choose the type of surgery by taking into account the type of arthritis, its severity, and the patient's physical condition. Surgical procedures include
- removal of the diseased or damaged joint lining;
- realignment of the joints;
- total joint replacement; and
- fusion, which permanently places the bone ends of a joint together to prevent joint motion.

Is there a cure for arthritis?

At present, most types of arthritis cannot be cured. Researchers continue to make progress in finding the underlying causes for the major types of arthritis. Often, when a cause is found for a disease, preventive measures can be undertaken. In the meantime, orthopaedists, working with other physicians and scientists, have developed effective treatments for arthritis.

In most cases, persons with arthritis can continue to perform the activities of daily living. Weight reduction for obese persons with osteoarthritis, physical therapy, and anti-inflammatory drugs are measures that can be taken to reduce pain and stiffness.

In persons with severe cases of arthritis, ortho-paedic surgery can provide dramatic pain relief and restore lost joint function. A total joint replacement, for example, can enable a person with severe arthritis in the hip or the knee to walk without pain or stiffness.

Some types of arthritis, especially the rheumatoid form, are best treated by a team of health care professionals with special abilities. These professionals include rheumatologists (nonsurgeons who are arthritis specialists), physical and occupational therapists, social workers, as well as orthopaedic surgeons.

MODEL 3-3   Commentary

This narrative describes a combat mission during Allied Force, a NATO operation in Bosnia in 1999. Major Jeff Hubbard, a U.S. Air Force pilot with the 22nd Fighter Squadron, flew 37 combat missions for Allied Force. Here, he reports one incident. His remarks appeared in *CODE ONE,* a magazine published by Lockheed Martin Tactical Aircraft Systems. The magazine is distributed to U.S. Air Force, U.S. Navy, and foreign air bases that use the F-16 and F-111 aircraft. The magazine also goes to selected members of the military, aerospace industry, media, and academic institutions.

Discussion

1. Discuss the language Major Hubbard uses. How specialized is it? How much can you understand? What terms are you unsure of?

2. Discuss the types of readers this magazine reaches. Are they all reading for the same purpose? How would you characterize the readers' levels of technical knowledge?

3. In groups, summarize Major Hubbard's experience in nontechnical language. Compare your results with those of the other groups. Does the removal of technical language make any difference in the impact of the narrative? Why, or why not?

Major Jeff Hubbard
F-16 Block 50 Pilot
22nd Fighter Squadron
Spangdahlem Air Base, Germany

Another Memorable Mission

I will remember another mission the rest of my life. I call it the Easter SAM dance. We were in the Belgrade area that night protecting F-117s. We turn cold in the CAP and turn in hot again. I see a signal I had seen in the same exact place during the last circuit. I look again and see nothing there. The signal looks like a tanker on my sensors. But it can't be a tanker, so I don't say anything. The third time I turn the corner and come around and the same thing happens again. I roll inverted and take another look. That's when I see a salvo of three SA-3s taking a belly shot on me. The middle SAM looks like a flashlight with a black center. Like a doughnut of fire. I say to myself, "This is it. My number is up." I figure from my last encounter with SAMs that I have about seven seconds to live. I start a defensive maneuver. I put the SAMs off my right wing, light the afterburner, and start to dive at the ground.

I am not even thinking about the ground at this point. I am watching the missiles. I don't punch off my tanks because I just simply forget. It's hard to find the jettison button at night under such circumstances. As I am going down, I realize the missiles are arcing over. As they hit the horizon, I pull back on the stick and put the missiles at the top of my canopy and pull at them. I've rolled and I'm starting to do an orthogonal roll, what we call a last-ditch maneuver. At this point, the first missile swings by. Then the second one goes by and I lose track of the third one. All I see is a huge light out of the back of the canopy. I realize I am upside down at this point and I see nothing but AAA below me. I roll out the aircraft and realize that my burner has been lit the whole time. The AAA gunners are shooting at my afterburner, which lights me up against the night sky. Gunners from the east and west are all shooting at my afterburner. I'm at 18,000 feet and the AAA is getting closer and closer to me. So I keep the burner lit and just climb. The engagement is over.

MODEL 3-3

MODEL 3-4   Commentary

This model shows the opening page of a report by research chemists for the U.S. Bureau of Mines. The report is an investigative study of the thermodynamics of certain chemical compounds. Readers are other research scientists interested in this experiment and the results that might influence their own research. The abstract is written for expert readers and does not provide a summary for nonexpert readers. You probably cannot understand the language or the content of this document.

The report language is highly specialized. None of the technical terms or chemical symbols is defined because the writer expects only experts to read the report.

Discussion

1. Compare the language, kinds of information, and formats of Models 3-1, 3-2, and 3-3, and discuss how each document is appropriate for its readers.

2. Find a set of instructions used in your workplace. Bring a copy to class. In groups, exchange instructions, and read them from the perspective of a new employee. Discuss what items might confuse a reader and what information may need to be added to the instructions to help the reader.

Enthalpy of Formation of 2CdO • CdSO$_4$

by H. C. Ko[1] and R. R. Brown[1]

ABSTRACT

The Bureau of Mines maintains an active program in thermochemistry to provide thermodynamic data for the advancement of mineral science and technology. As part of this effort, the standard enthalpy of formation at 298.15 K, $\Delta Hf°$, for 2CdO · CdSO$_4$ was determined by HCl acid solution calorimetry to be –345.69 ±0.61 kcal/mol.

INTRODUCTION

Cadmium oxysulfate (2CdO · CdSO$_4$) is known as an intermediate compound in the decomposition process of cadmium sulfate (CdSO$_4$). There is only one reported value ([4])[2] for the enthalpy of formation at 2CdO · CdSO$_4$, and that value was determined by a static manometric method. The objective of this investigation was to establish the standard enthalpy of formation for this compound by HCl solution calorimetry, which is inherently a more accurate method. This investigation was conducted as part of the Bureau of Mines effort to provide thermodynamic data for the advancement of mineral technology.

MATERIALS

Cadmium oxide: Baker[3] analyzed reagent-grade CdO was dried at 500°C for 2 hr. X-ray diffraction analysis indicated that the pattern matched the one given by PDF card 5–640 ([2]). Spectrographic analysis indicated 0.05 pct Pb to be present as the only significant impurity.

Cadmium oxysulfate: 2CdO · CdSO$_4$ was prepared by reacting stoichiometric quantities of anhydrous CdSO$_4$ and CdO in a sealed, evacuated Vycor tube according to the following procedure. The materials were blended and transferred to a Vycor reaction tube. The reaction tube was evacuated and heated to 330°C for 17 hr to remove traces of water that may have been introduced during blending and transferring. The tube was then sealed and heated to 600°C for 19 hr. The temperature was. . . .

---

[1]Research chemist, Albany Research Center, Bureau of Mines, Albany, Oreg.
[2]Underlined numbers in parentheses refer to items in the list of references at the end of this report.
[3]Reference to specific trade names does not imply endorsement by the Bureau of Mines.

MODEL 3-4

Chapter 3   Exercises

1. Select a document used by students on your campus (e.g., parking regulations, information about majors, instructions for using the library, career center information), *or* select consumer instructions. Interview an international student about what changes in language or graphics would be helpful before the document is used in that student's country. Discuss your results with the class.

2. Read "Prepare, Listen, Follow up" by Messmer in Part 2. Assume you are going to interview someone who currently holds the kind of job you would like some day. Prepare a set of 8 to 12 questions you would use in an interview with that person. Consider such topics as major responsibilities; interaction with coworkers, clients, the public, or regulatory agencies; work experience and education; and advice for students. Report your questions for class discussion, and submit a written set of questions to your instructor.

COLLABORATIVE EXERCISES

3. Melanoma, a potentially fatal skin cancer, now ranks as a leading cause of death among people aged 25 to 29 years. In groups, search the Internet for information about melanoma. Discuss how to present the information most effectively for teenagers and how they are likely to react to this information. Consider how to present tips for avoiding melanoma in a way that will interest readers. As a group, design an information sheet on melanoma for teenage readers. Present your design to the class and, explain your audience analysis of the potential teenage readers.

4. In groups, select a brochure or an Internet information site. Develop five questions to test a reader's understanding of the information or Web site. Follow the procedures described under "Readers Answering Questions," and test the usability of the brochure or Web site on members of another group. Then, your group will take the test devised by the other group on their material. Analyze the results of your test, and report to the class about any difficulties readers had with the brochure or Web site. *Or,* if your instructor prefers, write a group short report on your findings. Short reports are covered in Chapter 12.

5. The following announcement was sent by email to all 750 people planning to attend the All Sports Conference in four days. The attendees are college sports coaches and athletic directors, and they will be staying at Hotel Illinois in Chicago. The conference sessions are held in the adjoining Conference Center. Discuss the problems the readers might have with this message. In groups, revise the announcement for clarity. Add any details you wish to help clarify the following message:

> A special session has been added to the program for Thursday night. This event is not listed in your printed program. Carter Sullivan will be speaking on "Close

Encounters in Climbing." Mr. Sullivan is well known for his book, *Big Rocks and Ice: Climbing in the Midwest*. He will be signing book copies after his speech. On Friday, he will be available for photographs at a luncheon. Reservations are required.

INTERNET EXERCISES

6. Study your school's Web site. Imagine you are a potential transfer student looking for a new school. Analyze the usefulness of the information available and how easy it is to locate information about various majors and directions about applying for admission. In groups, compare your analyses. Identify two or three features of the Web site you would change in some way.

7. Select three countries each from a different continent. Search the Internet for information about each country's customs for (a) business appointments, (b) business dress, and (c) public behavior. Discuss your results with your classmates. *Or,* if your instructor prefers, write a short report of your findings. Short reports are covered in Chapter 12.

8. Select a product (e.g., tennis racquets, cameras) that you already know something about. Find the Web sites of three companies that make the product. Compare the Web sites' usability for the average consumer who is trying to decide what to buy. Report your findings in class discussion.

9. Find a city government Web site and a state government Web site. Select a city and state that is not your own, so you can consider the information available from the perspective of a tourist planning a trip. Evaluate the sites for usability by general readers. Report your results during class discussion.

10. Find three Web sites that offer information on a particular topic relevant to your major. Evaluate what audience the sites are aimed at. Compare the kinds of information available at the sites and the level of expertise needed to understand the information. Report your findings in class discussion.

CHAPTER 4

Organization

SORTING INFORMATION

The organization of a document has a strong influence on how well the reader understands and is able to use the information. If a document is well organized with headings, lists, and key words that trigger a reader's memory of prior knowledge of the subject, a reader can learn and use new information effectively. Reading studies show that if adults are given poorly organized information, they tend to reorganize it themselves and to omit information that does not clearly fit into their prior knowledge.[1] As a writer you must organize documents to help readers learn and remember new information.

In some writing situations, documents have predetermined organizational patterns. Social workers at a child welfare agency may be directed to present information in all case reports under the major topics of living conditions, parental attitudes, previous agency contact, and suspected child abuse. Because these major topics appear in all the agency case reports, readers who use them regularly, such as attorneys and judges, know where to look for the specific details they need. Such consistency is useful to readers if the documents have a standard format—only the details change from situation to situation.

If you are writing a document that does not have standard categories for information, however, your first step will be to organize the information into major topics and identify appropriate subtopics.

Select Major Topics

Begin organizing your document by sorting the information into major topics. Select the topics based on your analysis of your readers' interest in and need for information. If you are writing a report about columns or pillars, you should group your information into major topics that represent what your readers want to know about the subject. If your readers are interested in Greek and Roman column design, you can group your information according to classic column type—Doric, Ionic, and Corinthian. Readers interested in learning about column construction may prefer an organization based on construction materials, such as brick, marble, wood, stone, and metal. Readers interested in the architectural history of columns may prefer an organization based on time period, such as Ancient, Medieval, and Renaissance. Readers interested in the cultural differences in column design may prefer the information grouped according to geographic region, such as Asian, Mediterranean, and North American.

As you sort your information into major topics, remember that the topics should be similar in type and relatively equal in importance. You should not have one major topic based on column design and one based on time period. Such a mix of topics would confuse your readers. You should also be certain that the major topics you select are distinct. You would not want to use time periods that overlap, such as the Renaissance and the fifteenth century, or column designs that overlap, such as Doric and Tuscan. Consider carefully which major topics represent appropriate groups of information for your readers.

Identify Subtopics

After you determine your major topics, consider how to sort the information into appropriate subtopics. Do not simply lump facts together. Instead, think about what specific information you have that will support each major topic. If your major topics are based on classic column types, your readers may be interested in such subtopics as Greek design, Roman design, changes in design over time, and how the Greeks and Romans used the columns. These subtopics, or specific pieces of information, should appear under each of the major topics because your readers want to know how these details relate to each column type. When you have identified your major topics and have selected appropriate subtopics, you are ready to begin outlining.

CONSTRUCTING OUTLINES

Think of an outline as a map that identifies the major topics and shows their location in a document. Outlines provide three advantages to a writer:

1. A writer can "see" the structure of a document before beginning a rough draft in the same way that a traveler can follow the route of a journey on a road map before getting into a car.

2. Having a tentative organizational plan helps a writer concentrate on presenting and explaining the information in the rough draft rather than on writing and organizing at the same time.

3. A writer can keep track of organizational decisions no matter how many interruptions occur in the writing process.

An outline usually begins as a short, informal list of topics and grows into a detailed list that includes all major topics and subtopics and groups them into chunks of information most appropriate for the reader. Consider this situation: Al Martinez's first project as a summer intern in the Personnel Department of Tri-Tech Chemical Company is to write a bulletin explaining the company rules on travel and business expenses. Al's readers will be both the managers who travel and the secretaries who handle the paperwork. At present, the pertinent information is scattered through more than a dozen memos written over the last three years. After reading all these memos and noting appropriate items, Al begins to organize his information. He begins with an informal outline.

Informal Outlines

An *informal outline* is a list of main topics in the order the writer expects to present them in the rough draft. Computers have made outlining faster and easier than it used to be. With word processing, a writer can quickly (1) add or delete items and rearrange them without recopying sections that do not change and (2) keep the outline up to date by changing it to match revisions in rough drafts. In developing an informal outline, remember these guidelines:

1. List all the relevant topics in any order at first. Al listed the main topics for his bulletin in the order in which he found them discussed in previous memos in company files.

2. Identify the major groups of related information from your list. Looking at his list of items, Al saw immediately that most of his information centered around types of expenses and reimbursement and approval procedures. Al, therefore, clustered his information into groups representing these topics.

3. Arrange the information groups in an order that will best serve the readers' need to know. Al decided to present reimbursement procedures first because readers would need to know about those before traveling. He then organized groups representing types of expenses from most common to least common so that the majority of readers could find the information they needed as quickly as possible.

Model 4-1 shows Al's informal outline for his bulletin. Writers usually use informal outlines to group and order information in the planning stage. Often, an informal outline is all that a writer needs. For some documents, such as a manual, however, a writer must prepare a formal outline that will be part of

---

1. Travel Authorization and Reimbursement Rules
2. Meals
3. Hotels
4. Transportation
5. Daily Incidental Expenses
6. Spouse's Travel
7. Entertainment
8. Gifts
9. Loss or Damage
10. Nonreimbursed Items

MODEL 4-1   Al's Informal Outline

the document. Also, some writers prefer to work with formal outlines in the planning stage.

Formal Outlines

A *formal outline* uses a special numbering system and includes subtopics under each major section. A formal outline is a more detailed map of the organization plan for a document. Writers usually use one of two numbering systems for outlines: (1) Roman and Arabic numerals or (2) decimals (also called the *military numbering system*). Choose whichever numbering system you prefer unless your outline will be part of the final document. In that case, use the numbering system most familiar to your readers or the one your company prefers. Model 4-2 shows a set of topics organized by both numbering systems.

Roman–Arabic Numbering System
I. Costs
   A. Equipment
      1. Purchase
      2. Maintenance
         a. Weekly
         b. Breakdown
   B. Employees
      1. Salaries
      2. Benefits
II. Locations
   A. European
      1. France

Decimal Numbering System
1.0 Costs
   1.1 Equipment
      1.1.1. Purchase
      1.1.2. Maintenance
         1.1.2.1. Weekly
         1.1.2.2. Breakdown
   1.2 Employees
      1.2.1. Salaries
      1.2.2. Benefits
2.0 Locations
   2.1 European
      2.1.1. France

MODEL 4-2   Formal Outline Numbering Systems

Al Martinez decided to develop his informal outline into a formal one because he wanted a detailed plan before he began his rough draft.

Topic Outlines

A *topic outline* lists all the major topics and all the subtopics in a document by key words. Like the major topics, all subtopics should be in an order that the readers will find useful. Al developed his formal topic outline by adding the items he knew he had to explain under each major topic and using the Roman–Arabic numbering system. Model 4-3 shows Al's formal topic outline.

In drafting his formal topic outline, Al knew that he wanted to begin his bulletin by explaining the process of applying for and receiving reimbursement for travel expenses because his readers needed to understand that system before they began collecting travel-expense information and preparing expense reports. He organized expense topics in the order in which they appeared on the forms his readers would have to submit and then listed specific items under each main topic. While redrafting his outline, Al revised some of his earlier organizational decisions. First, he decided that the topic "Spouse's Travel" was not really a separate item and ought to be covered under each travel expense item, such as "Hotels." He also decided that the topic "Gifts" was really a form of entertainment for clients, so he listed gifts under "Entertainment" in his formal topic outline.

Sentence Outlines

A *sentence outline* develops a topic outline by stating each point in a full sentence. When a project is long and complicated or the writer is not confident about the subject, a sentence outline helps clarify content and organization before drafting begins. Some writers prefer sentence outlines because they can use the sentences in their first drafts, thereby speeding up the initial drafting. Al Martinez also decided to expand his topic outline into a sentence outline as a way of moving closer to his first draft. Model 4-4 shows a portion of Al's sentence outline.

DEVELOPING EFFECTIVE PARAGRAPHS

Each paragraph in a document is a unit of sentences that focuses on one idea and acts as a visual element to break up the text into manageable chunks of information. Effective paragraphs guide readers by

- Introducing individually distinct but related topics
- Emphasizing key points

I. Travel Authorization and Reimbursement
  A. Approval
  B. Forms
    1. Form 881
    2. Form 669
    3. Form 40-A
    4. Form 1389
  C. Submission
II. Meals
  A. Breakfast
  B. Lunch
  C. Dinner
III. Hotels
  A. Chain Guaranteed Rates
  B. Suites
IV. Transportation
  A. Automobile
    1. Personal Cars
      a. Mileage
      b. Maintenance
    2. Rental Cars
      a. Approved Companies
      b. Actual Costs Only
  B. Airplane
    1. Tri-Tech Aircraft
    2. Other Companies' Aircraft
    3. Commercial Aircraft
  C. Railroad
  D. Taxi, Limousine, Bus
V. Daily Incidental Expenses
  A. Parking and Tolls
  B. Tips
  C. Laundry and Telephone
VI. Entertainment
  A. Parties
    1. Home
    2. Commercial
  B. Meals
  C. Gifts
    1. Tickets
    2. Objects
VII. Loss or Damage
VIII. Nonreimbursed Items
  A. Personal Items
  B. Legal, Insurance Expenses

MODEL 4-3 Al's Formal Topic Outline

I. Travel Authorization and Reimbursement. Tri-Tech will pay travel expenses directly or reimburse an employee for costs of traveling on company business.
  A. All official travel for Tri-Tech must be approved in advance by supervisors.
  B. Several company forms must be completed and approved.
    1. Form 881, "Travel Authorization," is used for trip approval if travel reservations are needed.
    2. Form 669, "Petty Cash Requisition," is used if no travel reservations are needed.
    3. Form 40-A, "Domestic Travel Expenses," is used to report reimbursable expenses in the United States.
    4. Form 1389, "Foreign Expense Payment," is used to report reimbursable expenses outside the United States.
  C. Travel authorization forms should be submitted to the employee's supervisor, but posttravel forms should be sent directly to the Travel Coordinator in the Personnel Department.
II. Meals. Tri-Tech will reimburse employees for three meals a day while traveling.
  A. Breakfast expenses are reimbursed if the employee is away overnight or leaves home before 6:00 a.m.

MODEL 4-4   A Portion of Al's Formal Sentence Outline

- Showing relationships between major points
- Providing visual breaks in pages to ease reading

For effective paragraphs, writers should consider unity, coherence, development, and an organizational pattern for presenting information.

Unity

Unity in a paragraph means concentration on a single topic. One sentence in the paragraph is the *summary sentence* (also called the *topic sentence*), and it establishes the main point. The other sentences explain or expand the main point. If a writer introduces several major points into one paragraph without developing them fully, the mix of ideas violates the unity of the paragraph and leaves the readers confused. Here is a paragraph that lacks unity:

Many factors influence the selection of roofing material. Only project managers and engineers have the authority to change the recommendations, and the changes should be in writing. The proper installation of roofing material is important, and contractors should be trained carefully. Roof incline, roof deck construction, and climatic conditions must be considered in selecting roofing material.

The first sentence appears to be a topic sentence, indicating that the paragraph will focus on factors in selecting roofing. Readers, therefore, would expect to find individual factors enumerated and explained. However, the second and third sentences are unrelated to the opening and, instead, introduce two new topics: (1) the authority to select roofing and (2) the training of contractors. The fourth sentence returns to the topic of selecting roofing material. Thus, the paragraph actually introduces three topics, none of which is explained. The paragraph fails to fulfill the expectations of the reader. Here is a revision to achieve unity:

> Three factors influence the selection of roofing material. One factor is roof incline. A minimum of ¼ in. per foot incline is recommended by all suppliers. A second factor is roof deck construction. Before installing roofing material, a contractor must cover a metal deck with rigid installation, prime a concrete deck to level depressions, or cover a wooden deck with a base sheet. Finally, climatic conditions also affect roofing selection. Roofing in very wet climates must be combined with an all-weather aluminum roof coating.

In this version, the paragraph is unified under the topic of which factors influence the selection of roofing materials. The opening sentence establishes this main topic; then, the sentences that follow identify the three factors and describe how contractors should handle them. Remember, readers are better able to process information if they are presented one major point at a time.

Coherence

A paragraph is *coherent* when the sentences proceed in a sequence that supports one point at a time. Transitional, or connecting, words and phrases help coherence by showing the relationships between ideas and by creating a smooth flow of sentences. Here are the most common ways writers achieve transition:

- Repeat key words from sentence to sentence.
- Use a pronoun for a preceding key term.
- Use a synonym for a preceding key term.
- Use demonstrative adjectives (e.g., *this* report, *that* plan, *these* systems, *those* experiments).
- Use connecting words (e.g., *however, therefore, also, nevertheless, before, after, consequently, moreover, likewise, meanwhile*) or connecting

phrases (e.g., *for example, as a result, in other words, in addition, in the same manner*).

- Use simple enumeration (e.g., *first, second, finally, next*).

Here is a paragraph lacking clear transitions and with sentences not in the best sequence to explain the two taxes under discussion:

> Married couples should think about two taxes. When a person dies, an estate tax is levied against the value of the assets that pass on to the heirs by will or automatic transfer. Income tax on capital gains is affected by whatever form of ownership of property is chosen. Under current law, property that one spouse inherits from the other is exempt from estate tax. Married couples who remain together until death do not need to consider estate taxes in ownership of their homes. The basic rule for capital gains tax is that when property is inherited by one person from another, the financial basis of the property begins anew. For future capital gains computations, it is treated as though it were purchased at the market value at the time of inheritance. When the property is sold, tax is due on the appreciation in value since the time it was inherited. No tax is due on the increase in value from actual purchase to time of inheritance. If the property is owned jointly, one-half gets the financial new start.[2]

This paragraph attempts to describe the impact of two federal taxes on married couples who own their homes jointly. The writer begins with the topic of estate tax, but he or she quickly jumps to the capital gains tax, then returns to the estate tax, and finally concludes with the capital gains tax. Since readers expect information to be grouped according to topic, they must sort out the sentences for themselves to use the information efficiently. The lack of transition also forces readers to move from point to point without any clear indication of the relationships between ideas. Here is a revision of the paragraph with the transitions highlighted:

> For financial planning, married couples must think about two taxes. **First,** is the estate tax. When an individual dies, **this tax** is levied against the value of the assets that pass on to the heirs by will or automatic transfer. Under current law, any amount of property that one spouse inherits from the other is exempt from **estate tax. Therefore,** married couples who remain together until death do not need to consider **estate taxes** in planning ownership of their home. The **second** relevant tax is capital gains. Unlike **estate tax, capital gains tax** is affected by what form of ownership of property a couple chooses. The basic rule for **capital gains tax** is that when property is inherited by one person from another, it begins anew for tax purposes. **That is,** for future **capital gains** computation, the **property** is treated as though it were purchased at the market value at the time of inheritance. When the **property** is sold, tax is due on the appreciation in value since the time it was inherited. No tax, **however,** is due on the increase in value from actual purchase to time of inheritance. If the **property** is owned jointly, one-half begins anew for tax purposes.

In this revision, the sentences are rearranged so that the writer gives information first about the estate tax and then about the capital gains tax. Transitions come from repetition of key words, pronouns, demonstrative adjectives, and connecting words or phrases.

Coherence in paragraphs is essential if readers are to use a document effectively. When a reader stumbles through a paragraph trying to sort out the information or to decide the relationships between items, the reader's understanding and patience diminish rapidly.

STRATEGIES—International Readers

Nonnative English speakers often have difficulty understanding the meaning of phrasal verbs (a verb plus a preposition)—for example, *pay for, turn in, fill out, join in*. These phrases usually are idioms—an expression that native English speakers understand, but the phrase itself does not indicate that meaning. Whether you are writing for print or for the Web, if your audience is likely to be international, use the standard verb—for example, *purchase, submit, complete, participate*.

Development

Develop your paragraphs by including enough details so that your reader understands the main point. Generally, one- and two-sentence paragraphs are not fully developed unless they are used for purposes of transition or emphasis or are quotations. Except in these circumstances, a paragraph should contain a summary or topic sentence and details that support the topic.

Here is a poorly developed paragraph from a brochure that explains diabetes to patients who must learn new eating habits:

> People with insulin-dependent diabetes need to plan meals for consistency. Insulin reactions can occur if meals are not balanced.

This sample paragraph is inadequately developed for a new patient who knows very little about the disease. The opening sentence suggests that the paragraph will describe meal planning for diabetics. Instead, the second sentence tells what will happen without meal planning. Either sentence could be the true topic sentence, and neither point is developed. The reader is left, therefore, with an incomplete understanding of the subject and no way to begin meal planning. Here is a revision:

> People with insulin-dependent diabetes need to plan meals for consistency. To control blood sugar levels, schedule meals for the same time every day. In addition, eat about the same amounts of carbohydrates, protein, and fat every day in the same combination and at the same times. Your doctor will tell you the exact amounts appropriate for you. This consistency in eating is important because your insulin dose is based on a set food intake. If your meal plan is not balanced, an insulin reaction may occur.

In this revision, the writer develops the summary sentence by giving examples of what consistency means. The paragraph is developed further by an explanation of the consequences if patients do not plan balanced meals. The point about an insulin reaction is now connected to the main topic of the paragraph—meal planning.

In developing paragraphs, think first about the main idea, and then determine what information the reader needs to understand that idea. Here are some ways to develop your paragraphs:

- Provide examples of the topic.

- Include facts, statistics, evidence, details, or precedents that confirm the topic.

- Quote, paraphrase, or summarize the evidence of other people on the topic.

- Describe an event that has some influence on the topic.

- Define terms connected with the topic.

- Explain how equipment operates.

- Describe the physical appearance of an object, area, or person.

In developing your paragraphs, remember that long, unbroken sections of detailed information may overwhelm readers, especially those without expert knowledge of the subject, and may interfere with the ability of readers to use the information.

Patterns for Presenting Information

The following patterns for presenting information can be effective ways to organize paragraphs or entire sections of a document. Several organizational patterns may appear in a single document. A writer preparing a manual may use one pattern to give instructions, another to explain a new concept, and a third to help readers visualize the differences between two procedures. Select the pattern that will best help your readers understand and use the information.

Ascending or Descending Order of Importance Pattern

To discuss information in order of importance to the reader, use the ascending (lowest-to-highest) or descending (highest-to-lowest) pattern. If you are describing the degrees of hazard of several procedures or the seriousness of several production problems, you may want to use the descending pattern to alert your readers to the matter most in need of attention. In technical writing, the descending order usually is preferred by busy executives because they want to know the most important facts about a subject first and may even stop reading once they understand the main points. However, the ascending order can be effective if you are building a persuasive case on why, for example, a distribution system should be changed.

This excerpt from an advertisement for a mutual fund illustrates the descending order of importance in a list of benefits for potential investors in the fund:

> Experienced and successful investors select the Davies-Meredith Equity Fund because
>
> 1. The Fund has outperformed 98% of all mutual funds in its category for the past 18 months.
> 2. The Fund charges no investment fees or commissions.
> 3. A 24-hour toll-free number is available for transfers, withdrawals, or account information.
> 4. Each investor receives the free monthly newsletter, "Investing for Your Future."

Although a reader may be persuaded to invest in the fund by any of these reasons, most potential investors would agree that the first item is most important and the last item least important. The writer uses the descending order of importance to attract the reader's attention.

The writer, however, uses the ascending order of importance to illustrate the gains in an investment over time:

> Long-term investment brings the greatest gains. An initial investment of $1000 would have nearly doubled in six years:
>
> | 1 year | $1045 |
> |--------|-------|
> | 2 years | $1211 |
> | 4 years | $1571 |
> | 6 years | $1987 |

The ascending-order pattern here emphasizes to the investor the benefits of keeping money in the fund for several years.

Cause-and-Effect Pattern

The cause-and-effect pattern is useful when a writer wants to show readers the relationship between specific events. If you are writing a report about equipment problems, the cause-and-effect pattern can help readers see how one breakdown led to another and interrupted production. Be sure to present a clear relationship between cause and effect, and give readers evidence of that relationship. If you are merely speculating about causes, make this clear:

> The probable cause of the gas leak was a blown gasket in the transformer.

Writers sometimes choose to describe events from effects back to cause. Lengthy research reports often establish the results first and then explain the causes of those results.

This excerpt from a pamphlet for dental patients uses cause and effect to explain how periodontal disease develops:

> Improper brushing or lack of flossing and regular professional tooth cleaning allows the normal bacteria present in the mouth to form a bacterial plaque. This plaque creates spaces or pockets between the gums and roots of the teeth. Chronic inflammation of the gums then develops. Left untreated, the inflammation erodes the bone in which the teeth are anchored, causing the teeth to loosen or migrate. Advanced periodontal disease requires surgery to reconstruct the supporting structures of the teeth.

Chronological Pattern

The chronological pattern is used to present material in stages or steps from first to last when readers need to understand a sequence of events or follow specific steps to perform a task. This excerpt from consumer instructions for a popcorn maker illustrates the chronological pattern:

1. Before using, wash cover, butter-measuring cup, and popping chamber in hot water. Rinse and dry. *Do not immerse housing in water.*
2. Place popcorn container under chute.
3. Preheat the unit for 3 minutes.
4. Using the butter-measuring cup, pour ½ cup kernels into popping chamber.

Instructions should always be written in the chronological pattern. This pattern also is appropriate when the writer needs to describe a test or a process, so the reader can visualize it, as in this description of starting up a test engine for a jet:

> An engineer brings the JSF119-611 test engine to life by turning a valve that sends compressed air to an air starter. The air starter is attached to a gearbox on the underside of the engine. The gearbox turns the engine's high-pressure compressor and high-pressure turbine. The various stages of the engine spin faster and faster. When the rotation reaches about 3500 rpm, fuel pumps send JP8 to fuel injectors in the combustion chamber. An igniter spark sets off a continuous reaction of air and fuel. The air starter shuts down. The engine is running.[3]

Classification Pattern

The classification pattern involves grouping items in terms of certain characteristics and showing your readers the similarities within each group. The basis for classification should be the one most useful for your purpose and your readers. You might classify foods as protein, carbohydrate, or fat for a report to dietitians. In a report on foods for a culinary society, however, you might classify them as beverages, appetizers, and desserts. Classification is useful when you have many items to discuss, but if your categories are too broad, your readers will have trouble understanding the distinctiveness of each class. Classifying all edibles under "food" for either of the two reports just mentioned would not be useful because you would have no basis for distinguishing the individual items. This paragraph from a student report uses classification to explain types of sugars:

> When analyzing a patient's nutritional needs, dietitians generally sort sugars into five types. Sucrose comes from sugar cane and sugar beets and is found in table sugar. Dextrose is derived from corn and is used in many commercially produced foods. Lactose is found in milk and milk products, and some people find it difficult to digest. Fructose is in fruits and is somewhat sweeter than the other sugars. Lastly, glucose is found in a variety of fruits, honeys, and vegetables.

Partition Pattern

The partition pattern, in contrast to the classification pattern, involves separating a topic or system into its individual features. This division allows readers to master information about one aspect of the topic before going on to the next. Instructions are always divided into steps so that readers can perform one step at a time. The readers' purpose in needing the information should guide selection of the basis for partition. For example, a skier on the World Cup circuit would be interested in a discussion of skis based on their use in certain types of races, such as downhill or slalom. A manufacturer interested in producing skis might want the same discussion based on materials used in production. A sales representative might want the discussion based on costs of the different types of skis. This excerpt from a weather information brochure for consumers uses partition to explain Earth's atmosphere:

> Earth's atmosphere is divided into five layers. The first layer, the troposphere, is next to Earth's surface and extends about 7 miles up. It contains most of the clouds and weather activity. The next layer is the stratosphere extending to about 30 miles up and containing relatively little water or dust. The third layer is the mesosphere extending 30 to 50 miles up. The mesosphere is very cold, dropping to about minus 100 Fahrenheit. The fourth layer, the thermosphere, is also called the ionosphere and extends to about 250 miles up. The thermosphere has rapidly increasing temperatures because of solar radiation. The final layer is the exosphere in which the traces of atmosphere fade into space.

Comparison and Contrast Pattern

The comparison and contrast pattern focuses on the similarities (comparison) or differences (contrast) between subjects. Writers often find this pattern useful because they can explain a complex topic by comparing or contrasting it with another, familiar topic. For this pattern, the writer has to set up the basis for comparison according to what readers want to know. In evaluating the suitability of two locations for a new restaurant, a writer could compare the two sites according to accessibility, neighborhood competition, and costs. There are two ways to organize the comparison and contrast pattern for this situation. In one method (topical), the writer could compare and contrast location A with location B under specific topics, such as

Accessibility
1. Location A
2. Location B

Neighborhood competition
1. Location A
2. Location B

Costs
1. Location A
2. Location B

In the other method (complete subject), the writer could present an overall comparison by discussing all the features of location A and then all the features of location B, keeping the discussion in parallel order, such as

Location A
1. Accessibility
2. Neighborhood competition
3. Costs

Location B
1. Accessibility
2. Neighborhood competition
3. Costs

In choosing one method over another, consider the readers' needs. For the report on two potential restaurant locations, readers might prefer a topical comparison to judge the suitability of each location in terms of the three vital factors. The topical method does not force readers to move back and forth between major sections looking for one particular item, as the complete subject method does. In comparing two computers, however, a writer may decide that readers need an overall description of the both to determine which seems to satisfy more office requirements. In this case, the writer would present all the features of computer A and then all those of computer B.

This excerpt from a column in a boating magazine contrasts steam fog with advection fog:

> Steam fog is formed when the air temperature drops below the water temperature in light winds. In contrast, advection fog, or sea fog, forms when moist air is transported, or advected, over a cold surface, like the ocean or a lake. If the water temperature is less than the dew-point temperature of the air, moisture in the air can condense, forming a cloud on the ground (fog). Advection fog is also distinguished from steam fog in that the wind is usually blowing, unlike the calm conditions during steam fog.[4]

The following excerpt from a superintendent report to a school board uses comparison to show that two history programs are similar:

> The Winfield Academy history program is similar to ours in three important areas. First, the initial year of the Winfield program covers American colonial history through the Gilded Age. Second, the Winfield program in the second year concentrates on the twentieth century. Third, the twentieth century material includes American and world history. This sequence matches the one we have been using. Merging the Winfield students with those in Central High for a joint history program will not disrupt either school's curriculum.

Writers sometimes use an *analogy*, a comparison of two objects or processes that are not truly similar but share important qualities that help the reader to understand the less familiar object or process. The following excerpt from a book about weather uses an analogy to explain the vertical movement of air:

> Imagine putting in a wood screw. Turn the screwdriver clockwise and the screw goes down. Turn the screwdriver counterclockwise, the screw comes up. This will help you remember that in the Northern Hemisphere, air goes clockwise and down around high pressure. Air goes counterclockwise and up around low pressure.[5]

Definition Pattern

The purpose of definition is to explain the meaning of a term that refers to a concept, process, or object. Chapter 8 discusses writing definitions for objects or processes. Definition also can be part of any other organizational pattern when a writer decides that a particular term will not be clear to readers. An *informal definition* is a simple statement using familiar terms:

> A drizzle is a light rainfall.

A *formal definition* places the term into a group and then explains the term's special features that distinguish it from the group:

> term   group   special features
>
> A pronator is a muscle in the forearm that turns the palm downward.

Writers use *expanded definitions* to identify terms and explain individual features when they believe readers need more than a sentence definition. Definitions can be expanded by adding examples, using an analogy, or employing one of the organization patterns, such as comparison and contrast or partition.

This excerpt from a student's biology report uses the cause-and-effect pattern to expand the definition of arteriosclerosis:

> Arteriosclerosis is a disease of the arteries, commonly known as hardening of the arteries. As we age, plaques of fatty deposits, called atheromas, form in the blood and cling to the walls of the arteries. These deposits build up and narrow the artery passage, interfering with the flow of blood. Fragments of the plaque, called emboli, may break away and block the arteries, causing a sudden blockage and a stroke. Even if no plaque fragments break away, eventually the artery will lose elasticity, blood pressure will rise, and blood flow will be sufficiently reduced to cause a stroke or a heart attack.

Spatial Pattern

In the spatial pattern, information is grouped according to the physical arrangement of the subject. A writer may describe a machine part by part from top to bottom so that readers can visualize how the parts fit together. The spatial pattern creates a path for readers to follow. Features can be described from top to bottom, side to side, inside to outside, north to south, or in any order that fits the way readers need to "see" the topic.

This excerpt is from an architect's report to a civic restoration committee about an old theater scheduled to be restored:

> The inside of the main doorway is surrounded by a Castilian castle facade and includes a red-tiled parapet at the top of the castle roof. A series of parallel brass railings just beyond the doorway creates corridors for arriving movie patrons. Along the side walls are ornamented white marble columns behind which the walls are covered with 12-ft-high mirrors. The white marble floor sweeps across the lobby to the wall opposite the entrance, where a broad split staircase curves up from both sides of the lobby to the triple doors at the mezzanine level. The top of the staircase at the mezzanine level is decorated with a life-size black marble lion on each side.

CHAPTER SUMMARY

This chapter discusses organizing information by sorting it into major topics and subtopics, constructing outlines, and developing effective paragraphs. Remember:

- Information should be grouped into major topics and subtopics based on the way the readers want to learn about the subject.

- An informal outline is a list of major topics.

- A formal outline uses a special numbering system, usually Roman and Arabic numerals or decimals, and lists all major topics and subtopics.

- A topic outline lists all the major topics and subtopics by key words.

- A sentence outline states each main topic and subtopic in a full sentence.

- Effective paragraphs need to be unified, coherent, and fully developed.

- Organizational patterns for effective paragraphs include ascending or descending order of importance, cause and effect, chronological, classification, partition, comparison and contrast, definition, and spatial.

SUPPLEMENTAL READINGS IN PART 2

Bagin, C. B., and Van Doren, J. "How to Avoid Costly Proofreading Errors," *Simply Stated*, p. 480.

Garhan, A. "ePublishing," *Writer's Digest*, p. 495.

McAdams, M. "It's All in the Links: Readying Publications for the Web," *The Editorial Eye*, p. 512.

Nielsen, J. "Top Ten Guidelines for Homepage Usability," *Alertbox*, p. 523.

■ *ENDNOTES*

1. Ann Mill Duin, "Factors That Influence How Readers Learn from Text: Guidelines for Structuring Technical Documents," *Technical Communication* 36.1 (February 1989): 97–101.

2. Adapted from Allen Bernstein, *1998 Tax Guide for College Teachers* (Washington, DC: Academic Information Service, 1997): 186.

3. Eric Hehs, "Propulsion System Testing," *CODE ONE* 14.2 (April 1999): 18.

4. David Schultz, "Advection (Sea) Fog," *Canoe & Kayak* 29.2 (May 2001): 23.

5. Jack Williams, *The Weather Book* (New York: Vintage Books, 1992), p. 34.

MODEL 4-5   Commentary

The following two Web pages show different methods of organizing information for users. The first is the home page of the U.S. Department of Agriculture. The second is the home page of the U.S. Department of Health & Human Services.

Discussion

1. Discuss the obvious differences in design between the two Web pages. How well do you think these differences fit the purposes of the two Web sites?

2. Discuss the links on the margins of the U.S. Department of Agriculture Web page and how helpful they would be to someone seeking information.

3. Discuss the organization of the Web page from the U.S. Department of Health & Human Services. Notice the different placement of "News" on both pages. Discuss why officials may have decided they wanted this placement.

4. What political emphasis do you see in these Web pages?

5. In groups, find the home pages for the U.S. Department of Labor and the U.S. Department of State. Discuss their organization and design. Decide which Web pages are most useful to Web readers.

MODEL 4-5   U.S. Department of Agriculture Home Page

United States Department of
Health & Human Services
Leading America to Better Health, Safety and Well-Being

Skip Navigation

- **HHS Home**
- **Questions?**
- **Contact HHS**
- **Site Map**

• Diseases & Conditions

- Heart Disease, Cancer, HIV/AIDS, Diabetes...
- Mental Health
- Treatment, Prevention, Genetics
- Clinical Trials
- Addictions, Substance Abuse

• Safety & Wellness

- Eating right
- Exercise, Fitness
- Safety Tips and Programs
- Smoking, Drinking
- Traveler's Health

• Drug & Food Information

- Drugs, Dietary Supplements
- Food Safety
- Recalls & Safety Alerts
- Medical Devices

• Disasters & Emergencies

- Bioterrorism
- Homeland Security
- Natural Disasters

• Grants & Funding

- Research & Program Funding
- Scholarships, Internships, Financial Aid

• Reference Collections

- Dictionaries, Libraries, Databases
- Publications, Fact Sheets
- Statistics, Reports

• Families & Children

- Medicaid, other health insurance
- Child Support, Child Care, Adoption
- Domestic Violence, Child Abuse
- Vaccines

• Aging

- Medicare
- Health Issues
- Coping and Caring

• Specific Populations

- Women, Men, Children, Seniors
- Disabilities
- Racial and Ethnic Minorities
- Homeless

• Resource Locators

- Nursing Homes
- Physicians, other Healthcare Providers
- Health Care Facilities

• Policies & Regulations

- Policies, Guidelines
- Laws, Regulations
- Testimony

• About HHS

- Plans and Budget
- HHS Agencies and Offices
- Employment, Training Opportunities

 News   1:12 PM Wed, Aug 04

August 3, 2004 — President Announces $100 Million in Grants to Support Substance Abuse Treatment

(Full Story)

August 3, 2004 — President Announces $43 Million in Grants from Compassion Capital Fund

(Full Story)

August 3, 2004 — President Announces Mentoring Grants for Children of Prisoners

(Full Story)

August 3, 2004 — HHS to Provide New Interactive Book of Health Information to Women of Afghanistan and Their Families

(Full Story)

• All HHS News

Features
- **Health IT Strategic Framework**
- **Surgeon General**
- **Medicare-Approved Drug Discount Cards**
- **Small Steps to Better Health**
- **Privacy of Health Information/HIPAA**
- **"Gift of Life" Organ Donation Initiative**
- **Faith-Based & Community Initiatives**

U.S. Department of Health & Human Services Home Page

MODEL 4-6   Commentary

This letter report was written by an engineer to a client who had requested an analysis of suitable roadway improvement processes and a recommendation for future work. The engineer reviews the current conditions of the city's roads, describes the two available types of recycled asphalt processes, and concludes with the firm's recommendation. The writer also offers to discuss the subject further and to provide copies of the reports he mentions in his letter.

Discussion

1. Discuss the headings the writer uses in his letter report. How helpful are they to the reader? Are there any revisions you would make in the headings?

2. In his "Recommendation" section, the writer mentions environmental concerns. How much impact does this mention of the environment have in this section? What revisions in content or organization would you make to increase the emphasis on the environmental reasons for using recycled asphalt?

3. Identify the organizational strategies the writer uses throughout his report.

MANNING-HORNSBY ENGINEERING CONSULTANTS
3200 Farmington Drive
Kansas City, Missouri 64132
(816) 555–7070

November 23, 2005

Mr. Marvin O. Thompson
City Manager
Municipal Building, Suite 360
Ridgewood, MO 64062

Dear Mr. Thompson:

This report covers the Manning-Hornsby review of the current roadway
conditions in Ridgewood, Missouri, analysis of the available recycled
asphalt processes, and our recommendation for the appropriate process
for future Ridgewood roadway improvement projects.

Current Ridgewood Maintenance Program

The City of Ridgewood has 216 miles of roadways. Half of these have
curbs and gutters, while the other half have bermed shoulders with
drainage ditches. About 95 percent of the roadways are of asphalt
composition. The remaining 5 percent are of reinforced concrete
construction or are unimproved.

Condition: Over the last five years, 40 miles of existing roadways have
been improved. Improvements generally consist of crack sealing and
removal and replacement of minor sections of pavement followed by a
resurfacing overlay. Roadways with curbs and gutters are typically
overlayed with 3–4 inch layers of hot-mix asphalt. Roadways with
berms and drainage ditches are typically overlayed with 3–4 inch layers
of cold-mix asphalt, also known as "chip and seal." No form of recycled
asphalt has ever been used on these roadways.

Both these processes are appropriate for Ridgewood roadways. Hot-mix
asphalt is a bituminous pavement, mixed hot at an asphalt plant,
delivered hot to the site, and rolled to a smooth compacted state. The
cold-mix asphalt is mixed and placed at lower temperatures with higher
asphalt contents. Cold-mix asphalt is usually covered with bitumen and
stone chips. Generally, the hot mix provides a smoother surface than
the cold mix but is more susceptible to future cracking. The cold mix is
more resilient than the hot mix and will reform itself in the summer
heat and high temperatures.

MODEL 4-6

Mr. Marvin O. Thompson   -2-   November 23, 2005

Problem: One of the problems associated with the current maintenance process is that roadways are being overlayed without the removal of existing asphalt. Therefore, the roadway grades are increasing. In some cases, the curbs are starting to disappear. Drainage problems are occurring due to the flattening road surface, permanent cross slopes, and grade changes. These conditions cause low spots that do not drain to existing catchbasins.

So that the curb and drainage problems noted above are eliminated, future projects should include the removal of existing layers of asphalt before resurfacing begins. Plans should call for milling off a depth of existing asphalt that is equal to or greater than the proposed overlay thickness. The milling process will make it possible to achieve curb heights, pavement cross slopes, and pavement grades that are similar to the originally constructed roadway. The City must then decide what it wants to do with the milled asphalt.

Recycled Asphalt

Recycling products has become a popular operation in today's environmentally conscious world. One waste material already being used extensively is asphalt. Most experts agree that using recycled asphalt in roadways does not detract from performance. Using recycled asphalt becomes advantageous specifically in cases where existing asphalt removal is necessary on projects because past overlays are affecting curb heights and creating drainage problems. In these cases, using recycled asphalt creates two main benefits. First, by reusing existing asphalt, the municipality takes another step toward solving its environmental problems by reducing solid waste in landfills. Second, using recycled asphalt reduces project costs as noted below. There are two types of recycled asphalt appropriate for roadways such as those in Ridgewood.

Hot In-Place Recycled Asphalt: Hot in-place asphalt recycling has been used throughout the United States with satisfactory results. Hot in-place recycling consists of milling off existing asphalt to a specified depth, mixing the recycled asphalt with virgin asphalt at high temperatures, laying the hot asphalt, and rolling it to a highly compacted density. Mixing is performed either at the asphalt plant or in a mixing vehicle that follows directly behind the milling equipment and directly in front of the paver. Reports from other municipalities and states indicate using hot in-place recycled asphalt can save 10 to 30 percent in road costs.

Cold In-Place Recycled Asphalt: Cold in-place asphalt recycling usually consists of milling off existing asphalt to a specified depth, mixing that asphalt with virgin materials and/or rejuvenating liquids, then laying it back in place. The mixing is done either at the plant or in a paving machine directly behind the milling machine. The advantage of this procedure is the ability to use existing material to correct road profile problems. During in-place recycling, a new roadway crown can be established. The cold mix is commonly used for base or intermediate layers on low-volume roads. The main benefit here is the cold mix's ability to prevent old cracks from reflecting through the overlay. The majority of users of recycled cold mix top it with virgin asphalt or chip and seal. The latest reports indicate that using cold in-place recycled asphalt can save local governments from 6 to 67 percent in road costs.

Recommendation

The existing roadways in Ridgewood have undergone years of overlays in road maintenance. The future improvement projects will have to reestablish road grades for proper drainage and for standard curb heights. Increasing evidence shows that cost savings can be achieved by using both hot and cold in-place recycled asphalt. In addition, these processes provide an environmentally sound alternative to wasting asphalt in landfills. Manning-Hornsby recommends that the City of Ridgewood adopt this recycled asphalt process in future roadway improvement projects.

I would be happy to discuss this recommendation with you in more detail, and I can provide copies of the reports from other areas about the use of recycled asphalt. I will call your office next week to set up a time convenient to you. In the meantime, please do not hesitate to contact me with any questions.

Sincerely,

Joseph F. Lennetti

Joseph F. Lennetti
Senior Engineer
MANNING-HORNSBY ENGINEERING CONSULTANTS

Chapter 4 Exercises

1. The following bulletin at a large construction company provides guidelines for workers involved in cleanup operations after flooding has occurred. Revise and reorganize the bulletin so it is easier to read and understand.

All workers should be vigilant about protecting themselves when engaged in cleanup operations at a site after a flood. Flooded areas can be dangerous because of downed wires or infectious disease. It is usually difficult to maintain good hygiene during cleanup operations, but workers should be conscious of the need to wash their hands with soap and running water before meals and at the end of the shift. Supervisors should be sure to check on whether flooded areas and surrounding areas have been cleared by local authorities for safety. The signs of water-borne bacteria are nausea, vomiting, abdominal cramps, and fever. Tetanus can be acquired from contaminated water or soil entering the body through cuts and puncture wounds. Supervisors will provide bottled water if there is any danger in the local water supply. Cleanup workers also need to wear protective clothing—goggles, rubber gloves, boots, and other outerwear—as necessary. Pools of standing water are homes for mosquitoes. Some floodwaters have collected chemical spills that cause dizziness, skin rashes, nausea, weakness, and fatigue. Workers should have up-to-date tetanus shots. Supervisors should call the local fire or police department for removal of chemical hazards, such as propane tanks.

2. The following section from a company's administrative manual describes the job functions of the company's Security Director. Reorganize and revise this section into readable paragraphs.

The Company Security Director must report accidents or fatalities. He must investigate the physical circumstances surrounding the event. A monthly report of injuries and illnesses is required by the Occupational Safety and Health Act regulations. The Director is responsible for developing, coordinating, and administering the local safety program. He shall cooperate with, give counsel, and render assistance to all levels of management regarding safety. The Director is the liaison with local police, fire, and municipal authorities. In case of national or regional emergency, the Director shall coordinate company functions under the direction of the FBI or other authorities as necessary. He also provides the company management with the necessary information and assistance so that the management responsible for the safety of the employees and equipment can fulfill those responsibilities. Security service to the management and employees must be the first priority. Reports must be maintained accurately. The Company Security Director must also inspect the company physical facilities to insure compliance with established standards and rules of operation and to identify any potentially unsafe conditions or practices that require additional security. He maintains the security system checks and procedures for entering and exiting the buildings. The Director must be knowledgeable about security legislation and regulations and maintain an information exchange with the medical and Worker's Compensation representatives. Those practices are the Director's paramount responsibility.

COLLABORATIVE EXERCISES

3. The Camlann Village Chamber of Commerce wants a handout to inform visitors about the quaint marketplace. In groups, organize the following facts into an information sheet that would be an appropriate handout for the Camlann Hill Marketplace. *Or,* if your instructor prefers, draft an information sheet individually and then meet in groups to discuss the drafts and produce a final version.

> The Camlann Hill Marketplace was developed after World War II when enterprising veterans returning home decided to build up the Camlann Village economy by creating a market that would attract customers from the surrounding counties. One of the founders was the prominent playwright Brian Wright. Wright was born in Camlann Village and became a Pulitzer Prize winner. The marketplace is open Wednesday-Thursday-Friday-Saturday, 8 A.M. to 2 P.M. The food stalls are open from May 15 to October 31. Other shops are open all year. Brian Wright was honored by the Village after his death in 2003 with a statue in the community rose garden. Local farmers bring vegetables, fruits, and homemade goods, such as breads, jams, and baked goods for sale at the marketplace. The marketplace is closed the first Saturday in October for the annual blessing of the pets. The marketplace opens daily for business when a merchant, selected annually by the other merchants, blows the old whistle taken from the nineteenth-century mill. A nearby park has picnic tables where customers can eat their purchases from the food stalls. Shops in the marketplace sell antiques, copper cookery, fish and meat, gourmet spices and groceries, and china.

INTERNET EXERCISES

4. As the Assistant Manager of the Southwest Motorcycle Center in Phoenix, you have to prepare an information Web page about types of motorcycles (e.g., dirt bikes, pocket bikes). Search the Web for information about types of motorcycles. Your Web page should provide information about the purpose, sizes, and styles of the motorcycles. Include links to some sources for racing information or the history of the types of motorcycles.

5. Oceans are rising. Mountain ice caps are melting. Search the Web for information about global warming. Select a particular area to study (e.g., North America, the Arctic, the Alps). Now, assume you are an intern at the Global Weather Research Center, working for Dr. Neal Wong. Dr. Wong is preparing bulletins to be distributed to high-school science teachers. Write a report to Dr. Wong on the current global warming situation in your selected area. *Or,* if your instructor prefers, write the information bulletin for your area. Report style is covered in Chapter 12.

6. Assume you need to write a brief article for a hospital newsletter. Select one of the following surgeries: hip replacement, knee replacement, rotor cuff repair, or carpal tunnel release. Search the Web for the latest information about number of cases, treatments, and ongoing research. The newsletter readers are current hospital employees, donors, and residents of the community. Write a short article presenting the latest information. You are limited to 1,000 words or as your instructor directs.

7. Using the information in Exercises 1 and 2, create Web pages that offer readers the same information.

CHAPTER 5

Revision

CREATING A FINAL DRAFT

As you plan and draft a document, you probably will change your mind many times about content, organization, and style. In addition to these changes, however, you should consider revision as a separate stage in producing your final document. During drafting, you should concentrate primarily on developing your document from your initial outline. During revision, concentrate on making changes in content, organization, and style that will best serve your readers and their purpose. These final changes should result in a polished document that your readers can use efficiently.

Thinking about revision immediately after finishing a draft is difficult. Your own writing looks "perfect" to you, and every sentence seems to be the only way to state the information. Experienced writers allow some time between drafting and revising whenever possible. The longer the document, the more important it is to let your writing rest before final revisions so that you can look at the text and graphic aids with a fresh attitude and read as if you were the intended reader and had never seen the material before. Most writers divide revision into two separate stages—global, or overall document revision, and fine-tuning, or style and surface-accuracy revision.

Global Revision

The process of *global revision* involves evaluating your document for effective content and organization. At this level of revision, you may add or delete information; reorganize paragraphs, sections, or the total document; and redraft sections.

Fine-Tuning Revision

The process of *fine-tuning revision* involves the changes in sentence structure and language choice that writers make to ensure clarity and appropriate tone. This level also involves checking for surface accuracy, such as correct grammar, punctuation, spelling, and mechanics of technical style. Many organizations have internal style guides that include special conventions writers must check during fine-tuning revision.

Revising On-Screen

With computers, you can do both global and fine-tuning revisions quickly once you decide what you want. Depending on your software, you can revise on-screen using these features:

1. *Delete and insert.* You can omit or add words, sentences, and paragraphs anywhere in the text.

2. *Search for specific words.* You can search through the full text for specific words or phrases that you want to change. If you want to change the word *torpid* to *inactive* throughout the document, the computer will locate every place *torpid* appears so that you can delete it and insert *inactive.* Be cautious when using the "Replace All" function. If your reference list contains the word you are replacing in the text, your reference list will be incorrect at the end of editing.

3. *Moving and rearranging text.* You can take your text apart as if you were cutting up typed pages and arranging the pieces in a new order. You can move whole sections of text from one place to another and reorder paragraphs within sections. You also may decide that revised sentences should be reorganized within a paragraph.

4. *Reformat.* You can easily change margins, headlines, indentations, spacing, and other format elements.

5. *Check spelling and word choice.* Most software programs have spelling checkers that scan a text and flag misspelled words that are in the checker's dictionary. Spelling checkers, however, cannot find usage errors, such as *affect* for *effect,* or correctly spelled words that you have used incorrectly, such as *personal* for *personnel.* A thesaurus program will suggest synonyms for specific words. You can select a synonym and substitute it wherever the original word appears in the text. Add your field's specialized vocabulary to the spelling dictionary by clicking on "add" whenever the specialized terms come up on the spell check.

6. *Check grammar and punctuation.* Most grammar and punctuation checkers are not reliable. They miss problems, such as dangling modifiers, and they flag "errors" that are not really errors. The checkers cannot identify punctuation errors and almost never identify apostrophe requirements correctly. Do not automatically accept a suggested revision from a grammar and punctuation checker. As a careful writer, reread your text slowly, consult a handbook, and make appropriate decisions about correctness.

7. *Customize the screen.* Look for options on your software that enhance your revision process. Perhaps you can adjust the contrast on the screen for easier reading. You might change to a larger font if you are tired. Use "View" selections to review how an entire page looks.

Revision on a computer is so convenient that some people cannot stop making changes in drafts. Do not change words or rearrange sentences simply for the sake of change. Revise until your document fulfills your readers' needs, and then stop.

MAKING GLOBAL REVISIONS

For global revisions, rethink your reader and purpose, and then review the content and organizational decisions you made in the planning and drafting stages. Consider changes that will help your readers use the information in the document efficiently.

Content

Evaluate the content of your document and consider whether you have enough information or too little. Review details, definitions, and emphasis:

- *Details*—Have you included enough details for your readers to understand the general principles, theories, situations, or actions? Have you included details that do not fit your readers' purpose and that they do not need? Do the details serve both primary and secondary readers? Are all the facts, such as dates, amounts, names, and places, correct? Are your graphic aids appropriate for your readers? Do you need more? Have you included examples to help your readers visualize the situation or action?

- *Definitions*—Have you included definitions of terms that your readers may not know? If your primary readers are experts and your secondary readers are not, have you included a glossary for secondary readers?

- *Emphasis*—Do the major sections stress information suitable for the readers' main purpose? Do the conclusions logically result from the information provided? Is the urgency of the situation clear to readers? Are deadlines explained? Do headings highlight major topics of interest to readers? Can readers easily find specific topics?

After evaluating how well the document serves your readers and their need for information, you may discover that you need to gather more information or that you must omit information you have included in the document. Reaching the revision stage and finding that you need to gather more information can be frustrating, but remember, readers can only work with the information they are given. The writer's job is to provide everything readers need to use the document efficiently for a specific purpose.

Organization

Evaluate the overall organization of your document as well as the organization of individual sections and paragraphs. Review these elements:

- *Overall organization*—Is the information grouped into major topics that are helpful and relevant to your readers? Are the major topics presented in a logical order that will help readers understand the subject? Are the major sections divided into subsections that will help readers understand the information and that highlight important subtopics? Do some topics need to be combined or further divided to highlight information or to make the information easier to understand? Do readers of this document expect or prefer a specific organization, and is that organization in place?

- *Introductions and conclusions*—Consider the introductions and conclusions both for the total document and for major sections. Does the introduction to the full document orient readers to the purpose and major topics covered? Do the section introductions establish the topics covered in each section? Does the conclusion to the full document summarize major facts, results, recommendations, future actions, or the overall situation? Do the section conclusions summarize the major points and show why these are important to readers and purpose?

- *Paragraphs*—Do paragraphs focus on one major point? Are paragraphs sufficiently developed to support that major point with details, examples, and explanations? Is each paragraph organized according to a pattern that will help readers understand the information, such as chronological or spatial?

- *Headings*—Are the individual topics marked with major headings and subheadings? Do the headings identify the key topics? Are there several pages in a row without headings or is there a heading for every sentence, and therefore, do headings need to be added or omitted?

- *Format*—Are there enough highlighting devices, such as headings, lists, boxes, white space, and boldface type, to lead readers through the information and help them see relationships among topics? Can readers quickly scan a page for key information?

After evaluating the organization of the document and of individual sections, make necessary changes. You may find that you need to write more introductory or concluding material or reorder sections and paragraphs. Remember that content alone is not enough to help readers if they have to hunt through a document for relevant facts. Effective organization is a key element in helping readers use a document quickly and efficiently.

MAKING FINE-TUNING REVISIONS

When you are satisfied that your global revisions have produced a complete and well-organized draft, go on to the fine-tuning revision stage, checking for sentence structure, word choice, grammar, punctuation, spelling, and mechanics of technical style. This level of revision can be difficult because, by this time, you probably have the document or sections of it memorized. As a result, you may read the words that should be on the page, rather than the words that are on the page. For fine-tuning revisions you need to slow your reading pace so that you do not overlook problems in the text. You also should stop thinking about content and organization and think instead about individual sentences and words. Experienced writers sometimes use these techniques to help them find trouble spots:

- *Read aloud*. Reading a text aloud will help you notice awkward or overly long sentences, insufficient transitions between sentences or paragraphs, and inappropriate language. Reading aloud slows the reading pace and helps you hear problems in the text that your eye could easily skip over.

- *Focus on one point at a time*. You cannot effectively check for sentence structure, word choice, correctness, and mechanical style all at the same time, particularly in a long document. Experienced writers usually revise in steps, checking for one or two items at a time. If you check for every problem at once, you are likely to overlook items.

- *Use a ruler*. By placing a ruler under each line as you read it and moving the ruler from line to line, you will focus on the words rather than on the content of the document. Using a ruler also slows your reading pace, *so* you are more likely to notice problems in correctness and technical style.

- *Read backward*. Some writers read a text from the bottom of the page up to concentrate only on the typed lines. Reading from the bottom up will help you focus on words and find typographic errors, but it will not help you find problems in sentence structure or grammar because sentence meaning will not be clear as you move backward through the text.

You may find a combination of these methods useful. You may use a ruler, for example, and focus on one point at a time as you look for sentence structure, word choice, and grammar problems. As a final check when you proofread, you may read backward to look for typographic errors.

The following items are those most writers check for during fine-tuning revision. For a discussion of grammar, punctuation, and mechanical style, see Appendix A.

Overloaded Sentences

An overloaded sentence includes so much information that readers cannot easily understand it. This sentence is from an announcement to all residents in a city district:

> You are hereby notified, in accordance with Section 101–27 of the Corinth City Code, that a public hearing will be held by a committee of the Common Council on the date and time, and at the place listed below, to determine whether or not to designate a residential permit parking district in the area bounded by E. Edgewood Avenue on the north, E. Belleview Place on the south, the Corinth River on the west, and Lake McCormick on the east, except the area south of E. Riverside Place and west of the north/south alley between N. Oakland Avenue and N. Bartlett Avenue and the properties fronting N. Downer Avenue south of E. Park Place.

No reader could understand and remember all the information packed into this sentence. Here is a revision:

> As provided in City Code Section 101–27, a committee of the Common Council of Corinth will meet at the time and place shown below. The purpose of the meeting is to designate a residential permit parking district for one area between the Corinth River on the west and Lake McCormick on the east. The area is also bounded by E. Edgewood Avenue on the north and E. Belleview Place on the south. Not included is the area south of E. Riverside Place and west of the north/south alley between N. Oakland Avenue and N. Bartlett Avenue and the properties fronting N. Downer Avenue south of E. Park Place.

To revise the original overloaded sentence, the writer separated the pieces of information and emphasized the individual points in shorter sentences. The first sentence in the revision announces the meeting. The next two sentences identify the meeting's purpose and the streets included in the parking regulations. A separate sentence then identifies streets not included. Readers will find the shorter sentences easier to understand and, therefore, the information easier to remember. Because so many streets are involved, readers also will need a graphic aid—a map with shading that shows the area in question. Avoid writing sentences that contain so many pieces of information that readers must mentally separate and sort the information as they read.

Even short sentences can contain too much information for some readers. Here is a much shorter sentence than the preceding one, yet if the reader were a nonexpert interested in learning about cameras, the sentence holds more technical detail than he or she could remember easily:

> The WY–30X camera is housed in a compact but rugged diecast aluminum body, including a 14X zoom lens and 1.5-in. viewfinder, and has a signal-to-noise ratio of 56 dB, resolution of more than 620 lines on all channels, and registration of 0.6% in all zones.

In drafting sentences, consider carefully how much information your readers are likely to understand and remember at one time.

Clipped Sentences

Sometimes when writers revise overloaded sentences, the resulting sentences become so short they sound like the clipped phrases used in telegrams. Do not cut these items from your sentences just to make them shorter: (1) articles (*a, an, the*), (2) pronouns (*I, you, that, which*), (3) prepositions (*of, at, on, by*), and (4) linking verbs (*is, seems, has*). Here are some clipped sentences and their revisions:

Clipped: Attached material for insertion in Administration Manual.

Revised: The attached material is for insertion in the Administration Manual.

Clipped: Questioned Mr. Hill about compensated injuries full-time employees.

Revised: I questioned Mr. Hill about the compensated injuries of the full-time employees.

Clipped: Investigator disturbed individual merits of case.

Revised: The investigator seems disturbed by the individual merits of the case.

Do not eliminate important words to shorten sentences. The reader will try to supply the missing words and may not do it correctly, resulting in ambiguity and confusion.

Lengthy Sentences

Longer sentences demand that readers process more information, whereas shorter sentences give readers time to pause and absorb facts. A series of short sentences, however, can be monotonous, and too many long sentences in a row can overwhelm readers, especially those with nonexpert knowledge of the subject. Usually, short sentences emphasize a particular fact, and long sentences show relationships among several facts. With varying sentence lengths you can avoid the tiresome reading pace that results when writers repeat sentence style over and over. This paragraph from a letter to bank customers uses both long and short sentences effectively:

> Your accounts will be automatically transferred to the Smith Road office at the close of business on December 16, 2005. You do not need to open new accounts or order new checks. If for some reason another Federal location is more convenient, please stop by 2900 W. Manchester Road and pick up a special customer courtesy card to introduce you to the office of your choice. The Smith Road office is a full-service banking facility. Ample lobby teller stations, auto teller

service, 24-hour automated banking, night depository, as well as a complete line of consumer and commercial loans, are all available from your new Federal office. You will continue to enjoy the same fine banking service at the Smith Road office, and we look forward to serving you. If you have questions, please call me at 555–8876.

In this paragraph the writer uses short sentences to emphasize the fact that customers do not have to take action, as well as that the new bank office is a full-service branch. Longer sentences explain the services and how to use them. The offer of help also is in a short sentence for emphasis.

Passive Voice

Active voice and passive voice indicate whether the subject of a sentence performs the main action or receives it. Here are two sentences illustrating the differences:

Active: The bridge operators noticed severe rusting at the stone abutments.

Passive: Severe rusting at the stone abutments was noticed by the bridge operators.

The active-voice sentence has a subject (*bridge operators*) that performs the main action (*noticed*). The passive-voice sentence has a subject (*rusting*) that receives the main action (*was noticed*). Passive voice is also characterized by a form of the verb *to be* and a prepositional phrase beginning with *by* that identifies the performer of the main action—usually quite late in the sentence. Readers generally prefer active voice because it creates a faster pace and is more direct. Passive voice also may create several problems for readers:

1. Passive voice requires more words than active voice because it includes the extra verb (e.g., *is, are, was, were, will be*) and the prepositional phrase that introduces the performer of the action. The preceding passive-voice sentence has two more words than the active-voice sentence. This may not seem excessive, but if every sentence in a ten-page report has two extra words, the document will be unnecessarily long.

2. Writers often omit the *by* phrase in passive voice, decreasing the number of words in the sentence but also possibly concealing valuable information from readers. Notice how these sentences offer incomplete information without the *by* phrase:

   Incomplete: At least $10,000 was invested in custom software.

   Complete: At least $10,000 was invested in custom software by the vice president for sales.

   Incomplete: The effect of welded defects must be addressed.

   Complete: The effect of welded defects must be addressed by the consulting engineer.

3. Writers often create dangling modifiers when using passive voice. In the active-voice sentence below, the subject of the sentence (*financial analyst*) also is the subject of the introductory phrase.

    Active: Checking the overseas reports, the company financial analyst estimated tanker capacity at 2% over the previous year.

    Passive: Checking the overseas reports, tanker capacity was estimated at 2% over the previous year.

    In the second sentence, the opening phrase (*Checking . . .*) cannot modify *tanker capacity* and is a dangling modifier.

4. Passive voice is confusing in instructions because readers cannot tell *who* is to perform the action. Always write instructions and direct orders in active voice.

    Unclear: The cement should be applied to both surfaces.

    Revised: Apply the cement to both surfaces.

    Unclear: The telephone should be answered within three rings.

    Revised: Please answer the telephone within three rings.

In technical writing, passive voice is sometimes necessary. Use passive voice when

1. You do not know who or what performed the action:

    The fire was reported at 6:05 A.M.

2. Your readers are not interested in who or what performed the action:

    Ethan McClosky was elected district attorney.

3. You are describing a process performed by one person and naming that person in every sentence would be monotonous. Identify the performer in the opening, and then use passive voice:

    The carpenter began assembling the base by gluing each long apron to two legs. The alignment strips were then cut for the tray bottoms to finished dimensions. The tabletop was turned upside down and . . .

4. For some reason, perhaps courtesy, you do not want to identify the person responsible for an action:

    The copy machine jammed because a large sheaf of papers was inserted upside down.

Jargon

Jargon refers to the technical language and abbreviated terms used by people in one particular field, company, department, or unit. People in every workplace and every occupation use some jargon. A bookstore manager may tell the clerks to put out the "shelf talkers" (printed description cards under a book display). A public relations executive may ask for the "glossies" (glossy photographs). A psychologist may remark to another psychologist that the average woman has a "24F on the PAQ" (a femininity score of 24 on the Personal Attributes Questionnaire). In some cases, jargon may be so narrow that employees in one department cannot understand the jargon used in another department. Use jargon in professional communications only when you are certain your readers understand it.

Sexist Language

Sexist language refers to words or phrases that indicate a bias against women in terms of importance or competence. Most people would never use biased language to refer to ethnic, religious, or racial groups, but sexist language often goes unnoticed because many common terms and phrases that are sexist in nature have been part of our casual language for decades. People now realize that sexist language influences our expectations about what women can accomplish and also relegates women to an inferior status. Using such language is no longer acceptable. In revising documents, check for these slips into sexist language:

1. Demeaning or condescending terms. Avoid using casual or slang terms for women.

   Sexist: The girls in the Records Department will prepare the reports.
   Revised: The clerks in the Records Department will prepare the reports.

   Sexist: The annual division picnic will be June 12. Bring the little woman.
   Revised: The annual division picnic will be June 12. Please bring your spouse.

2. Descriptions for women based on standards different from those for men. Avoid referring to women in one way and men in another.

   Sexist: The consultants were Dr. Dennis Tonelli, Mr. Robert Lavery, and Debbie Roberts.
   Revised: The consultants were Dr. Dennis Tonelli, Mr. Robert Lavery, and Ms. Debra Roberts.

   Sexist: There were three doctors and two lady lawyers at the meeting.
   Revised: There were three doctors and two lawyers at the meeting.

   Sexist: The new mechanical engineers are Douglas Ranson, a graduate of MIT and a Rhodes scholar, and Marcia Kane, an attractive redhead with a B.S. from the University of Michigan and an M.S. from the University of Wisconsin.
   Revised: The new mechanical engineers are Douglas Ranson, a graduate of MIT and a Rhodes scholar, and Marcia Kane, with a B.S. from the University of Michigan and an M.S. from the University of Wisconsin.

3.   Occupational stereotypes. Do not imply that all employees in a particular job are the same sex.

Sexist: An experienced pilot is needed. He must . . .

Revised: An experienced pilot is needed. This person must . . .

Sexist: The new tax laws affect all businessmen.

Revised: The new tax laws affect all businesspeople.

Although some gender-specific occupational terms remain common, such as *actor/actress* and *host/hostess*, most such terms have been changed to neutral job titles.

Sexist: The policeman should fill out a damage report.

Revised: The police officer should fill out a damage report.

Sexist: The stewardess reported the safety problem.

Revised: The flight attendant reported the safety problem.

4.   Generic *he* to refer to all people. Using the pronoun *he* to refer to unnamed people or to stand for a group of people is correct grammatically in English, but it might be offensive to some people. Avoid the generic *he* whenever possible, particularly in job descriptions that could apply to either sex. Change a singular pronoun to a plural:

Sexist: Each technical writer has his own office at Tower Industries.

Revised: All technical writers have their own offices at Tower Industries.

Eliminate the pronoun completely:

Sexist: The average real estate developer plans to start his construction projects in April.

Revised: The average real estate developer plans to start construction projects in April.

Use *he* or *she* and *his* or *hers* very sparingly:

Sexist: The X-ray technician must log his hours daily.

Revised: The X-ray technician must log his or her hours daily.

Better: All X-ray technicians must log their hours daily.

Concrete versus Abstract Words

Concrete words refer to specific items, such as objects, statistics, locations, dimensions, and actions that can be observed by some means. Here is a sentence using concrete words:

The 48M printer has a maximum continuous speed of 4000 lines per minute, with burst speeds up to 5500 lines per minute.

Abstract words refer generally to ideas, conditions, and qualities. Here is a sentence from a quarterly report to stockholders that uses abstract words:

> The sales position of Collins, Inc., was favorable as a result of our attention to products and our response to customers.

This sentence contains little information for readers who want to know about the company activities. Here is a revision using more concrete language:

> The 6% rise in domestic sales in the third quarter of 2005 resulted from our addition of a safety catch on the Collins Washer in response to over 10,000 requests by our customers.

This revision in concrete language clarifies exactly what the company did and what the result was. Using concrete or abstract words, as always, depends on what you believe your readers need. However, in most professional situations, readers want and need the precise information provided by concrete words.

The paragraphs that follow are from two reports submitted by management trainees in a large bank after a visit to the retail division of the main branch. The trainees wrote reports to their supervisors, describing their responses to the procedures they observed. The paragraph from the first report uses primarily abstract, nonspecific language:

> The first department I visited was very interesting. The way in which the men buy and investigate the paper was very interesting and informative. This area, along with the other areas, helped me where I am right now—doing loans. Again, the ways in which the men deal with the dealers was quite informative. They were friendly, but stern, and did not let the dealers control the transaction.

The paragraph from the second report uses more concrete language:

> The manual system of entering sales draft data on the charge system seems inefficient. It is clear that seasonal surges in charge activity (for example, Christmas) or an employee illness would create a serious bottleneck in processing. My talk with Eva Lockridge confirmed my impression that this operation is outdated and needs to be redesigned.

The first writer did not identify any specific item of interest or explain exactly what was informative about the visit. The second writer identified both the system that seemed outdated and the person with whom she talked. The second writer is providing her reader with concrete information rather than generalities.

STRATEGIES—Standard American English

To ensure clarity, especially for international readers, use standard American English in all documents and on Web sites.

Avoid **regionalisms** (expressions used only in certain parts of the country).

No: The courthouse had a **bubbler** on every floor.

Yes: The courthouse had a **drinking fountain** on every floor.

Do not use **dialect** constructions that are specific to certain regions or groups and that violate standard grammatical form.

No: The boiler **needs repaired.**

Yes: The boiler **needs to be repaired.**

No: The engine **be running** too fast.

Yes: The engine **is running** too fast.

Do not use **colloquialisms** (informal terms or slang).

No: The merger collapse was a **tough break.**

Yes: The merger collapse was **disappointing.**

Gobbledygook

Gobbledygook is writing that is indirect, vague, pompous, or longer and more difficult to read than necessary. Writers of gobbledygook do not care if their readers can use their documents because they are interested only in demonstrating how many large words they can cram into convoluted sentences. Such writers are concerned primarily with impressing or intimidating readers rather than with helping them use information efficiently. Gobbledygook is characterized by these elements:

1.   Using jargon the readers cannot understand.

2.   Making words longer than necessary by adding suffixes or prefixes to short, well-known words (e.g., *finalization, marginalization*). Another way to produce gobbledygook is to use a longer synonym for a short, well-known word. A good example is the word *utilize* and its variations to substitute for *use. Utilize* is an unnecessary word because *use* can be substituted in any sentence and mean the same thing. Notice the revisions that follow these sentences:

     Gobbledygook: The project will utilize the Acme trucks.

     Revision: The project will use the Acme trucks.

Gobbledygook:	By utilizing the gravel already on the site, we can save up to $5000.
Revision:	By using the gravel already on the site, we can save up to $5000.
Gobbledygook:	Through the utilization of this high-speed press, we can meet the deadline.
Revision:	Through the use of this high-speed press, we can meet the deadline.

The English language has many short, precise words. Use them.

3. Writing more elaborate words than are necessary for document purpose and readers. The writer who says "deciduous perennial, genus *Ulmus*" instead of "elm tree" probably is engaged in gobbledygook unless his or her readers are biologists.

This paragraph from a memo sent by a unit supervisor to a division manager is supposed to explain why the unit needs more architects:

> In the absence of available adequate applied man-hours, and/or elapsed calendar time, the quality of the architectural result (as delineated by the drawings) has been and will continue to be put ahead of the graphic quality or detailed completeness of the documents. In order to be efficient, work should be consecutive not simultaneous, and a schedule must show design and design development as scheduled separately from working drawings on each project, so that a tighter but realistic schedule can be maintained, optimizing the productive efforts of each person.

You will not be surprised to learn that the division manager routinely threw away any memos he received from this employee without reading them.

Wordiness

Wordiness refers to writing that includes superfluous words that add no information to a document. Wordiness stems from several causes:

1. *Doubling terms.* Using two or more similar terms to make one point creates wordiness. These doubled phrases can be reduced to one of the words:

final ultimate result (result)

the month of July (July)

finished and completed (finished or completed)

unique innovation (innovation)

2. *Long phrases for short words.* Wordiness results when writers use long phrases instead of one simple word:

   at this point in time (now)

   in the near future (soon)

   due to the fact that (because)

   in the event that (if)

3. *Unneeded repetition.* Repeating words or phrases increases wordiness and disturbs sentence clarity:

   The Marshman employees will follow the scheduled established hours established by their supervisors, who will establish the times required to support production.

   This sentence is difficult to read because *established* appears three times, once as an adjective and twice as a verb. Here is a revision:

   The Marshman employees will follow the schedule established by their supervisors, who will set times required to support production.

4. *Empty sentence openings.* These sentences are correct grammatically but their structure contributes to wordiness:

   It was determined that the home office sent income statements to all pension clients.

   It is necessary that the loader valves (No. 3C126) be scrapped because of damage.

   It might be possible to open two runways within an hour.

   There will be times when the videotapes are not available.

   There are three new checking account options available from Southwest National Bank.

   These sentences have empty openings because the first words (*There will be/There are/It is/It was/It might be*) offer the reader no information, and the reader must reach the last half of the sentence to find the subject. Here are revisions that eliminate the empty openings:

   The home office sent income statements to all pension clients.

   The loader valves (No. 3C126) must be scrapped because of damage.

   Two runways might open within an hour.

   Sometimes the videotapes will not be available.

   Southwest National Bank is offering three new checking account options.

Remember that wordiness does not refer to the length of a document, but to excessive words in the text. A 40-page manual containing none of the constructions discussed here is not wordy, but a half-page memo full of them is.

REVISING FOR THE WEB

Use the principles of clear and effective writing as key revision tools for both print and Web revising. The Web does present new challenges for revising text. Words tend to dominate the message on a printed page. However, a Web page is more visual, and the visual elements can detract from the clarity if they are overused.[1] In preparing text for a Web page, remember that Web pages are likely to have more casual readers and more international readers than a printed document.[2] As you revise text for Web pages, use the principles in this chapter and consider the following guidelines:

- Because you will use several Web pages for one overall document, think of each page as a paragraph focused on one main idea. Repeat key words on each page to ensure that readers feel they are reading a consistent document.

- Traditional phrases that move readers through a document (e.g., "as you have just learned") do not work on Web pages because readers may not be reading the pages in the order you would prefer.

- In print, you might use variety in word choice, such as referring to "international readers" one time and "worldwide readers" another time. On Web pages, avoid such variety and use the same key term on all pages.

- Avoid any humor and culturally specific content (e.g., a reference to the Dallas Cowboys or George Washington cutting down the cherry tree) unless these references are the subjects of the Web page.

- Label all icons to be sure readers understand them. Too often, Web designers assume that icons are easy to interpret because they are not bound to a specific language. What an icon signifies is determined by the culture, and the icon rarely indicates the action connected to the object. If an icon shows a knife, fork, and spoon, does that mean a restaurant or a store selling silverware? Identify the icon in words to make sure the readers understand it.[3]

CHAPTER SUMMARY

This chapter discusses two stages of revision—global revision and fine-tuning revision. Remember:

- Revision consists of two distinct stages: (1) global revision, which focuses on content and organization, and (2) fine-tuning revision, which focuses on sentence structure, word choice, correctness, and the mechanics of technical style.

- For global revision, writers check content to see if there are enough details, definitions, and emphasis on the major points.

- For global revision, writers also check organization of the full document and of individual sections for (1) overall organization, (2) introductions and conclusions, (3) paragraphs, (4) headings, and (5) format.

- For fine-tuning revision, writers check sentence structure for overloaded sentences, clipped sentences, overly long sentences, and overuse of passive voice.

- For fine-tuning revision, writers also check word choice for unnecessary jargon, sexist language, overly abstract words, gobbledygook, and general wordiness.

SUPPLEMENTAL READINGS IN PART 2

Garhan, A. "ePublishing," *Writer's Digest*, p. 495.

McAdams, M. "It's All in the Links: Readying Publications for the Web," *The Editorial Eye,* p. 512.

ENDNOTES

1. Steven L. Anderson, Charles P. Campbell, Nancy Hindle, Jonathan Price, and Randall Scasny, "Editing a Web Site: Extending the Levels of Edit," *IEEE Transactions on Professional Communication* 41.1 (March 1998): 47–57.

2. Jan H. Spyridakis, "Guidelines for Authoring Comprehensive Web Pages and Evaluating Their Success," *Technical Communication* 47.3 (August 2000): 359–382.

3. Thomas R. Williams, "Guidelines for Designing and Evaluating the Display of Information on the Web," *Technical Communication* 47.3 (August 2000): 383–396.

MODEL 5-1   Commentary

This model shows three drafts of a company bulletin outlining procedures for security guards who discover a fire while on patrol. The bulletin will be included in the company's "Security Procedures Manual," which contains general procedures for potential security problems, such as fire, theft, physical fights, and trespassing. The bulletins are not instructions to be used on the job. Rather, the guards are expected to read the bulletins, become familiar with the procedures, and follow them if a security situation arises.

Discussion

1. In the first draft, identify the major topics the writer covers, and discuss how well the general organization fits the purpose of the readers.

2. Compare the first and second drafts, and discuss the changes the writer made in organization, sentence structure, and detail. How will these changes help the readers understand the information?

3. Compare the third draft with the first two, and discuss how the specific changes will improve the usefulness of the bulletin.

4. In groups, draft a fourth version of this bulletin, making further organization changes to increase clarity. *Or,* if your instructor prefers, rewrite one section of the third draft, and make further fine-tuning revisions according to the guidelines in this chapter.

BULLETIN 46-RC   FIRE EMERGENCIES—PLANT SECURITY PATROLS

Following are procedures for fire emergencies discovered during routine security patrols. All security personnel should be familiar with these procedures.

General Inspection

Perimeter fence lines, parking lots, and yard areas should be observed by security personnel at least twice per shift. Special attention should be given to outside areas during the dark hours and nonoperating periods. It is preferable that the inspection of the yard area and parking lots be made in a security patrol car equipped with a spot light and two-way radio communications. Special attention should be given to the condition of fencing and gates and to yard lighting to assure that all necessary lights are turned on during the dark hours and that the system is fully operative. Where yard lights are noted as burned out, the guard should report these problems immediately to the Plant Engineer for corrective action and maintain follow-up until the yard lighting is in full service. Occasional roof spot checks should be made by security patrols to observe for improper use of roof areas and fire hazards, particularly around ventilating equipment.

Discovery

When a guard discovers a fire during an in-plant security patrol, he should immediately turn in an alarm. This should be done before any attempt is made to fight the fire because all too frequently guards think they can put out the fire with the equipment at hand, and large losses have resulted.

Whenever possible, the guard should turn in the alarm by using the alarm box nearest the scene of the fire. If it is necessary to report the fire on the plant telephone, he should identify the location accurately so that he may give this information to the plant fire department or to the guard on duty at the Security Office, who will summon the employee fire brigade and possibly the city fire department.

Once he has turned in the alarm, the guard should decide whether he can effectively use the available fire protection equipment to fight the fire. If the fire is beyond his control, he should proceed to the main aisle of approach where he can make contact with the fire brigade or firemen and direct them to the scene of the fire. If automatic sprinklers have been engaged, the men in charge of the fire fighting crew will make the decision to turn off the system.

MODEL 5-1   First Draft

BULLETIN 46-RC   FIRE EMERGENCIES—PLANT SECURITY PATROLS

All security guards should be familiar with the following procedures for fire emergencies that occur during routine security patrols.

Inspection Areas

Inspection of the yard area and parking lots should be made in a security patrol car equipped with a spot light and two-way radio communications. At least twice per shift, the guard should inspect (1) fence lines, (2) parking lots, and (3) yard areas. These outside areas require special attention during the dark hours and nonoperating periods. The guard should inspect carefully the condition of the fencing and gate closures. The yard lighting should be checked to ensure that all necessary lights are turned on during the dark hours and that the system is fully operative. If the guard notes that yard lights are burned out, he should report these problems to the Plant Engineer for corrective action and maintain follow-up until the yard lighting is in full service.

Roof Checks

Occasional roof spot checks should be made by the guard to observe improper use of roof areas and any fire hazards, particularly around ventilating equipment.

Fires

When a guard discovers a fire during an in-plant security patrol, he should immediately turn in an alarm before he makes any attempt to fight the fire. In the past, attempts to fight the fire without sounding an alarm have resulted in costly damage and larger fires than necessary.

Whenever possible, the guard should turn in the alarm at the alarm box nearest the scene of the fire. If he must use a plant telephone, he should identify the location accurately. This information helps the plant fire department or the guard on duty at the Security Office, who must summon the employee fire brigade or the city fire department.

Once he has turned in the alarm, the guard should decide whether he can fight the fire with the available equipment. If the fire is beyond his control, he should go to the main entrance of the area so that he can direct the fire fighters to the scene of the fire. If the automatic sprinklers are on, the supervisor of the fire fighting crew will decide when to turn off the system.

Second Draft

BULLETIN 46-RC   FIRES—PLANT SECURITY PATROLS

All security guards should be familiar with the following procedures for fires that occur during routine security patrols.

Inspection Areas

The guard should inspect the outside areas in a security patrol car equipped with a spot light and two-way radio. The outside areas require special attention after dark and during nonoperating hours. At least twice per shift, the following should be inspected:

Fence Lines—The guard should check fences for gaps or breaks in the chain and gate closures for tight links and hinges.

Parking Lots and Yards—The guard should monitor the lighting in all parking lots and open yards to ensure that all necessary lights are on after dark and that the system is fully operative. If any lights are burned out, the guard should report the locations to the Plant Engineer and maintain follow-up until the lighting is repaired.

Occasional spot checks during a shift should be made of the following:

Roofs—The guard should check for improper use of the roof areas and any fire hazards, particularly around ventilating equipment.

Fires

Discovery—Upon discovering a fire during an in-plant security patrol, the guard must turn in an alarm before attempting to fight the fire. In the past, attempts to fight the fire before turning in an alarm have resulted in costly damage and larger fires than necessary.

Alarms—Whenever possible, the guard should turn in the alarm at the alarm box nearest the scene of the fire. If a plant telephone is more convenient or safer, the guard should be sure to identify the fire location accurately. This information is needed by the plant fire department and the guard at the Security Office, who must summon the employee fire brigade or the city fire department.

Fire Fighting—Once the alarm is in, the guard should decide whether to fight the fire with the available equipment. If the fire is beyond control, the guard should go to the main entrance to the fire area and direct the fire brigade or city fire fighters to the scene of the fire. If the automatic sprinklers are on, the supervisor of the fire brigade or of the fire department crew will decide when to turn off the system.

Third Draft

Chapter 5   Exercises

1. Find a "real-world" letter, memo, bulletin, pamphlet, or report. Analyze the writing style according to the guidelines in this chapter. Identify specific style problems. Write a one-page evaluation of the style and clarity of the document, and identify specific areas where revision would help readers. *Or,* if your instructor prefers, rewrite the original to improve the style according to chapter guidelines. Hand in both your revision and the original document.

2. The following memo was sent by the supervisor to the staff. Identify the style problems in the memo, and rewrite it in a more effective style. Compare your results with those of your classmates.

> Pursuant to my recent memo, I must insist that the telephonic communication for personal problems be reduced immediately. I know you gals have to take calls from your kids when they are out of control, but the general conversations about everyone's surgical operations needs to cease. There seems to have developed a general problem with attending to our work tasks on a regular and timely basis. I provide the coffee, but that doesn't mean I'm encouraging a general milling around the coffeemaker for 20 minutes every morning. In short, it is acknowledged that a certain amount of verbal exchange is typical for every office and is useful for maintaining a relaxed atmosphere in the office surroundings, but it can all interfere with normal working schedules and inhibit the accomplishment of the work we are gathered to perform

COLLABORATIVE EXERCISES

3. The following is a bulletin issued at a shipbuilding facility. In groups, discuss the readability problems and rewrite in a more effective style. Use the organization principles in Chapter 4 to revise.

> It is important that all employees who work near or pass through or under areas where hot work is being performed be aware of the potential hazards associated with this work. In shipbuilding and repair, burns are among the most common and prevalent injuries to workers performing hot work. It is imperative, therefore, that all workers be cautious about coming into close proximity to surfaces that have recently had or are having hot work performed. It has been determined by surveys conducted by OSHA that burn injuries usually occur when hot sparks of molten slag become trapped in footwear or apparel. It was further determined that inexperienced workers

tend to get into danger by coming in contact with hot surfaces that remain hot even after hot work is completed. Personal protective equipment (PPE) should be issued to all workers in shipbuilding and repair. Shipyard operations often require that workers perform many different and varied jobs that cause them to come into close contact with hot surfaces that have had hot work performed on them. Regulations require that these surfaces be labeled "Hot." PPE consists of a hood, face shield, goggles with filter lenses, protective sleeves, and flame-resistant gloves. The gloves should be gauntlet type. Skullcaps of leather or flame-resistant materials should be available to all workers because it was decided that they would be important for protecting against head burns. When walking under staging, platforms, or deck grating, hot work also causes burns to workers. When the average worker is asked, he usually says he needs training for types of welding hazards, electrical safety, and ventilation hazards. Shipyards that do not operate welding schools and training seminars for fire hazards may wish to avail themselves of training offered by insurance companies and state agencies.

4. In groups, discuss this segment of a report by a consultant to a fitness center. Revise for more readability, and compare your results with those of other groups.

It has been determined by various studies that most Americans do not obtain enough exercise either at their workplaces or at their leisure. Exercise is required in only a few jobs today, such as lumberjacking or construction work, but not usually in any service jobs or any office jobs, or education jobs. Europeans walk much more than Americans, who tend to drive their cars when they want to perform errands. Europeans walk more and ride more bicycles on a daily basis. Exercising harder or faster will not increase the calories spent but exercising longer will improve calories spent. Cross-country skiing uses about 700 calories an hour. People who exercise regularly have found that the exercise makes them feel more energetic and helps in coping with stress. Jogging at the pace of 7 miles per hour uses up 920 calories in an hour. Surveys of Americans have been done and have found that many people believe doing chores around the house constitutes exercise of the type that studies have found to be beneficial. This is not necessarily true. Walking at 3 miles per hour uses up approximately 320 calories. Exercise by and large improves someone's ability to fall asleep quickly and sleep well.

INTERNET EXERCISES

5. Find three Web sites that contain information about medical equipment. Evaluate the readability of the information, and report your findings during class discussion.

CHAPTER 6

Document Design

UNDERSTANDING DESIGN FEATURES

Document design refers to the physical appearance of a document. Because the written text and its presentation work together to provide readers with the information they need, think about the design of your documents during the planning stage—even before you select appropriate information and organize it.

Readers do not read only the printed words on a page; they also "read" the visual presentation of the text, just as a television viewer pays attention not only to the main actor and the words he or she speaks but also to the background action, noises, music, and other actors' movements. In an effective document, as in an effective television scene, the words and visuals support each other.

Desktop publishing has expanded the writer's role in producing technical documents. A writer with sophisticated computer equipment and design software packages can create finished documents with most of the design features discussed in this chapter. Desktop publishing makes it easy to produce newsletters, catalogs, press releases, brochures, training materials, reports, and proposals—all with graphic aids, columns, headlines, mastheads, bullets, and any design feature needed by readers.

Rapidly developing computer technology increases the capabilities of desktop publishing with each new scanner, printer, and software package. It is not an exaggeration to say that desktop publishing has launched a printing revolution as more companies produce more of their own documents. No matter whether a document is produced on centuries-old movable type or on the latest computer technology, however, the writer's goal is the same—to provide readers with the information they need in a form they can use.

Purpose of Design Features

Some documents, such as business letters, have well-known, conventional formats, but letters and other documents also benefit from additional design features—graphic aids and the format elements of written cues, white space, and typographic devices. These design features increase the usefulness of documents in several ways:

1. They guide readers through the text by directing attention to individual topics and increasing the ability of readers to remember the important, highlighted sections.

2. They increase reader interest in the document. Unbroken blocks of type have a numbing effect on most readers, but eye-catching graphic aids and attention-getting format devices keep readers focused on the information they need.

3. They create a document that reflects the image you wish readers to have. A conservative law firm may want to project a solid, traditional, no-nonsense image with its documents, whereas a video equipment company may prefer to project a trendy, dramatic image. Both images can be enhanced by specific design features.

Design Principles

The principles of design are qualities important to any visual presentation regardless of topic or audience. Experienced designers use the principles of design to create the "look" they want for a document. The general principles most designers consider in all documents are balance, proportion, sequence, and consistency.

Balance

Page balance refers to having comparable visual "weight" on both sides of a page or on opposing pages in a longer document. A page in a manual filled with text and photographs followed by a page with only a single paragraph in the center would probably jar the reader. If this unsettling effect is not your intent, avoid such imbalance in page design.

Think of page balance as similar to the scales held by the figure of Justice, which you have seen so often. Formal balance on a page would be the same as two evenly filled scales hanging at the same level. Informal balance, which is used more often by experienced page designers, would be represented by a heavily weighted scale on one side balanced by two smaller scales equal to the total weight of the larger scale on the other side.

One large section of a page, then, can be balanced by two smaller sections. Every time an element is added to or removed from a page, however, the balance shifts. A photograph that is dominant on one page may not be dominant on another page. Remember these points about visual "weight":

- Big weighs more than small.

- Dark weighs more than light.

- Color weighs more than black and white.

- Unusual shapes weigh more than simple circles or squares.

Proportion

Proportion in page design refers to the size and placement of text, graphic aids, and format elements on the page. Experienced designers rarely use an equal amount of space for text and graphics page after page. Not only would this be monotonous for readers, it would interfere with the readers' ability to use the document. Reserving the same amount of space for one heading called "Labor" and another called "Budgetary Considerations" would result in the long heading looking cramped and the short heading looking lost in the space available. In a similar fashion, you would not want every drawing in a parts manual to be the same size regardless of the object it depicts. Each design feature should be the size that is helpful to the readers and appropriate for the subject.

Sequence

Sequence refers to the arrangement of design features so that readers see them in the best order for their use of the document. Readers usually begin reading a page at the top left corner and end at the bottom right corner. In between these two points, readers tend to scan from left to right and up to down. Readers also tend to notice the features with the most "weight" first. Effective design draws readers through the page from important point to important point.

Consistency

Consistency refers to presenting similar features in a similar style. Keep these elements consistent throughout a document:

- *Margins*—Keep uniform margins on all pages of a document.

- *Typeface*—Use the same size and style of type for similar headings and similar kinds of information.

- *Indentations*—Keep uniform indentations for such items as paragraphs, quotations, and lists.

Do not mistake consistent format for boring format. Consistency helps readers by emphasizing similar types of information and their similar importance. A brochure published by the Ohio Department of Health to alert college students to the dangers of AIDS is designed as a series of questions. Each question (e.g., "What Is AIDS?" and "How Do You Get AIDS?") is printed in all capital letters in light blue. The answers are printed in black. This consistency in design helps readers quickly find the answers to their questions.

CREATING GRAPHIC AIDS

Graphic aids, called *figures* and *tables,* are not merely decorative additions to documents or oral presentations. Graphic aids often are essential in helping readers understand and use the information in a document. Instructions may be easier to use when they have graphics that illustrate some steps, such as directions for gripping a tennis racquet properly or for performing artificial respiration.

Purpose of Graphic Aids

Think about which graphic aids would be appropriate for your document during the planning stage. Graphic aids are important in technical documents in these ways:

- Graphic aids provide quick access to complicated information, especially numerical data. For example, a reader can more quickly see the highs and lows of a production trend from a line graph than from a long narrative explanation.

- Graphic aids isolate the main topics in complex data and appeal particularly to general and nonexpert readers. Newspapers, for instance, usually report government statistics with graphic aids so that their readers can easily see the scope of the information.

- Graphic aids help readers see relationships among several sets of data. Two pie graphs side by side will illustrate more easily than a narrative can the differences between how the federal government spent a tax dollar in 1980 and in 2000.

- Graphic aids, such as detailed statistical tables, can offer expert readers quick access to complicated data that would take pages to explain in written text.

Readers of some documents expect graphic aids. Scientists reading a research report expect to see tables and formulas that show the experimental method and results. Do not, however, rely solely on graphic aids to explain important data. Some readers are more comfortable with written text than with graphics, and for these readers, your text should thoroughly analyze the facts, their impact on the situation or on the future, their relevance for decision making or for direct action, and their relation to other data. For consistency and reader convenience, follow these guidelines for all graphic aids:

1. Identify each graphic aid with a specific title, such as "2000–2004 Crime Rates" or "Differences in Patient Response to Analgesics."

2. Number each graphic aid. All graphics that are not tables with words or numbers in columns are called figures. Number each table or figure consecutively throughout the document, and refer to it by number in the text:

   The results shown in Table 1 are from the first survey. The second survey is illustrated in Figure 2.

3. Place each graphic aid in the text as near after the first reference to it as possible. If, however, the document has many tables and charts or

only some of the readers are interested in them, you may decide to put them all in an appendix to avoid breaking up the text too frequently.

The guidelines in this chapter for creating various types of graphic aids provide general advice for such illustrations. Effective graphic aids, however, can violate standard guidelines and still be useful to the readers who need them. The daily newspaper *USA Today* contains many innovative illustrations of data for its readers who are interested in a quick understanding of the latest statistical or technical information. The key to creating effective graphic aids is to analyze purpose and reader before selecting graphic formats. While an industrial psychologist may prefer a detailed statistical table of the results of a survey of construction workers, a general reader may want—and need—a simple pie chart showing the key points.

Tables

A table shows numerical or topical data in rows and columns, thus providing readers with quick access to quantitative information and allowing them to make comparisons among items easily. Model 6-1 shows a table that presents a comparison of data from company service centers over a two-year period.

TABLE 1   Shipments from Service Centers

	2003	2004	% Change
Liberty, MO	4,167	4,012	− 3.72
Saukville, WI	4,857	4,118	− 15.22
Franklin, TN[a]	623	5,106	+719.58
Medina, OH	3,624	3,598	− 0.72
Salem, OR	3,432	3,768	+ 9.79
Lafayette, LA	2,740	3,231	+ 17.92
Tucson, AZ	5,024	5,630	+ 12.06
Sweetbriar, KY[b]	2,616	3,967	+ 51.64
TOTALS	27,083	33,430	+ 23.44

[a]Opened in June 2003.
[b]Opened in March 2003.

MODEL 6-1   A Typical Table Comparing Data from Two Different Years

Here are general guidelines for setting up tables:

1. Provide a heading for each column that identifies the items in the column.

2. Use footnotes to explain specific items in columns. Footnotes for specific numbers or columns require lowercase superscript letters (e.g., [a], [b], [c]). If the footnote is for a specific number, place the letter directly after the number (e.g., 432[a]). If the footnote applies to an entire column, place the letter directly after the column heading (e.g., Payroll[b]). List all footnotes at the left margin of the table, directly below the data.

3. Space columns sufficiently so that the data do not run together.

4. Give the source of your data below the table. If you have multiple sources and have compiled the information into a table, explain this in the text if necessary.

5. Use decimals and round off figures to the nearest whole number.

6. Indicate in the column heading if you are using a particular measure for units, such as "millions of dollars" or "per 5000 barrels."

Figures

All graphic aids that are not tables are considered figures. Information in tables often can be presented effectively in figures as well, and sometimes readers can use figures more readily than tables. A geologist seeking information about groundwater levels may want a table giving specific quantities in specific geologic locations. Someone reading a report in the morning newspaper on groundwater levels across the United States, however, may be better served by a set of bars of varying lengths representing groundwater levels in regions of the country, such as the Pacific Northwest. Before selecting graphic aids for your document, consider your readers and their ability to interpret quantitative information and, particularly, their need for specific or general data. The most common types of graphic aids are bar graphs, pictographs, line graphs, pie graphs, organization charts, flowcharts, line drawings, cutaway drawings, exploded drawings, maps, photographs, and screen shots.

Bar Graphs

A *bar graph* uses bars of equal width in varying lengths to represent (1) a comparison of items at one particular time, (2) a comparison of items over time, (3) changes in one item over time, or (4) a comparison of portions of a single

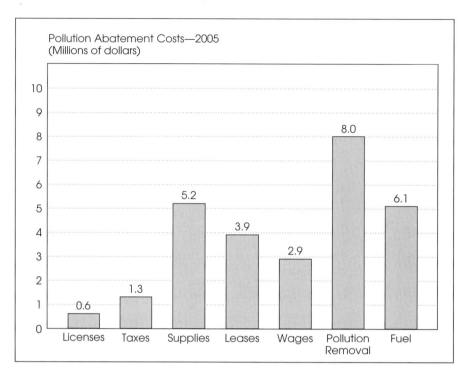

Pollution Abatement Costs—2005
(Millions of dollars)

MODEL 6-2   A Typical Vertical Bar Graph Showing Costs for Different Items

item. The horizontal and vertical axes represent the two elements being illustrated, such as time and quantity. Model 6-2 is a typical vertical bar graph in which bars represent different types of pollution-reduction costs incurred by a company.

Bars can extend in either a vertical direction, as in Model 6-2, or in a horizontal direction, as in Model 6-3. In Model 6-3, the bars emphasize the increasing sales over the five-year period. Notice that in both Models 6-2 and 6-3, a specific figure indicating a dollar amount is printed at the top or the end of the bar.

Bars also can appear on both sides of the axis to indicate positive and negative quantities. Model 6-4 uses bars on both sides of the horizontal axis, indicating positive and negative quantities. Notice that the zero point in the vertical axis is about one-third above the horizontal axis and that the quantities are labeled positive above the zero point and negative below it.

Bar graphs cannot represent exact quantities or provide comparisons of quantities as precisely as tables can, but bar graphs generally are useful for readers who want to understand overall trends and comparisons.

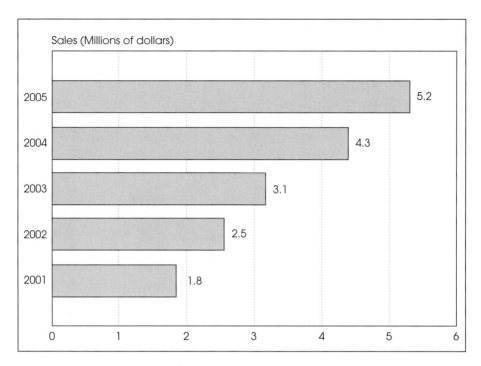

MODEL 6-3   A Typical Horizontal Bar Graph Showing Sales Growth

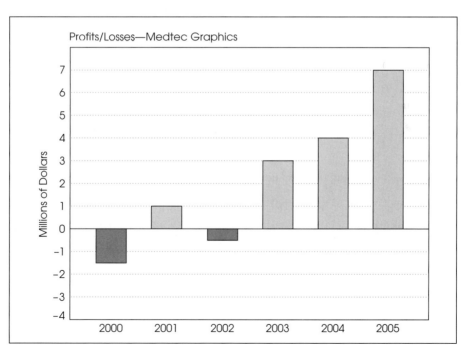

MODEL 6-4   A Typical Bar Graph Presenting Positive and Negative Data for Profits and Losses

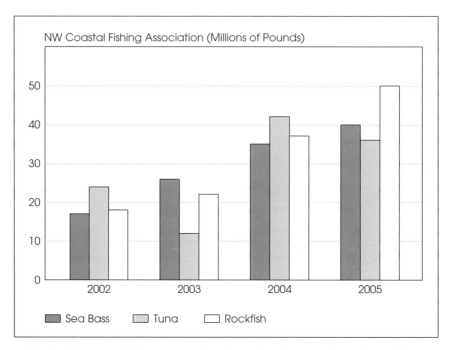

MODEL 6-5   A Typical Bar Graph with Multiple Bars Showing Differences in Three Items

Depending on the size of the graph and the shading or color distinctions, up to four bars can represent different items at any point on an axis. Label each bar, or provide a key to distinguish among shadings or colors. Model 6-5 shows a bar graph with multiple bars representing three different kinds of fish.

Instead of multiple bars on an axis, writers sometimes use a stacked or subdivided bar. The purpose is to show the comparative amounts of selected items. If the bar components represent 100% of the total, the bar is called a 100% bar graph. The stacked bar shows the reader a quick comparison of items, but it is not useful if the reader needs the specific amounts of the items. To distinguish the sections on the bar if there is no room for labeling, writers can use colors or shading. Model 6-6 shows a stacked bar graph representing a comparison of the U.S. adult overweight population on two different dates.

Pictographs

A *pictograph* is a variation of a bar graph that uses symbols instead of bars to illustrate specific quantities of items. The symbols should realistically correspond to the items, such as, for example, a cow representing milk production. Pictographs provide novelty and eye-catching appeal, particularly in documents intended for consumers. Pictographs are limited, however, because

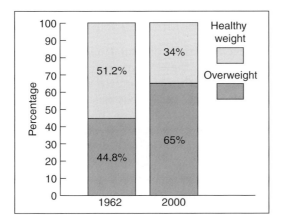

MODEL 6-6   A Typical Stacked Bar Graph Comparing Two Items

symbols cannot adequately represent exact figures or fractions. When using a pictograph, (1) make all symbols the same size, (2) space the symbols equally on the axis, (3) show increased quantity by increasing the number of symbols rather than the size of the symbol, (4) round off the quantities represented instead of using a portion of a symbol to represent a portion of a unit, and (5) include a key indicating the quantity represented by a symbol. Model 6-7 uses houses to represent the trend in housing sales during a six-month period.

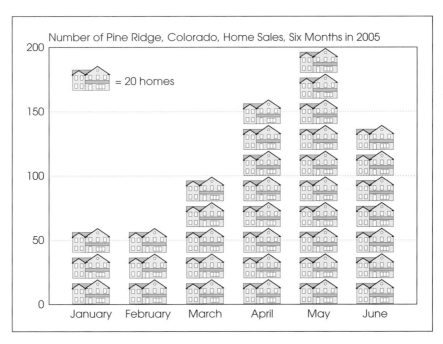

MODEL 6-7   A Typical Pictograph Showing Housing Sales

Line Graphs

A *line graph* uses a line between the horizontal and vertical axes to show changes in the relationship between the elements represented by the two axes. The line connects points on the graph that represent a quantity at a particular time or in relation to a specific topic. Line graphs usually plot changes in quantity or in position and are particularly useful for illustrating trends. Three or four lines representing different items can appear on the same graph for comparison. The lines must be distinguished by color or design, and a key must identify them. Label both vertical and horizontal axes, and be sure that the value segments on the axes are equidistant. Model 6-8 uses three lines, each representing sales for a specific corporation. The lines plot changes in sales (vertical axis) during specific years (horizontal axis). The amounts indicated are not exact, but readers can readily see differences in the corporate sales over time.

Model 6-9 uses three lines, each representing energy consumption by a different type of user. Notice that the lines are distinguished by design and are labeled in the graph rather than by a key, as in Model 6-8.

Pie Graphs

A *pie graph* is a circle representing a whole unit, with the segments of the circle, or pie, representing portions of the whole. Pie graphs are useful if the

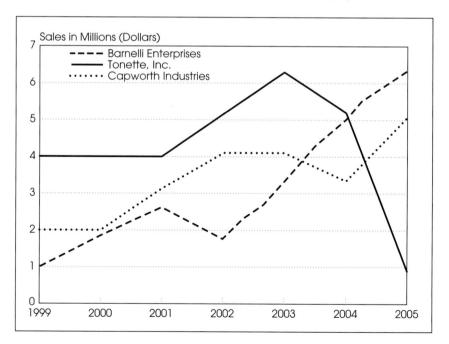

MODEL 6-8   A Typical Line Graph Comparing Sales of Three Companies

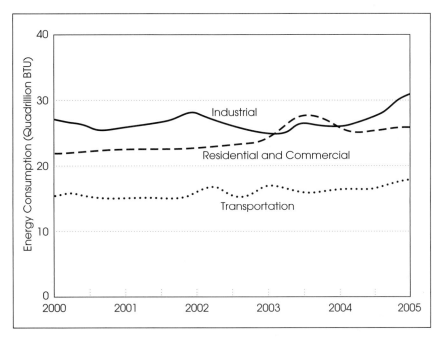

MODEL 6-9   A Typical Line Graph Comparing Energy Consumption by Three Types of Users

whole unit has between three and ten segments. Use colors and shadings to highlight segments of special importance or separate one segment from the pie for emphasis. In preparing a pie graph, start the largest segment at the 12 o'clock position and follow clockwise with the remaining segments in descending order of size. If one segment is "Miscellaneous," it should be the last. Label the segments, and be sure their values add up to 100% of the total. Model 6-10 is a pie graph with six segments. The segments represent four different companies—those heating and cooling contractors with the largest share of the Chicago-area business. Notice that the section called "Wisconsin based" is larger (8%) than the Richland section (3%). The Wisconsin-based section follows Richland because it represents a group and not a single company. "Other" appears last because it is the least-specific group. One section of a pie chart may be separated from the pie to emphasize the item or quantity represented by that section, as in Model 6-11. In this case, the writer wants to emphasize the percentage spent on Hondas. Notice that the Honda section of the pie is marked by stripes, further distinguishing it for the reader.

Pie charts and other figures often have call-outs to identify specific parts of the illustration. Call-outs consist of lines leading from a specific part of the

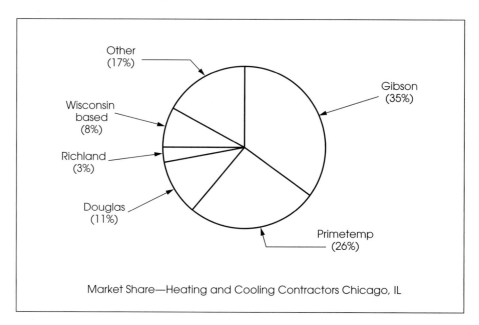

Market Share—Heating and Cooling Contractors Chicago, IL

MODEL 6-10   A Typical Pie Graph Showing Market Share

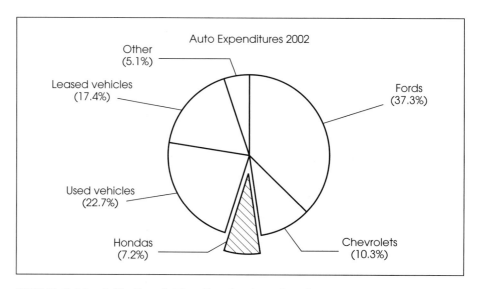

MODEL 6-11   A Pie Graph That Emphasizes One Segment

illustration to a number, often circled, which corresponds to an item in a separate list. The call-out line also may lead to a written term. Notice that the call-outs in Model 6-10 are small arrows that connect a specific section of the pie to the identification of the contractors and percentages represented. The call-out lines in Model 6-11 lead to the type of expenditure and the percentage. Call-out lines should always touch the edge of the section or part being identified.

Organization Charts

An *organization chart* illustrates the individual units in a company or any group and their relationships to each other. Organization charts most often are used to illustrate the chain of authority—the position with the most authority at the top and all other positions leading to it in some way. Organization charts also indicate the lines of authority between units or positions and which positions are on the same level of authority. Rectangles or ovals usually represent the positions in an organizational chart. Label each clearly. Model 6-12 is an organization chart showing the committees at a charitable foundation.

Flowcharts

A *flowchart* illustrates the sequence of steps in a process. A flowchart can represent an entire process or only a specific portion of it. As a supplement to a written description of a process, a flowchart is useful in enabling readers to

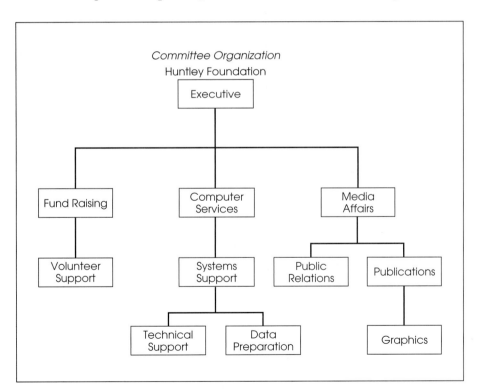

MODEL 6-12   A Typical Organization Chart

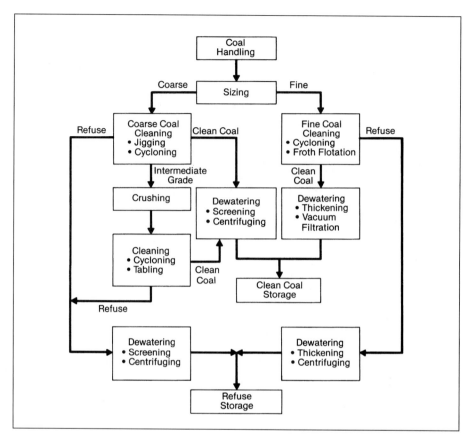

MODEL 6-13   A Typical Flowchart: Coal-Cleaning Operations

visualize the progression of steps. An open-system flowchart shows a process that begins at one point and ends at another. A closed-system flowchart shows a circular process that ends where it began. Use rectangles, circles, or symbols to represent the steps of the process, and label each clearly. Model 6-13 uses rectangles to show the steps of coal-cleaning operations. Model 6-14 uses drawings to show the stages of mass burning of refuse. The truck dumps refuse into storage pits. The refuse then moves through the hopper onto stoker grates and finally emerges as ash at the end of the process. The simple drawings guide the reader through the process.

Line Drawings

A *line drawing* is a simple illustration of the structure of an object or the position of a person involved in some action. The drawing may not show all the details but, instead, highlights certain areas or positions that are important to the discussion. If details of the interior of an object are important, a cutaway

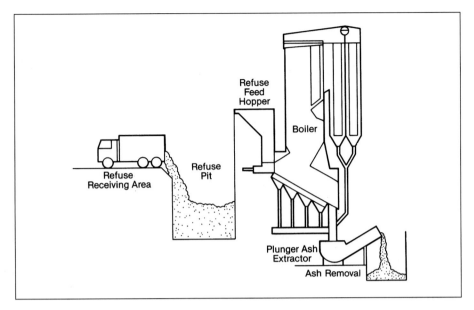

MODEL 6-14   A Typical Flowchart Using Drawings: Mass Burning

drawing is more appropriate. If the relationship of all the components is important, an exploded drawing is more appropriate. Model 6-15 is a line drawing of a police officer conducting a driver sobriety test. The drawing shows the customary positions of the driver and police officer as the driver tries to walk a straight line.

Cutaway Drawings

A *cutaway drawing* shows an interior section of an object or an object relative to a location and other objects so that readers can see a cross-section below the surface. Cutaway drawings show the location, relative size, and relationships of interior components. Use both horizontal and vertical cutaway views if your readers need a complete perspective of an interior. Model 6-16 is a cutaway drawing of a suction dredge revealing its parts and its position in the water.

Model 6-17 shows another cutaway drawing, this time showing the interior of a specialized furnace used in testing the transmission qualities of non-silicon-based glass.

Exploded Drawings

An *exploded drawing* shows the individual components of an object as separate, but in the sequence and location they have when put together. Most often used in manuals and instruction booklets, exploded drawings help

MODEL 6-15   A Line Drawing of a Police Officer Conducting a Driver Sobriety Test

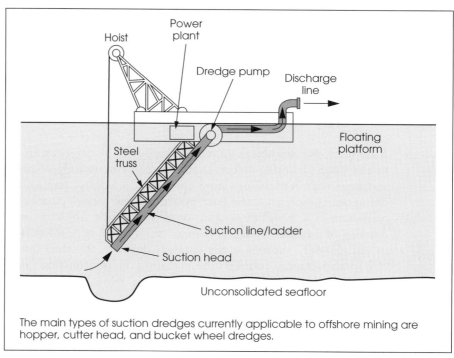

The main types of suction dredges currently applicable to offshore mining are hopper, cutter head, and bucket wheel dredges.

MODEL 6-16   A Typical Cutaway Drawing Showing Components of a Suction Dredge

150

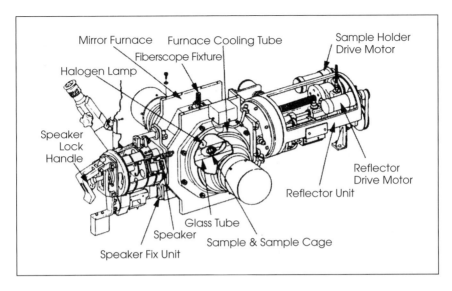

MODEL 6-17   A Cutaway Drawing Showing an Acoustic Levitation Furnace

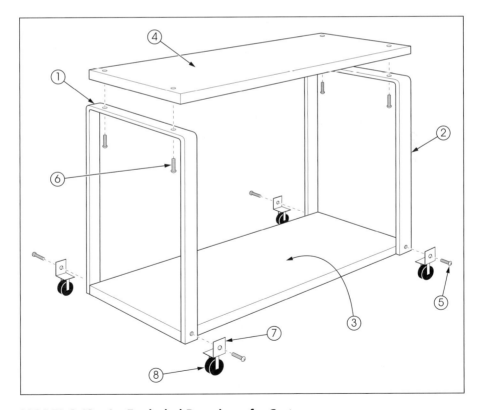

MODEL 6-18   An Exploded Drawing of a Cart

readers visualize how the exterior and interior parts fit together. Model 6-18 is an exploded drawing used to illustrate assembly instructions of a cart for a consumer.

Maps

A *map* shows (1) geographic data, such as the location of rivers and highways, or (2) demographic and topical data, such as density and distribution of population or production. If demographic or topical data are featured, eliminate unnecessary geographic elements, and use dots, shadings, or symbols to show distribution or density. Include (1) a key if more than one topical or demographic item is illustrated, (2) a scale of miles if distance is important, and (3) significant boundaries separating regions. Model 6-19 is a map of the U.S. census regions available on the Web site of the U.S. Census Bureau. Notice that the map is exploded to emphasize the separate regions. Alaska and Hawaii are represented in inserts at the left of the map.

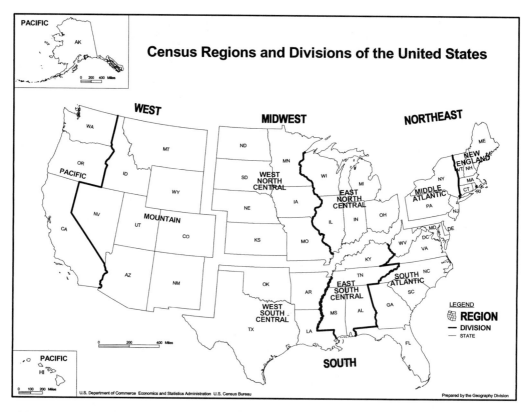

MODEL 6-19   A Map Showing U.S. Census Regions

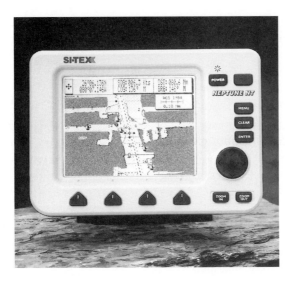

MODEL 6-20   A Product Photograph

Photographs

A *photograph* provides a surface view of an object or event. To be effective, photographs must be clear and focus on the pertinent item. Eliminate distracting backgrounds or other objects from the photograph. If the size of an object is important, include a person or well-known item in the photograph to illustrate the comparative size of the object. Model 6-20 is a typical product photograph.

Screen Shots

A *screen shot,* also called a *screen capture* or *screen dump,* is the image of the text and graphics visible on a computer screen at a particular moment. These most often are used to illustrate computer instructions, as in steps for users to follow as they complete a computer task, such as registering online.

USING FORMAT ELEMENTS

For many internal documents, such as reports, company procedures, and bulletins, the writer is responsible for determining the most effective format elements. Just as you use graphic aids to provide readers with easy access to complicated data, use format elements to help readers move through the document, finding and retaining important information. For printed documents, the writer often does not have the final word on which format elements to use. However, always consider these elements, and be prepared to consult with and offer suggestions to the technical editor and art director about document design. The following guidelines cover four types of format elements: (1) written cues, (2) white space, (3) color, and (4) typographic devices.

Written Cues

Written cues help readers find specific information quickly. The most frequently used written cues are headings, headers and footers, jumplines, icons, and company logos.

Headings

Headings are organizational cues that alert readers to the sequence of information in a document. Use headings

- To help readers find specific data

- To provide an outline that helps readers see the hierarchical relationship of sections

- To call attention to specific topics

- To show where changes in topics occur

- To break up a page so that readers are not confronted with line after line of unbroken text

Include headings for major sections and for topics within the sections. If one heading is followed by several pages of text, divide that section into subsections with more headings. If, on the other hand, every heading is followed by only two or three sentences, regroup your information into fewer topics. Headings are distinguished by level and style. Model 6-21 illustrates the placement and style of the four levels of headings most commonly used in technical documents.

Placement.   *First-level* headings represent major sections or full chapters. Center first-level headings, and use all capital letters. Begin the text two lines below the heading. *Second-level* headings represent major sections. Center second-level headings, and use upper- and lowercase letters. Begin the text two lines below the heading. *Third-level* headings are flush with the left margins and in upper- and lowercase letters. Begin the text two lines below the heading. *Fourth-level* headings are flush with the left margin, in upper- and lowercase letters, followed by a period. Begin the text on the same line.

SAFETY IN SHIP BUILDING

This bulletin provides information about worker safety. . . .

Deck Openings and Edges

When workers are near unguarded hatches or edges, they . . .

Edges of Decks

Guard rails must be in place at edges more than 5 ft above a solid surface . . .

Personal Flotation Devices.   Workers near unguarded edges on a vessel afloat must wear . . .

MODEL 6-21   Heading Placement and Style

You do not have to use all four levels of headings in any one document. A two-page report, for example, may have only third-level headings. Never use only one heading at any level. Divide your text so that you have at least two or more headings at whatever level you have selected. Always include at least one or two lines of text between headings. Most writers prefer boldface for headings because it stands out more than underlining or italics.

Style.   Use nouns, phrases, or sentences for headings. All headings of the same level, however, should be the same part of speech. If you want to use a noun for a heading, all headings of that level should be nouns. If one heading is a phrase, all headings of that level should be phrases. Here are three types of headings:

Nouns: **Protein**

            Fat

            Carbohydrate

Phrases: **Use of Tests**

            Development of Standards

            Rise in Cost

Sentences: **Accidents Are Rising.**

            Support Is Not Available.

            Reductions Are Planned.

Nouns and phrases are the most commonly used headings. Be sure also that your headings clearly indicate the topic of the section they head so that readers do not waste time reading through unneeded information. Keep sentence headings short. Headings such as "Miscellaneous" or "General Principles" provide no guide to the information covered in the section. Write headings with specific terms and informative phrases so that readers can scan the document for the sections of most interest.

Headers and Footers

A *header* or *footer* is a notation on each page of a document identifying it for the reader and including one or more of the following: (1) document date, (2) title, (3) document number, (4) topic, (5) author, or (6) title of the larger publication of which the document is part. Place headers at the tops of pages and footers at the bottoms flush with either the right or left margin. Here is an example:

> Interactive Systems Check, No. 24-X,
> May 1, 2005

Jumplines

A *jumpline* is a notation indicating that a section is continued in another part of a document. Place a jumpline below the last line of text on the page. Here is an example:

> Continued on page 39

Logos

A *company logo* is a specific design used as the symbol of a company. It may be a written cue or a visual one or a combination. Examples include the red cross of the American Red Cross, the golden arches of McDonald's, and the line drawing of a bell for the regional telephone companies. A logo identifies an organization for readers, and it also may unify the document by appearing at the beginning of major sections or on each page.

Icons

An *icon* is a drawing or visual symbol of a system or object, such as a mailbox representing email or a suitcase representing the airport baggage pickup. Icons often are used on public signs, as links on Web pages, or as guides in documents, directing readers to certain sections. Computer clip art supplies many common icons, but use them for specific purposes, not just as decoration. For international readers, check whether an icon has the same meaning in their culture as in yours. Avoid using national flags or religious symbols as icons.

White Space

White space is the term for areas on a page that have no text or graphics. Far from being just a wasted area on a page, white space helps readers process the text efficiently. In documents with complicated data and lots of detail, white space rests readers' eyes and directs them to important information. Think of white space as having a definite shape on the page, and use it to create balance and proportion in your documents. White space commonly appears in margins, heading areas, columns, and indentations.

Margins

Margins of white space provide a frame for the text and graphics (called the "live area") on a page. To avoid monotony, experienced designers usually vary the size of margins in printed documents. The bottom margin of a page is generally the widest, and the top margin is slightly smaller. If a document, such as a manual, is bound with facing pages, the inside margin is usually the narrowest, and the outside margin is slightly wider than the top margin.

Heading Areas

White space provides the background and surrounding area to set off headings. Readers need sufficient white space around headings to be able to separate them from the rest of the text.

Columns

In documents with columns of text, the white space between the columns both frames and separates the columns. Too much or too little space will upset the balance and proportion of the page. Generally, the wider the columns, the more space that is needed between columns to help readers see them as separate sections of text.

Indentations

The white space at the beginning of an indented sentence and after the last word of a paragraph sets off a unit of text for readers, as does the white space between lines, words, and single-spaced paragraphs. Add extra space between paragraphs to increase readability. Youhaveonlytotrytoreadalinewithoutwhite spacebetweenwordstoseehowimportantthistinywhitespaceis.

White space is actually the most important format element because, without it, most readers would quickly give up trying to get information from a document.

Web Pages

White space may not always be white. Solid black, a textured pattern, or a pale shade can all function as white space on a Web page.[1] The available reading space on a computer screen is limited because toolbars and status bars automatically take up certain areas. Many Web pages are overloaded with images, drop-down announcements, columns, boxes, and moving text, so an uncluttered Web page can have great appeal to readers. The white space creates a path that guides the reader from heading to text or images.

STRATEGIES—Voice Mail

Be sure that you have a clear and understandable voice mail message. Follow these guidelines:

Do identify yourself and your subject immediately.

Do cover one point at a time in your message.

Do be specific about times and dates.

Do speak clearly and slowly when leaving a telephone number.

Do close by starting what kind of response you want.

Do not put confidential information or bad news on voice mail.

Do not take up time with small talk.

Do not use voice mail to avoid talking to someone directly.

Color

Color in a document or on a Web site is eye-catching and appealing to readers. It also creates an image and helps a reader move through a document and find specific kinds of information. Color can distinguish among levels of information and can code information as to purpose or importance. When using color in printed documents or on Web sites, you should consider how color creates style and supports usability.

Style

Consider the message you want your document or Web site to convey. Choose colors that reinforce your message and appeal specifically to your intended audience. Here are some suggestions:

- Use pale or muted shades to create a conservative, stable image. Bright colors express more activity or a more experimental style.

- Select bright colors for younger audiences and softer colors for older audiences.

- Research international audiences for cultural differences in interpreting colors.

- Avoid following the latest trends in color preferences. You want your document or Web site to stand out, not look like dozens of others.

Usability

Color should enhance usability. Do not rely on color to convey the entire message, and do not use color as a coding mechanism because color-blind users will be confused. Always use text or another visual in addition to color.[2] Also, readers vary in their ability to detect differences in shades. Use the following strategies for color:

- For the highest possible contrast, use black on white. If using other combinations, select the darkest possible color on the lightest background.

- Limit the number of colors you use in one document or on one Web page. Too much color reduces readability and distracts readers.

- Draw readers' attention to similar types of information by using the same shade as the background.

- Establish a consistent color for specific navigation buttons and headings.

- Use consistent colors for frames around similar kinds of information or for linked illustrations.

Typographic Devices

Typographic devices are used to highlight specific details or specific sections of a document. Highlighting involves selecting different typefaces or using boldface type, lists, and boxes.

Typefaces

Type is either *serif* or *sans serif*. Serif type has small projections at the top or bottom of the major vertical strokes of a letter. Sans serif type does not have these little projections.

This sentence is printed in serif type.

This sentence is printed in sans serif type.

A *typeface* is a particular design for the type on a page. Typeface should be appropriate to readers and purpose. An ornate, script typeface suitable for a holiday greeting card would be out of place in a technical manual.

Poor: *Feasibility Tests in Mining Ventures*

Better: **Feasibility Tests in Mining Ventures**

Follow these general guidelines if you are involved in selecting the typeface for a printed document:

1. Stay in one typeface group, such as Times New Roman, throughout each document. Most typefaces have enough variety in size and weight to suit the requirements of one document.

2. Use the fewest possible sizes and weights of typeface in one document. A multitude of type sizes and weights, even from the same typeface, distracts readers.

3. Use 12- to 14-point type because most readers find that size easy to read. If your document or Web site will be used by people with visual problems or by older readers, consider even larger type.[3]

4. Use both capital and lowercase letters for text. This combination is the most readable. Words in all capitals lose their distinctive shape, and that shape often works as a cue to meaning for many readers.

5. Use italics sparingly. Appendix A explains the correct use of italics. Italics on Web pages often are hard to read because the letters may be formed inconsistently.

Boldface

Boldface refers to type with extra weight or darkness. Use boldface type to add emphasis to (1) headings; (2) specific words, such as warnings; or

(3) significant topics. This example shows how boldface type sets off a word from a sentence:

> Depending on the type of incinerator, **gas temperatures** could vary more than 2000 degrees.

Lists

Lists highlight information and guide readers to the facts they specifically need. In general, lists are distinguished by numbers, bullets, and squares.

Numbers.   Traditionally, lists in which items are numbered imply either that the items are in descending order of importance or that the items are sequential stages in a procedure. Here is a sample from an operator's manual:

1. Clean accumulated dust and debris from the surface.
2. Run the engine, and check for abnormal noises and vibrations.
3. Observe three operating cycles.
4. Check operation of the brake.

Because numbered lists set off information so distinctly from the text, they often are used even when the items have only minor differences in importance. If a list has no chronology, you may use bullets to set off the items.

Bullets.   *Bullets* are small black circles or squares that appear before each item in a list just as a number would. On a typewriter or word processor without a bullet symbol, use the small o, and fill it in with black ink. Use bullets where there is no distinction in importance among items and where no sequential steps are involved. Here is an example from a company safety bulletin. The bullets are in the usual vertical style:

> A periodic survey should check for the following:
> - Gasoline and paint vapors
> - Alkaline and acid mists
> - Dust
> - Smoke

Squares.   Squares are used with checklists if readers are supposed to respond to questions or select items on the page itself. A small square precedes each item in the same position as a number or a bullet would appear. Here is an example from a magazine subscription form:

> ☐ Please bill me for the total cost.
> ☐ Please bill me in three separate installments.
> ☐ Please charge to my credit card.

Boxes

A box is a frame that separates specific information from the rest of the text. In addition to the box itself, designers often use a light color or shading to further set off a box. Here is an example of boxed information:

Boxes are used

- To add supplemental information that is related to the main subject but not part of the document's specific content
- To call reader attention to special items such as telephone numbers, dates, prices, and return coupons
- To highlight important terms or facts

Boxes are effective typographic devices, but overuse of them will result in a cluttered, unbalanced page. Use restraint in boxing information in your text.

CHAPTER SUMMARY

This chapter discusses document design features—graphic aids and format elements. Remember:

- Design features guide readers through the text, increase reader interest, and contribute to a document "image."

- Basic page design principles include balance, proportion, sequence, and consistency.

- Graphic aids (1) provide readers with quick access to information, (2) isolate main topics, (3) help readers see relationships among sets of data, and (4) offer expert readers quick access to complicated data.

- All graphic aids should be identified with a descriptive heading and a number and should be placed as near as possible to the reference in the text.

- The most frequently used graphic aids are tables, bar graphs, pictographs, line graphs, pie graphs, organization charts, flowcharts, line drawings, cutaway drawings, exploded drawings, maps, photographs, and screen shots.

- Written format elements, such as headings, headers and footers, jumplines, logos, and icons, help readers find special information.

- White space, the area without text or graphics, directs readers to information sections.

- Color can be appealing and enhance usability.

- Typographic devices, such as typefaces, boldface type, lists, and boxes, highlight specific sections or specific details.

SUPPLEMENTAL READINGS IN PART 2

Allen, L., and Voss, D. "Ethics in Technical Communication," *Ethics in Technical Communication,* p. 475.

Kostelnick, C., and Roberts, D. D. "Ethos," *Designing Visual Language,* p. 509.

McAdams, M. "It's All in the Links: Readying Publications for the Web," *The Editorial Eye,* p. 512.

Nielsen, J. "Top Ten Guidelines for Home Page Usability," *Alertbox,* p. 523.

ENDNOTES

1. Molly Holzschlag, "Give Me My Web Space," *Webtechniques Web site.* Retrieved May 22, 2001, from http://www.webtechniques.com/archives.

2. Jakob Nielsen, "Error Message Guidelines," *Alertbox,* Retrieved June 24, 2001, from http://www.useit.com.

3. Thomas R. Williams, "Guidelines for Designing and Evaluating the Display of Information on the Web," *Technical Communication* 47.3 (August 2000): 383–396.

Model 6-22   Commentary

This model shows two Web pages. The first is a page from the Web site of the U.S. Department of Labor, Occupational Safety and Health Administration. It uses graphics to illustrate the hazards in construction work. The second is from the Web site of the U.S. Department of the Interior, Bureau of Land Management. This Web page uses graphics to present the National Wild Horse and Burro Program.

Discussion

1. Discuss the intended readers of these pages. How expert are the readers expected to be?

2. Discuss the graphic aids used on these pages. How well do the graphics illustrate the text? How eye-catching are they for the intended readers?

3. In groups, discuss the graphics that would be effective on a Web page covering the following:

- An appearance at your school by a famous athlete who will be signing copies of his autobiography

- A track meet

- A swimming meet

- A motorcycle race

- A political rally in support of a candidate for governor of your state

Compare your designs with those of the other groups.

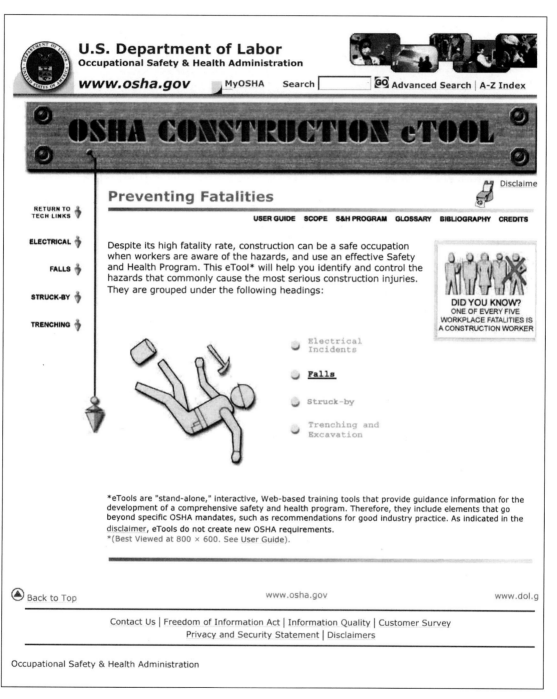

MODEL 6-22   U.S. Department of Labor Web Page

BUREAU OF LAND MANAGEMENT

National Wild Horse and Burro Program

Previous Page   WHB Home Page

Text Version

What We Do
Adoption
Events
News
Facts & Stats
Who We Are
Photo Gallery

For more information on the BLM's Wild Horse and Burro Program, please call toll-free 1-866-4MUSTANGS

State Programs

The Bureau of Land Management (BLM) is responsible for managing the nation's public lands. The wild horses and burros on the public rangelands are managed consistent with BLM's multiple use mission which takes into consideration natural resources such as wildlife and vegetation and other users such as livestock & recreationists.

Upcoming Wild Horse and Burro Adoptions

Check out the BLM Adoption Centers

In The Spotlight...

Wild Horse Workshop in California

Upcoming Adoption
Wheatland, WY
August 14, 2004

Adopted Wild Horses Honored

U.S. Bureau of Land Management Web Page

MODEL 6-23   Commentary

These pages are from a brochure prepared by Moen Incorporated. The brochure, written for consumers, presents the special features in a filtering faucet system.

Discussion

1. Identify and discuss the design elements used in these pages. In what ways do they assist the reader in understanding the information?

2. This brochure is written for nonexpert readers. Discuss how the headings direct a reader's attention to relevant topics.

3. Discuss the use of white space and how it helps or hinders the reader of these pages.

4. Discuss the text. How does the language appeal to consumers?

5. Bring a brochure for a consumer product to class. Identify the special design elements, and discuss how they help the reader understand the information in the brochure.

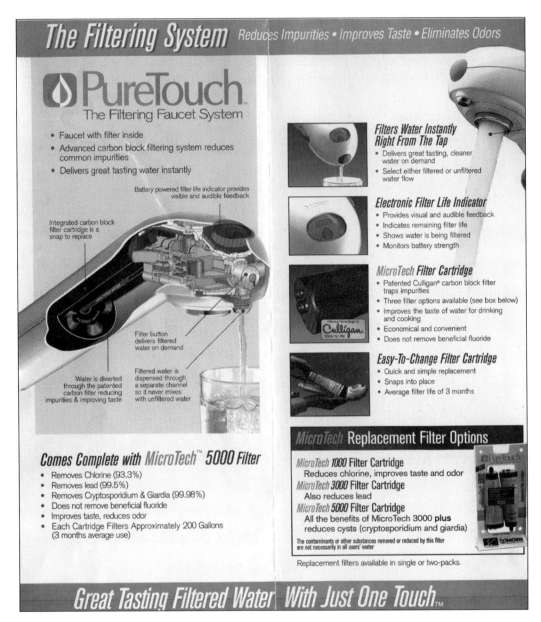

MODEL 6-23   Moen Pure Touch Filter

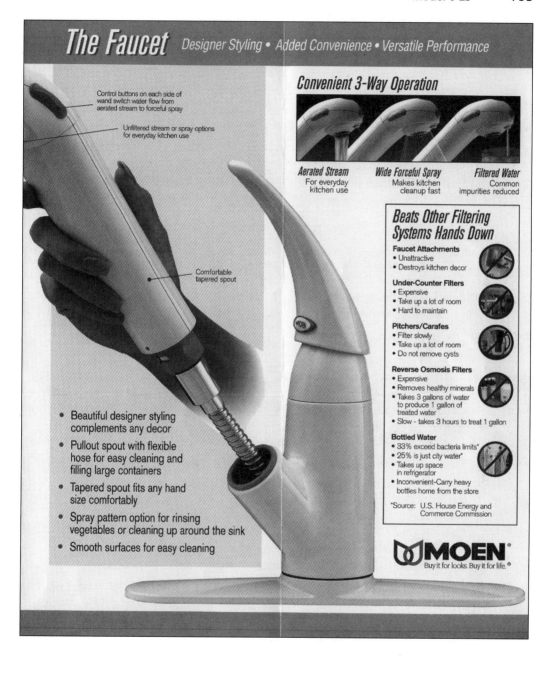

The Faucet *Designer Styling • Added Convenience • Versatile Performance*

Control buttons on each side of wand switch water flow from aerated stream to forceful spray

Unfiltered stream or spray options for everyday kitchen use

Comfortable tapered spout

- Beautiful designer styling complements any decor
- Pullout spout with flexible hose for easy cleaning and filling large containers
- Tapered spout fits any hand size comfortably
- Spray pattern option for rinsing vegetables or cleaning up around the sink
- Smooth surfaces for easy cleaning

Convenient 3-Way Operation

Aerated Stream
For everyday kitchen use

Wide Forceful Spray
Makes kitchen cleanup fast

Filtered Water
Common impurities reduced

Beats Other Filtering Systems Hands Down

Faucet Attachments
- Unattractive
- Destroys kitchen decor

Under-Counter Filters
- Expensive
- Take up a lot of room
- Hard to maintain

Pitchers/Carafes
- Filter slowly
- Take up a lot of room
- Do not remove cysts

Reverse Osmosis Filters
- Expensive
- Removes healthy minerals
- Takes 3 gallons of water to produce 1 gallon of treated water
- Slow - takes 3 hours to treat 1 gallon

Bottled Water
- 33% exceed bacteria limits*
- 25% is just city water*
- Takes up space in refrigerator
- Inconvenient-Carry heavy bottles home from the store

*Source: U.S. House Energy and Commerce Commission

MOEN
Buy it for looks. Buy it for life.

Chapter 6   Exercises

1. Find a technical document from a company, and make enough copies for the other students in class. Prepare to critique the document's design features. Explain how you would redesign the document for a more effective format.

2. In your new job with the city recycling center, you have been given the task of preparing the graphic illustrations about the current recycling efforts. Your supervisor, Charlane Brunner, is making a presentation to the city council in a few days and needs graphics for her handout to the council members. Ms. Brunner asks you to write an explanatory paragraph for each graphic aid. You gather the following statistics and get to work:

   a. The city recycling program last year recycled 12 million tons of trash. Percentages of types of trash included the following: 31% paper, 16% grass/leaves, 3% metals, 12% food, 6% wood, 16% plastics, 6% glass, 7% rubber/textiles, and 3% other. The U.S. national percentages were as follows: 35.7% paper, 12.2% grass/leaves, 11.4% food, 11.1% plastics, 7.9% metals, 7.1% rubber/textiles, 5.5% glass, 5.7% wood, and 3.4% other.

   b. The recycling rates for the city and for the county have risen over the last five years. In 2000, the city rate was 40%, and the county rate was 45%. In 2001, the city rate was 44%, and the county rate was 46%. In 2002, the city rate was 45%, and the county rate was 45%. In 2003, the city rate was 50%, and the county rate was 49%. In 2004, the city rate was 53%, and the county rate was 54%. In 2005, the city rate was 59%, and the county rate was 62%.

After finishing the graphics and the explanatory paragraphs, write a cover memo to Ms. Brunner to accompany the graphics and text. Chapter 14 explains memo writing.

3. In groups, assume you are working at a state health agency and need to create graphics that can be used in the annual agency report to the state legislature. Design an appropriate graphic for the following state statistics covering the last full year of reporting:

   • The most frequent health problems among 1,000 elderly people surveyed were arthritis (48%), cataracts (16%), diabetes (10%), hypertension (36%), heart disease (36%), visual impairment (10%), hearing impairment (32%), and orthopedic impairments (22%). Some elderly reported more than one health problem.

- Death rates in the state declined for heart disease (4%), cancer (2%), stroke (5%), accidents (2%), and flu/pneumonia (7%). Rates increased for kidney disease (4%), hypertension (3%), and Alzheimer's disease (5%).

- The number of on-the-job deaths declined to 591. The main causes of on-the-job deaths were highway crashes (36%), falls (14%), fires (3%), electrocutions (5.5%), homicides (11.5%), and other (29.5%).

- The number of on-the-job injuries increased to 4,660. The main types of injuries were broken bones (43%), burns (16%), cuts and bruises (20%), muscle and ligament strains (14%), and other (7%).

4. In groups, prepare graphic aids, illustrating the following financial information, to help the clients of your management company understand the importance of retirement planning:

- Most people who invest without professional help, place their retirement assets in bank certificates of deposit (41%), stocks (21%), money market funds (10%), bonds (16%), annuities (9%), and other (3%).

- Depending on the age and how long the client has until retirement, you recommend the following investments: 20 years or more to retirement—stocks 70%, cash 30%; 10 years to retirement—stocks 60%, bonds 20%, cash 20%; 1–2 years to retirement or newly retired—stocks 50%, bonds 30%, cash 20%; elderly retired—stocks 40%, bonds 40%, cash 20%.

- If the client had invested $10,000 in each of several areas 20 years ago, the investments would be worth the following today: stocks, $126,725; bonds, $100,131; real estate, $96,343; gold, $32,000; and cash, $72,925.

After creating the graphic aids, write brief paragraphs that explain the significance of each set of statistics.

INTERNET EXERCISES

5. Find a Web site you use frequently, and analyze how you use it. How easy is it to find the features you want to see? Has the home page changed since you began using it? Can you recall which features were redesigned? How consistent is the placement of navigation links on the pages you use? How informative are the headings and page titles? Are there any distracting elements? Are the graphics helpful or simply decorative? Are there moving graphics or text? If so, do they enhance the usefulness of the site? Write a memo to your instructor that analyzes the design of the Web site. Prepare to discuss your analysis in class. Chapter 14 covers memo format.

6. Find two Web sites concerned with one of the subjects you are studying, such as biology, history, accounting, or engineering. Print the home page from each site. Sketch on a blank piece of paper the placement of elements, such as boxes, banners, and navigation buttons for each page. Consider what you like about the home pages and what you would change if you were in charge of re-designing the pages. Picture a new user and what that person would find appealing or difficult. Sketch a redesigned page for each original home page, and prepare to discuss your design suggestions in class.

7. Find a Web site with information about a specific sport. Analyze the graphics presented for the user. How effective are they? What suggestions would you make for more or different graphics? Report your analysis to the class.

CHAPTER 7

Writing for
the Web

PLANNING A WEB SITE

Before thinking about links and graphics, consider the purpose of the Web site you are designing. If it is a company Web site, review the company's goals for the site, such as selling to customers, presenting new products, directing users to dealers, providing the company's history and annual reports, and emphasizing the company's success. If the Web site is for a charity or professional organization, focus on activities that serve clients or members. Charities will want to provide a system on their Web sites for users to donate money or volunteer services. A professional organization will want to attract new members through an appealing Web site.

A personal Web site will represent other goals. Freelance photographers use personal Web sites to present their qualifications for photo assignments and appeal to potential clients. Writers and actors have personal Web sites so that fans can access information about past successes and upcoming creative projects. A family Web site might help relatives in distant locations keep track of each other.

Just as with any document, however, you also must consider your intended readers/users. What are their goals in coming to your Web site? Anyone using the Internet can look at your site, but focus specifically on the users you want to attract and serve. If possible, talk to potential users before building the site to discover their preferences regarding types of information and design.

Research indicates that Web users generally scan Web pages looking for keywords that point to the information they are seeking. Results of surveys of Web users show that 79 to 81% of users report that they scan Web material.[1]

The basic principles of clarity and correctness for technical writing apply to the Web as well. The visual elements of the Internet, however, demand new considerations.

ORGANIZING WEB PAGES

Although users might enjoy the sound and graphics on a Web site, they usually come to a site for information. Therefore, content is the most important element of the site.

Home Page

The home page gives the user the overall impression of the Web site. Consider the following principles for home page design: identity, links, index, and security.

Identity

Identify your home page clearly, placing the name of the site, company, agency, or organization at the top with the appropriate logo. Avoid having a user enter the site on a home page that has no clear identification. Some sites open with only a graphic on a plain background. This design does not clearly guide users into the site. Mysterious designs confuse users—particularly inexperienced users.

Links

Show the site's major options on the home page so that the user can select the appropriate links.[2] Do not include a link to the home page on the home page, but do include a link to the home page on every other page. Use descriptive words for links so the user knows the topic before clicking on the link. Avoid links that do not tell the user where the link leads, such as "More News" or "Click Here."

Index

Provide an index of your Web site's topics if possible. Although creating an index can be time-consuming, an index quickly guides users to the information they need. Some Web sites offer a table of contents for each main section or long document. The contents should be at the top of the page so users can read the topics without scrolling. Each entry in the table of contents should be a link to the relevant section of the topic or document.

Security

Include links to any legal warnings, copyright statements, or privacy information on the home page. Also, provide a link for users to contact you or to find the contact information for others. For your own security, do not put your home address, home telephone number, Social Security number, detailed family information, or travel itineraries on your Web site. Maintain a secure distance between you and the Web site users. If you expect users to enter information on your site, you must protect their privacy with secure software and explain your policy in a privacy link.

Information Pages

Content pages are the most important pages on your Web site. You need to assess whether the information you have is relevant to the users and whether the users want and need that information. Once you decide the content you need to include on the Web pages, you can organize the Web pages based on several principles.

Information Chunks

Web pages use "chunks" of information, and "chunking" refers to the process of dividing the Web site content into specific topics. If you have a topic called "Cost," divide it into subtopics, so you can have separate chunks or pages for "Overhead Costs," "Labor Costs," and "Maintenance Costs." Internet users tend to be looking for very specific information, and the more specific the information chunks, the more likely users will find what they want quickly.

Headings

Write useful headings that direct the reader to specific information on separate pages. Instead of "Conference" as a heading, write "Conference Registration" so users know exactly what the link represents. Headings can be hyperlinks that take the user to the pages with appropriate information. If the content pages are designed for specific readers, use that key word in the heading. Many health information pages have separate links for "Patient Information" and "Physician Information." If the content segment extends to more than one page, repeat the heading on each page to be sure the users know where they are.

Graphics

Graphics should reinforce the specific content of the Web site and can act as a link if the connection is clear. A photo of the company president, for instance, can be a link to the person's biography or a letter to the stockholders.[3] Tables, charts, and graphs offer data as a concise visual, but you should include a written summary of the data for users as well.

Placement

Put the most important information at the top of the page. Also, arrange lists of links with the most important topic at the top and the others in descending order of importance.

Keep related information links in the same area on the page. If the Web site covers a discussion of wildlife issues, information might be grouped by fish, birds, mammals, and reptiles. Thus, the user could click on "Fish" and then select from the options on that page. Hunting through links that are not grouped by topic wastes the user's time. Be sure to keep logos, search buttons, and recurring graphics in the same position on all Web pages. Users tend to re-call the locations of certain kinds of links or buttons.

Print Options

Provide a link that allows users to print a complete document without the extra graphics or frames on the page. Many users prefer to read long docu-ments on paper, and the Internet is not permanent. If a user wants to keep a document, it must be printable.[4]

STRATEGIES—Cell Phones

As email became widespread, we developed email etiquette. With cell phones everywhere, we now need some etiquette rules to prevent rudeness:

- **Do** walk and drive safely. Step aside or pull over to make accident-free calls.
- **Do** turn off your cell phone during business meetings, presentations, or lunches.
- **Do** step outside to take or make calls. Avoid talking in the middle of a crowded room.
- **Do** turn off cell phones and all electronics in theaters, houses of wor-ship, other people's offices, and airplanes when taking off or landing.
- **Do not** discuss your personal life on a cell phone in public places, such as the grocery store.

WRITING EFFECTIVELY

Writing for the Web requires that you visualize how users will approach your content. Web users tend to scan content pages, looking for the information that interests them most. Also, testing has always indicated that reading on a monitor is more difficult and tiring than reading on paper.[5] In addition, the longer the pages, the less likely the users are to find what they are looking for. Follow these guidelines for effective writing on Web sites:

- Keep paragraphs short. Write short sentences. Use key words. Users should be able to scan quickly and find the specific information they want.

- Use standard language, and avoid jargon. Your Web pages are available to everyone on the Internet. You want nonexperts and nonnative speakers to be able to use your information as well. Even experts might not understand jargon that is local to an area or a company.

- Avoid using such terms as *above, aforementioned,* and *below* on Web pages, as in "This conclusion is based on the statistics above." These references in paper documents often are not clear either, but Web documents with separate chunks of information have no "above." Repeat key terms, or specifically send the users to another link, as in "See data in <u>sales reports</u>."

- Use word-based navigation aids. Users can interpret words more quickly than images, and words can link topics.

- Use numbered or bulleted lists. Lists are easier to read than long paragraphs, and they give users an immediate sense of the scope of a topic. Users can tell from a quick glance at a list how many steps are needed to complete an online task or how many subtopics an area has. Links also should be in rows or columns, so users do not overlook them.

- Use graphics to break up the text. A table, line drawing, or bar chart might convey information that would take several paragraphs to explain.

- Be cautious about using color to emphasize your text. Avoid dark backgrounds on Web pages; text is difficult to pick out against such backgrounds. Textured backgrounds can distort line graphs and bar charts. Rotating background colors make some sections of the text more readable than other sections. Users of Web sites need sufficient contrast between text and background.

- Revise printed material before putting it on the Web site. Do not simply add a long document to a Web page without breaking it into information chunks that stand on their own. Revise paragraphs to make them shorter and easier to read on the monitor. Include a summary of the document, so users can decide whether they need to read all the sections.

DESIGNING ELEMENTS FOR SPECIAL NEEDS

The Internet has become an important force in users' personal lives as well as in work-related tasks. Users now rely on the Internet for various personal activities, such as keeping in touch with friends and relatives through email, gathering information about medical conditions and prescriptions, shopping, banking, getting updates on investments, and checking on news events. People who are extremely knowledgeable about the Internet and computers design Web pages, and they sometimes assume that all users not only are equally knowledgeable but also are equally mobile and perceptive. Web page designers need to consider the needs of the growing senior population and those with various disabilities.

Senior Users

Seniors (age 65 years and older) are one of the fastest-growing groups of Web users. Even if these seniors are retired, they use the Web for information about the same diverse personal activities as younger users do. To make Web pages more accessible for senior users, consider the following design elements:

- *Font*—Use at least a 12-point font as the default. Younger users might be comfortable with a smaller font, but the larger font will fit the needs of most users regardless of age.

- *Moving interface elements*—Elements such as pull-down menus cause problems for users who may not be flexible with the mouse.[6] Avoid elements that require perfect maneuvering of the mouse.

- *Colors*—Studies indicate that our perception of color hues diminishes with age.[7] Seniors need strong color contrasts, so they can see the differences in values or understand the chronology or hierarchy of the text topics. If the colors do not have enough contrast, the users may not discern these distinctions.

- *Movement on screen*—Multiple and simultaneous movement on the screen frustrates senior users.[8] Younger users prefer quick movement and quick cuts, but seniors need slower and more consecutive movement. If the Web site is likely to have many senior users, the design should reduce movement.

Disabled Users

The U.S. Department of Health and Human Services estimates that 8% of Internet users have a disability that can interfere with full access to Web pages.[9] The elements listed previously that are a concern for seniors also can be troublesome for Web users with a variety of physical challenges, such as those of vision, hearing, and making exact movements. Generally, a Web designer wants to design Web sites that everyone, young or old, able-bodied or physically challenged, can use. If the designer believes that nontext items, such as animation, symbols, or audio files, are necessary for the majority of intended users, the best solution is to provide basic text information pages as well to ensure that all users can access the same information.[10] In addition to the elements for older users, consider the following:

- *Online forms*—Ensure that users with assistive technology can complete and submit online forms. Assistive technology includes many devices, such as voice-recognition software, large-print monitors, and braille printers. Usability studies will help you determine if the online forms are accessible to all users.

- *Frames*—Frames divide the screen into boxes that represent different but related topics. Provide a frame title for each one you use, so users can orient themselves and readily identify the topic in the frame.

- *Color*—Because color-blind people may not discern changes in color, do not rely on color as the only indication of critical information. Select color contrasts that color-blind users can distinguish. Sight-impaired users may have trouble identifying certain colors or shades.

All U.S. federal government Web sites must comply with the requirements in Section 508 of the Federal Accessibility Standards (http://www.section508.gov). If all Web sites complied with these requirements, users would have easy access to all information.

International Users

English is becoming the language of business and science throughout the world, but many Internet users do not read English well—or at all. Web pages for international companies usually offer a selection of languages. Web site designers also may decide that a translation alone is not adequate for helping international users.

In addition to translating Web content, Web designers may develop "localization"—that is, adjusting the content to reflect local market conditions, cultural preferences and traditions, and regional activities.[11] Adjusting content in this way usually requires a consultant who is familiar with the audience you are trying to serve. Even without the resources to localize the content completely, Web writers should expect the site to be translated, and translation will affect the Web design.

In general, consider the following elements when developing a Web site for international users:

- Translated text usually is longer than the English text. English has many short words, and it uses fewer words than most other languages to convey information. In German, for instance, the word *deodorant* becomes *desodorierendes mittel*. In this case, English requires nine letters; German uses 21 letters.

- Images may create confusion in international users. Cultures react differently to specific images.

- Check for vague wording. For example, "Recently" is not specific, but "October 2004" tells the user the time period in question.

- Check acronyms and initialisms.[12] Are they as clear to international readers as they are to domestic readers?

- Use informal definitions in online documents so international users do not have to stop and clarify a term before reading the whole document.

International Web users want the same things English-speaking users want—good content and clear writing.

EVALUATING YOUR WEB SITE

Give your Web site a final review for effective communication before considering it finished. The site has to support the goals you started with. Evaluate the following elements:

- *Purpose*—Is the purpose of your Web site clear to anyone who happens across it?

- *Audience*—Is the content appropriate for the intended audience? Have you considered users with special needs?

- *Content*—Do you have one topic per page? Are the topics grouped appropriately?

- *Links*—Do the headings lead users to relevant content? Are the links based on key words that identify content to the users?

- *Style*—Are the sentences and paragraphs short enough for easy reading?

- *Visuals*—Do graphics illustrate important information?

- *Page design*—Is there a link to the home page on every content page? Do the fonts, colors, and backgrounds enhance the user's ability to read and find content?

- *Repeat users*—Does the Web site give users a reason to return? You want users to bookmark your site for return. Sites that offer reviews, data updates, tutorials, advice columns, and links to other relevant sites will lure users back frequently.[13]

Users are easily frustrated by broken links and Web pages that have not been changed for years. Maintaining your Web site is a key to keeping your users interested.[14] Update and change the content regularly. Users will check the date at the bottom of the home page to see how current the site is. Check the links frequently. Other sites may move; links may not connect properly.

You want users to rely on your site for up-to-date information and connections. Redesign the site when technology or user expectations change, but do not redesign the site just to try out the latest technical elements.

CHAPTER SUMMARY

This chapter covers some basic principles of writing for the Web and designing your Web site. Remember:

- In planning Web sites, writers need to consider the goals of the site and of the potential users.

- Content is the most important element in a Web site.

- The home page should identify the owner, show the major options on the site, and provide links to legal warnings, contacts, copyright, and privacy information.

- Information pages require that the writer put content in information chunks, use headings with key words, and place important information at the top of the page.

- Writing style on the Web should include short paragraphs and sentences, standard language, numbered or bulleted lists, and graphics to illustrate information.

- Writers should be cautious about using color. Dark backgrounds or textured backgrounds can obscure information, and some users cannot distinguish colors.

- A print document must be revised for use on the Web. Writers must divide content into information chunks and rewrite for shorter paragraphs and sentences.

- A print document on a Web site needs a summary page so users can decide if they need to read the document.

- Web sites need to be accessible to seniors and users with special needs. Writers must check font size, moving interface elements, colors, multiple movement on screen, online forms, frames, and color contrasts.

- Users with problems in vision, hearing, or making precise movements may require all-text information pages.

- International Web users require translation and localization.

- Web writers should evaluate the Web pages based on purpose, audience, content, links, style, visuals, and a design that will keep users coming back. Web sites must be kept up to date.

SUPPLEMENTAL READINGS IN PART 2

Barefoot, D. K. "Ten Tips on Writing White Papers," *Intercom,* p. 483.

Garhan, A. "ePublishing," *Writer's Digest,* p. 495.

Kostelnick, C. and Roberts, D. D. "Ethos," *Designing Visual Language,* p. 509.

McAdams, M. "It's All in the Links: Readying Publications for the Web," *The Editorial Eye,* p. 512.

Nielsen, J. "Top Ten Guidelines for Homepage Usability," *Alertbox,* p. 523.

ENDNOTES

1. Janice C. Redish, "Letting Go of the Words," *Intercom* 51.6 (June 2004): 4–10.

2. "The Homepage." Retrieved July 22, 2004, from http://usability.gov/homepage/index.html.

3. Jakob Nielsen, "The Ten Most Violated Homepage Design Guidelines," *Alertbox.* Retrieved November 10, 2003, from http://www.useit.com.

4. "Optimizing the User Experience." Retrieved July 22, 2004, from http://usability.gov/optimizing/index.html.

5. "Methods for Designing Usable Web Sites." Retrieved July 22, 2004, from http://usability.gov/methods/collecting_writing.html.

6. Jakob Nielsen, "Usability for Senior Citizens," *Alertbox.* Retrieved April 28, 2002, from http://www.useit.com.

7. Zoe Strickler and Patricia Neafsey, "Visual Design of Interactive Software for Older Adults," *Visible Language* 36.1 (2002): 4–28.

8. Ibid.

9. "Accessibility." Retrieved July 22, 2004, from http://www.usability.gov/accessibility/index.html.

10. Ibid.

11. R. A. Bailie and J. Ryckborst, "Reaching Global Audiences: Doing More with Less," *Intercom* 49.6 (June 2002): 17–21.

12. Ibid.

13. Lynn Shuler, "Critiquing a Web Site," *Intercom* 46.9 (November 1999): 21–23.

14. Ibid.

MODEL 7-1   Commentary

This model shows four Web pages. The first is the home page of the U.S. Agency for International Development (http://www.usaid.gov). The second is the home page of the National Park Service (http://www.nps.gov). The third is an information page on the National Park Service site that describes the New Bedford Whaling National Historical Park (http://www.nps.gov/nebe). The fourth page is an information page on the U.S. Environmental Protection Agency Web site describing the Ocean Regulatory Programs (http://www.epa.gov/owow/oceans/regulatory/index.html).

Discussion

1. Discuss the U.S. Agency for International Development home page. How effective is the overall design? What features would be important in luring users back to the site? Is the purpose of the site obvious to a user? Which features would be most helpful to an interested user? Find this home page today, and discuss any changes that have been made since August 12, 2004, when this version was retrieved.

2. Discuss the home page of the National Park Service. Discuss the design differences between this home page and that of the U.S. Agency for International Development. Discuss possible differences in users and their goals in using the Web sites. Find this home page today, and discuss any changes that have been made since this version was retrieved on August 12, 2004. How appealing are the photos on these pages?

3. Discuss the information Web page about the New Bedford Whaling National Historical Park. Which design features seem to be most helpful? Are there any design features you would change, drop, or enhance? Find the National Park Service home page, and enter the following in the search box: (a) Antietam National Cemetery, (b) Grand Teton National Park, and (c) Stones River National Battlefield. Discuss the usefulness of the features on these Web pages for potential users.

4. Discuss the information page from the U.S. Environmental Protection Agency. Which features would be most helpful to users?

5. Identify features on any of these Web pages that would be helpful to seniors and users with any special needs.

6. In groups, design an information Web page for a building on your campus.

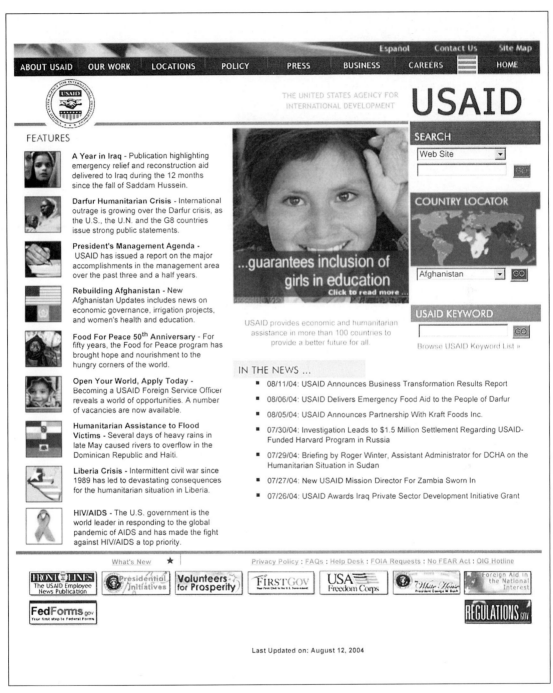

MODEL 7-1   U.S. Agency for International Development Web Page

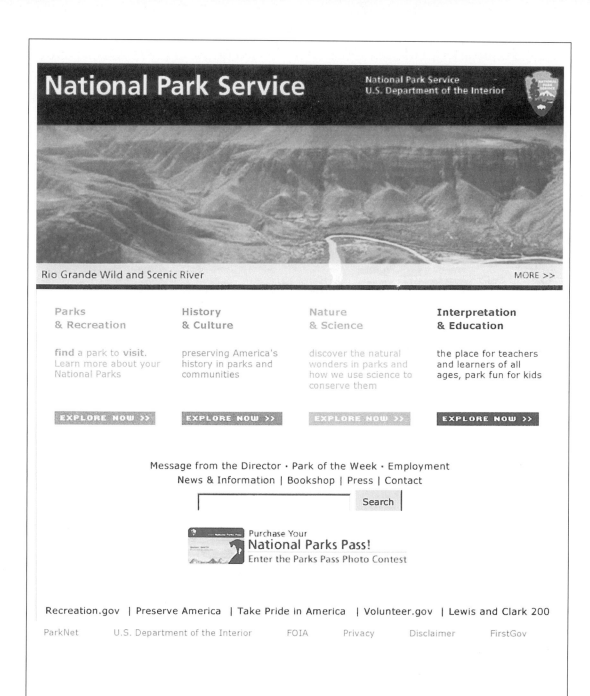

National Park Service Web Page

National Park Service

National Park Service
U.S. Department of the Interior

New Bedford
Whaling

National Historical Park
Massachusetts

Fee Information
View all Fees »

New Bedford Whaling National Historical Park commemorates the heritage of the world's preeminent whaling port during the 19th century. A variety of cultural landscapes, historic buildings, museum collections, and archives preserve this history and collectively recount the stories of a remarkable era. Whaling, a leading 19th century enterprise, contributed to America's economic and political vitality.

New Bedford Whaling National Historical Park was created in 1996. The park encompasses 34 acres spread over 13 city blocks and includes a visitor center, the New Bedford Whaling Museum,the Seamen's Bethel, the schooner Ernestina, and the Rotch-Jones-Duff House and Garden Museum.

The park's enabling legislation also established a legislative connection with the Inupiat Heritage Center in Barrow, Alaska, to commemorate the more than 2,000 whaling voyages from New Bedford to the Western Arctic.

Home
Accessibility
Activities
Facts
For Kids
In Depth
Management Docs
News
Plan Your Visit
Special Events

Bookstore »
Employment »
Volunteer »
Search »
Contact Us »

Designations

National Historic Landmark District - November 1966
National Historical Park - November 12, 1996

ParkNet   U.S. Department of the Interior   FOIA   Privacy   Disclaimer   FirstGov

New Bedford Whaling National Historical Park Web Page

Ocean Regulatory Programs

Contact Us | Print Version   Search: [   ] [GO]

EPA Home > Water > Wetlands, Oceans, & Watersheds > Oceans, Coasts, & Estuaries > Ocean Regulatory Programs

Ocean Dumping and Dredged Material Management

Marine and Ocean Discharges

Vessel Discharges

What You Can Do To Help

Estuaries for Kids

Oceans and Coasts for Children, Students and Teachers

Features

America's Living Oceans Charting a Course for Sea Change
[EXIT disclaimer ▶]

Keeping Our Oceans Clean & Safe

Marine Sanitation Devices

No Discharge Zones

Cruise Ships

Draft Ballast Water Report

National Dredging Team

Using Your Head to Help Protect Our Aquatic Resources

Federal regulation, in consultation with Federal, State, Tribal and local partners, is one of the fundamental tools the EPA uses to implement environmental policies designed to protect the environment and human health. Regulations are required by law and are necessary to interpret and implement laws intended to manage, protect, and restore water resources of the United States, including aquatic ecosystems of coastal watersheds and the oceans.

Ocean Dumping and Dredged Material Management
Ocean Dumping
Dredged Material Management
Site Monitoring
Partnerships

Marine and Ocean Discharges
In 1972, Congress passed the Federal Water Pollution Control Act, commonly known as the Clean Water Act (CWA). Under CWA section 402, any discharge of a pollutant from a point source (e.g., a municipal or industrial facility) to the navigable waters of the United States or beyond must obtain a National Pollutant Discharge Elimination System (NPDES) permit, which requires compliance with technology- and water quality-based treatment standards.

Vessel Discharges
Significant environmental impacts to coastal and ocean ecosystems occur via direct pollution from vessels, and as a vector for the invasion of non-indigenous species. Pollution from recreational, commercial, and military vessels emanates from a variety of sources, and include: gray water, bilgewater, blackwater (sewage), ballast water, anti-fouling paints (and their leachate), hazardous materials, and municipal and commercial garbage and other wastes.

Can't find what you want? Try our [A-Z Index] ☜

General Information & Resources   En Español

EPA Home | Privacy and Security Notice | Contact Us

U.S. Environmental Protection Agency Web Page

188

Chapter 7 Exercises

1. Check out one of the speeches available on the U.S. Environmental Protection Agency Web site (http://www.epa.gov). Print the speech and revise it into chunks suitable for information Web pages. Submit your revision to your instructor with a cover memo explaining why you made your decisions. Memo format is covered in Chapter 14.

2. Find the home pages for the following: (a) Ford Motor Company, (b) Chrysler, and (c) General Motors. Analyze the design features of these home pages in terms of users who are seniors or have problems making precise movements with the mouse. Report your findings to your instructor in a memo. Memo format is covered in Chapter 14.

3. Find two Web sites about one of your favorite sports or hobbies. Analyze which site is most likely to attract return visits. Present your analysis to the class.

COLLABORATIVE EXERCISES

4. In groups, find the Web page for the U.S. Bureau of Labor Statistics (http://www.bls.gov). Click on the link for "Kids' Page," and explore the information pages for the various careers. Analyze how well designed the pages are in terms of pictures, text, and links. Discuss the usefulness of each information page for potential users. Assume you are seeking career information for one of these careers. Do you want more information on any aspect? Discuss your analysis with the class.

5. As a group, draft a fact sheet about a campus activity. Then, create an information Web page based on the same material. Discuss the links to other pages you would want to add, and compare your results with those of other groups.

6. Individually, each group member should select a print ad about a product that seems to be interesting. Then, each should find the home page of the company advertising in the print ad. As a group, compare the design elements in the print ad with those on the Web page for the particular product. Prepare a group oral report on the central differences between the print ads and the company information pages for the products. *Or,* if your instructor prefers, write a group report on the differences. Report format is covered in Chapter 12.

INTERNET EXERCISES

7. Visit the Web site of the new National Museum of the American Indian. Search the Internet for two more museums that focus on specific ethnic or racial groups. Compare the design features on the home pages. Do you see any features that would be useful to users with special needs? How easy is it to find information about visiting these museums? Write a memo to your instructor presenting your comparisons. Memo format is covered in Chapter 14.

8. Find three of the Web sites listed in Appendix C under "Writing for the Web." Analyze the home pages for features that help someone navigate the site. Decide which Web site you would recommend to a beginning technical writer, and send a memo to your instructor recommending one of the sites. Memo format is covered in Chapter 14.

9. Select two of the job search Web sites listed in Appendix C under "Job Search." Which one is easier for a job seeker to use? Which features are most helpful for the job seeker? Discuss your conclusions with the class.

10. Compare the home pages for the following: (a) British Parliament (http://www.parliament.uk), (b) White House (http://whitehouse.gov), (c) Canadian Parliament (http://parl.gc.ca), and (d) Indian Parliament (http://www.parliamentofindia.nic.in). Analyze the features that would be helpful to those who are planning to visit these institutions. What features would you add or delete? Identify any differences in tone of the text on the Web sites. Are they equally welcoming to visitors? Write a short report to your instructor that explains your analysis. Short reports are covered in Chapter 12.

CHAPTER **8**

Definition

UNDERSTANDING DEFINITIONS

Definition is essential in good technical writing because of the specialized vocabulary in many documents. All definitions are meant to distinguish one object or procedure from any that are similar and to clarify them for readers by setting precise limits on each expression. In writing definitions, use vocabulary that your readers understand. Defining one word with others that are equally specialized will frustrate your readers; no one wants to consult a dictionary to understand an explanation that was supposed to make referring to a dictionary unnecessary.

Certain circumstances always call for definitions:

1. When technical information originally written for expert readers is revised for nonexpert readers, the writer must include definitions for all terms that are not common knowledge.

2. A document with readers from many disciplines or varied backgrounds must include definitions enabling readers with the lowest level of knowledge to understand the document.

3. All new or rare terms should be defined, even for readers who are experts in the subject. Change is so rapid in science and technology that no one can easily keep up with every new development.

4. When a term has multiple meanings, a writer must be clear about which meaning is being used in the document. The word *slate*, for example, can refer to a kind of rock, a color, a handheld chalkboard, or a list of candidates for election.

As discussed in Chapter 4, writers frequently use definitions in reports addressed to multiple readers because some of the readers may not be familiar with the technical terminology used. Definitions in reports may range from a simple phrase in a sentence to a complete appendix at the end of the report. In all cases, the writer decides just how much definition the reader needs to be able to use the report effectively. Definitions also may be complete documents in themselves; for example, entries in technical handbooks or science dictionaries are expanded definitions. Furthermore, individual sections of manuals, technical sales literature, and information pamphlets often are expanded definitions for readers who lack expert knowledge of the field but need to understand the subject.

The three types of definitions are (1) informal, (2) formal sentence, and (3) expanded. All three can appear in the same document, and your use of one or another depends on your analysis of the needs of your readers and your purpose in using a specific item of information.

WRITING INFORMAL DEFINITIONS

Informal definitions explain a term with a word or a phrase that has the same general meaning. Here are some examples of informal definitions:

Contrast is the difference between dark and light in a photograph.

Terra cotta—a hard, fired clay—is used for pottery and ornamental architectural detail.

Viscous (sticky) substances are used in manufacturing rayon.

Leucine, an amino acid, is essential for human nutrition.

The first definition is a complete sentence; the other definitions are words or phrases used to define a term within a sentence. Notice that you can set off the definition with dashes, parentheses, or commas. Place the definition immediately after the first reference to the term. Informal definitions are most helpful for nonexpert readers who need an introduction to an unfamiliar term. Be sure to use a well-known word or phrase to define a difficult term.

The advantage of informal definitions is that they do not significantly interrupt the flow of a sentence or the information. Readers do not have to stop thinking about the main idea to understand the term. The limitation of informal definitions is that they do not thoroughly explain the term and, in fact, are really identifications rather than definitions. If a reader needs more information than that provided by a simple identification, use a formal sentence definition or an expanded definition.

WRITING FORMAL SENTENCE DEFINITIONS

A *formal sentence definition* is more detailed and rigidly structured than an informal definition. The formal definition has three specific parts:

Term		*Group*	*Distinguishing features*
An *ace*	is	a tennis serve	that is successful because the opponent cannot reach the ball to return it.

The first part is the specific term you want to define, followed by the verb *is* or *are, was* or *were*. The second part is the group of objects or actions to which the term belongs. The third part consists of the distinguishing features that set off this term from others in the same group. In this example, the group part *a tennis serve* establishes both the sport and the type of action the term refers to. The distinguishing features part explains the specific quality that separates this tennis serve from others—the opponent cannot reach it.

These formal definitions illustrate how the group and the distinguishing features become more restrictive as a term becomes more specific:

Term		Group	Distinguishing features
A *firearm*	is	a weapon	from which a bullet or shell is discharged by gunpowder.
A *rifle*	is	a firearm	with spiral grooves in the inner surface of the gun barrel to give the bullet a rotary motion and increase its accuracy.
A *Winchester*	is	a rifle	first made about 1866, with a tubular magazine under the barrel that allows the user to fire a number of bullets without reloading.

In writing a formal sentence definition, be sure to place the term in as specific a group as possible. In the preceding samples, a rifle could be placed in the group *weapon*, but that group also includes clubs, swords, and nuclear missiles. Placing the rifle in the group *firearm* eliminates all weapons that are not also firearms. The distinguishing features part then concentrates on the characteristics that are special to rifles and not to other firearms. The distinguishing features part, however, cannot include every characteristic detail of a term. You will have to decide which features most effectively separate the term from others in the same group and which will best help your readers understand and use the information.

When writing formal sentence definitions, remember these tips:

1. Do not use the same key word in the distinguishing features part that you used in one or both of the other two units:

   Poor: A pump is a machine or device that pumps gas or liquid to a new level or position.

   Better: A pump is a machine or device that raises or moves gas or liquid to a new level or position.

   Poor: An odometer is a measuring instrument that measures the distance traveled by a vehicle.

   Better: An odometer is a measuring instrument that records the distance traveled by a vehicle.

In the first example, the distinguishing features section includes the word *pump,* which is also the term being defined. In the second example, the distinguishing features part includes the verb *measures,* which is used in the group part. In both cases, the writer must revise to avoid repetition that will send readers in a circle. Sometimes you can assume that your readers know what a general term means. If you are certain your readers understand the term *horse* and your purpose is to clarify the term *racehorse,* you may write, "A racehorse is a horse that is bred or trained to run in competition with other horses."

2. Do not use distinguishing features that are too general to adequately specify the meaning of the term:

Poor: Rugby is a sport that involves rough contact among players as they try to send a ball over the opponent's goal lines.

Better: Rugby is a team sport that involves 13 to 15 players on each side who try to send a ball across the opponent's goal line during two 40-minute halves.

Poor: A staple is a short piece of wire that is bent so as to hold papers together.

Better: A staple is a short piece of wire that is bent so both ends pierce several papers and fold inward, binding the papers together.

In both poor examples, the distinguishing features are not restrictive enough. Many sports, such as football and soccer, involve rough contact between players who try to send a ball over the opponent's goal line. The revision focuses on the number of players and the minutes played, features that distinguish rugby from similar team sports. Notice also that the group has been narrowed to *team sport* to further restrict the definition and help the reader understand it. In the second poor example, the distinguishing features could apply to a paper clip as well as a staple. Remember to restrict the group as much as possible and provide distinguishing features that isolate the term from its group.

3. Do not use distinguishing features that are too restrictive:

Poor: A tent is a portable shelter made of beige canvas in the shape of a pyramid, supported by poles.

Better: A tent is a portable shelter made of animal skins or a sturdy fabric and supported by poles.

Poor: A videotape is a recording device made of a magnetic ribbon of material ¾ in. wide and coated plastic that registers both audio and visual signals for reproduction.

Better: A videotape is a recording device made of a magnetic ribbon of material, usually coated plastic, that registers both audio and visual signals for reproduction.

In the first example, the distinguishing features are too restrictive because not all tents are pyramid-shaped or made of beige canvas. The second example establishes only one size for a videotape, but videotapes come in several sizes. Therefore, size is not an appropriate distinguishing feature. Do not restrict your definition to only one brand or one model if your term is meant to cover all models and brands of that particular object.

4. Do not use *is when, is where,* or *is what* in place of the group part in a formal definition:

Poor: A tongue depressor is what medical personnel use to hold down a patient's tongue during a throat examination.

Better: A tongue depressor is a flat, thin, wooden stick used by medical personnel to hold down a patient's tongue during a throat examination.

Poor: Genetic engineering is when scientists change the hereditary code on an organism's DNA.

Better: Genetic engineering is the set of biochemical techniques used by scientists to move fragments from the genes of one organism to the chromosomes of another to change the hereditary code on the DNA of the second organism.

In both examples, the writer initially neglects to place the term in a group before adding the distinguishing features. The group part is the first level of restriction, and it helps readers by eliminating other groups that may share some of the distinguishing features. The first example, for instance, could apply to any object shoved down a patient's throat. The poor definition of genetic engineering refers only to the result and does not clarify the term as applying to specific techniques that will produce that result.

STRATEGIES—Instant Messaging

Instant messaging (IM) software allows people to exchange messages even faster than by email. Many companies are adding IM to the workplace. Keep the following rules in mind when using IM:

- **Avoid** long or complicated messages; IM is for very short questions and answers.
- **Save** important information in another form for easy reference.
- **Avoid** conducting personal chats over IM on the job.

WRITING EXPANDED DEFINITIONS

Expanded definitions can range from one paragraph in a report or manual to an entry several pages long in a technical dictionary. Writers use expanded definitions in these circumstances:

1. A reader must fully understand a term to use the document successfully. A patient who has just been diagnosed with hypoglycemia probably will want a more detailed definition in the patient information booklet than just a formal sentence definition. Similarly, a decision

maker who is reading a feasibility study about bridge reconstruction will want expanded definitions of such terms as *cathodic protection* or *distributed anode system* to make an informed decision about the most appropriate method for the company project.

2. Specific terms, such as *economically disadvantaged* or *physical therapy,* refer to broad concepts and readers other than experts need to understand the scope and application of the terms.

3. The purpose of the document, such as a technical handbook or a science dictionary, is to provide expanded definitions to readers who need to understand the terms for a variety of reasons.

A careful writer usually begins an expanded definition with a formal sentence definition and then uses one or more of the following strategies to enlarge the definition with more detail and explanation. See also the examples of organization patterns in Chapter 4.

Cause and Effect

Writers use the cause-and-effect (or effect-and-cause) strategy to illustrate relationships among several events. This strategy is effective in expanded definitions of terms that refer to a process or a system. The following paragraph is from a student report about various types of exercises used in athletic training. The definition explains the effect aerobic exercise has on the body.

> Aerobic exercise is a sustained physical activity that increases the body's ability to obtain oxygen, thereby strengthening the heart and lungs. Aerobic effect begins when an exercise is vigorous enough to produce a sustained heart rate of at least 150 beats a minute. This exercise, if continued for at least 20 minutes, produces a change in a person's body. The lungs begin processing more air with less effort, while the heart grows stronger, pumping more blood with fewer strokes. Overall, the body's total blood volume increases and the blood supply to the muscles improves.

Classification

Classification is used in expanded definitions when a writer needs to break a term into types or categories and discuss the similarities and differences among the categories. The following expanded definition from a book about managing pain defines types of sleep. Notice that the writer does not include a formal definition of *sleep* because readers will already know what that term means.

There are two types of sleep—rapid eye movement (REM) and nonrapid eye movement (NREM). NREM is divided into three phases: light sleep, intermediate sleep, and deep sleep. . . .

Throughout the night you continually move from one phase of sleep to another. REM sleep is a period of increased activity. This is the phase of sleep during which you dream and your body functions, such as your heart rate, blood pressure, and breathing, increase. During NREM sleep, your brain activity decreases and these functions slow. Deep sleep is the most restful kind of sleep and lasts 30–40 minutes or less in each cycle. . . . Intermediate sleep helps refresh the body and most of the night is spent in this phase. . . . In light sleep, body movement decreases, and spontaneous awakening may occur.[1]

Comparison and Contrast

Writers use comparison and contrast to show readers the similarities and differences between the term being defined and another relevant term. This definition of carpal tunnel syndrome compares it to a similar condition called pronator syndrome:

Carpal tunnel syndrome causes numbness or tingling in the fingers and hands and may include a pain that shoots from the wrist into the palm or forearm. The carpal tunnel is a passageway through the wrist that protects nerves and tendons. The median nerve, which affects feeling in the thumb and all the fingers except the little finger, passes through the carpal tunnel. Carpal tunnel syndrome is similar to pronator syndrome because the median nerve is compressed, but the pain in the pronator syndrome centers in the wrist and forearm. The median nerve is compressed by the pronator muscle—a muscle that twists the forearm.[2]

When writing for nonexpert readers, develop your comparisons through analogy to help the readers understand the term you are defining. The following definition is from a book about Civil War military equipment written for general readers. The writer compares a specific style of tent to a hoop skirt and a teepee, objects most readers have seen in films or books. A third comparison occurs when the writer explains how the men arranged themselves to sleep, further illustrating the shape and size of the tent and creating an image that general readers will understand easily.

Much more popular and efficient was the Sibley tent, named for its inventor, Henry H. Sibley, now a brigadier general in the Confederate service. One Reb likened it to a "large hoop skirt standing by itself on the ground." Indeed, it resembled nothing so much as an Indian teepee, a tall cone of canvas supported by a center pole. Flaps on the sides could be opened for ventilation, and an iron replica of the tent cone called a Sibley stove heated the interior—sort of. Often more than twenty men inhabited a single tent, spread out like the spokes of a wheel, their heads at the outer rim and their feet at the center pole.[3]

Description

A detailed description of a term being defined will expand the definition. Reading about the physical properties of a term often helps readers visualize the concept or object and remember it more readily. Chapter 9 provides a detailed discussion of developing physical descriptions. This excerpt is from a NASA description of the sections of the Space Shuttle:

> The orbiter carries the crew and payload. It is 122 feet (37 meters) long and 57 feet (17 meters) high, has a wingspan of 78 feet (24 meters), and weighs from 168,000 to 175,000 pounds (76,000 to 79,000 kilograms) empty. It is about the size and general shape of a DC-9 commercial jet airplane. Orbiters may vary slightly from unit to unit.[4]

Development or History

Writers may expand definitions by describing how the subject has changed from its original form or purpose over time. When using development or history to expand a definition, you may include (1) discovery or invention of the concept or object, (2) changes in the components or design of the concept or object, or (3) changes in the use or function of the concept or object. The following excerpt from a science pamphlet for general readers expands the definition of the telegraph by briefly explaining some of its history:

> Although the term *telegraph* means a system or apparatus for sending messages over a long distance and, therefore, could include smoke and drum signals, the term now generally refers to the electric telegraph developed in the nineteenth century. In the late eighteenth century, Frenchman Claude Clappe designed a system of signals that relied on a vertical pole and movable crossbar with indicators. Using a telescope, operators in towers three miles apart read the signals and then passed the messages from tower to tower between principal French cities. The success of this system encouraged experimentation. In the early nineteenth century, researchers discovered that electric current through a wire could cause a needle to turn. The development of the electromagnet then allowed Samuel F. B. Morse to devise a practical system of transmitting and receiving electric signals over long distances, and he invented a message code of dots, dashes, and spaces. By the mid-nineteenth century, telegraph systems spread across the United States and Europe.

Etymology

Etymology is the study of the history of individual words. The term derives from the Greek *etymon* (true meaning) and *logos* (word). Writers rarely use etymology as the only strategy in an expanded definition, but they often include it with other strategies. The following definition of *ligament* is from a brochure for orthopedic patients:

In anatomy, a ligament is a band of white fibrous tissue that connects bones and supports organs. The word *ligament* comes from the Latin *ligare* (to tie or bind). Although a ligament is strong, it does not stretch, so if the bones are pulled apart, the ligament connecting them will tear.

Examples

Another way to expand a definition is to provide examples of the term. This strategy is particularly effective in expanding definitions for nonexpert readers who need to understand the variety included in one term. This definition of a combat medal is taken from a student examination on military history:

> A combat medal is a military award given to commemorate an individual's bravery under fire. Medals are awarded by all branches of the armed forces and by civilian legislative bodies. The Congressional Medal of Honor, for instance, is awarded by the President in the name of Congress to military personnel who have distinguished themselves in combat beyond the call of duty. The Purple Heart is awarded by the branches of the armed services to all military personnel who sustain wounds during combat. The Navy Cross is awarded to naval personnel for outstanding heroism against an enemy.

Method of Operation or Process

Another effective strategy for expanding a definition is to explain how the object represented by the term works, such as how a scanner reads images, stores them on disks, and prints them. Also, a writer may expand a definition of a system or natural process by describing the steps in the process, such as how a drug is produced or how a hurricane occurs. Chapter 10 covers developing process explanations.

This definition of a mountain-climbing device called a *jumar* defines the device by explaining how it operates:

> A jumar (also known as a mechanical ascender) is a wallet-sized device that grips the rope by means of a metal cam. The cam allows the jumar to slide upward without hindrance, but it pinches the rope securely when the device is weighted. Essentially ratcheting himself upward, a climber thereby ascends the rope.[5]

Negation

Writers occasionally expand a definition by explaining what a term *does not* include. This definition of *ocean* from a book about oceanography uses negation as part of the definition:

> The *ocean* is the big blue area on a globe that covers 72 percent of the earth and has a volume of 1.37 billion cubic kilometers. It does *not* include rivers, lakes, or

shallow, mostly landlocked bays and estuaries whose volume is insignificant by comparison.[6]

Partition

Separating a term into its parts and explaining each part individually can effectively expand a definition and help the reader visualize the object. In a biology report on blood circulation in humans, a student uses partition to expand her definition of *artery:*

> An artery is a blood vessel that carries blood from the heart to other parts of the body. All arteries have three main layers: the intima, the media, and the adventitia. The intima is the innermost layer and consists primarily of connective tissue and elastic fibers. The media, or middle layer, consists of smooth muscle fibers and connective tissue. The outermost layer is the adventitia, which is mainly connective tissue. Each of these layers consists of sublayers of muscle and elastic fibers. The thickness of all the layers depends on the size of the artery.

Stipulation

In some circumstances, writers need to restrict the general meaning of a term to use it in a particular context. Such stipulative definitions often are necessary in research reports when the reader needs to understand the limitations of the term as used by the writer. Without the stipulative definition, the reader probably would assume a broader meaning than the writer intends. The following is a stipulative definition from a sociologist's report of interviews with women in homeless shelters in a large city:

> For this report, the term *homeless women* applies to females over the age of 20 who, at the time of the study, had been in a recognized homeless shelter for at least one week. In addition, the women were unemployed when they were interviewed and had been on welfare assistance at least once before going to the shelter.

PLACING DEFINITIONS IN DOCUMENTS

If a definition is part of a longer document, you need to place it, so readers will find it both useful and nondisruptive. Include definitions (1) within the text, (2) in an appendix, (3) in footnotes, and/or (4) in a glossary.

Within the Text

Informal definitions or formal sentence definitions can be incorporated easily into the text of a document. If an expanded definition is crucial to the success of a document, it can also be included in the text. However, expanded definitions may interrupt a reader's concentration on the main topic. If the expanded definitions are not crucial to the main topic, consider placing them in an appendix or glossary.

In an Appendix

In a document intended for multiple readers, lengthy expanded definitions for nonexpert readers may be necessary. Rather than interrupt the text, place such expanded definitions in an appendix at the end of the document. Readers who do not need the definitions can ignore them, but readers who do can easily find them. Expanded definitions longer than one paragraph usually should be in an appendix unless they are essential in helping readers understand the information in the document.

In Footnotes

Writers often put expanded definitions in footnotes at the bottoms of pages or in endnotes listed at the conclusion of a long document. Such notes do not interrupt the text and are convenient for readers because they appear on the same page or within a few pages of the first reference to the term. If a definition is longer than one paragraph, however, it is best placed in an appendix.

In a Glossary

If your document requires both formal sentence definitions and expanded definitions, a glossary or list of definitions may be the most appropriate way to present them. Glossaries are convenient for readers because the terms are listed alphabetically and all definitions appear in the same location. Chapter 11 discusses preparation of glossaries.

CHAPTER SUMMARY

This chapter discusses how to write definitions of technical terms that readers need to understand. Remember:

- Definitions distinguish an object or concept from similar objects or concepts and clarify for readers the limits of the term.

- Definitions are necessary when (1) technical information for expert readers must be rewritten for nonexpert readers, (2) not all readers will understand the technical terms used in a document, (3) rare or new technical terms are used, and (4) a term has more than one meaning.

- Informal definitions explain a term with another word or a phrase that has the same general meaning.

- Formal sentence definitions explain a term by placing it in a group and identifying the features that distinguish it from other members of the same group.

- Expanded definitions explain the meaning of a term in a full paragraph or more than one paragraph.

- Writers expand definitions by using one or more of these strategies: cause and effect, classification, comparison and contrast, description, development or history, etymology, examples, method of operation or process, negation, partition, and stipulation.

- Definitions may be placed within the text or in an appendix, footnotes, or glossary, depending on the needs of readers.

SUPPLEMENTAL READINGS IN PART 2

Caher, J. M. "Technical Documentation and Legal Liability," *Journal of Technical Writing and Communication,* p. 486.

Garhan, A. "ePublishing," *Writer's Digest,* p. 495.

ENDNOTES

1. "Stages of Sleep," *Mayo Clinic on Chronic Pain* (Rochester, MN: Mayo Foundation for Medical Education and Research, 1999), p. 102.

2. K. Montgomery, *End Your Carpal Tunnel Pain without Surgery* (Nashville, TN: Rutledge Hill Press, 1998), p. 30.

3. W. C. Davis, *Rebels & Yankees: Fighting Men of the Civil War* (New York: Gallery Books, 1989), p. 132.

4. *Space Shuttle,* NASA, PMS 013-B (KSC), December 1991.

5. J. Krakauer, *Into Thin Air* (New York: Anchor Books, 1998), p. 161.

6. W. Bascom, *The Crest of the Wave: Adventures in Oceanography* (New York: Doubleday, 1988), p. 294.

MODEL 8-1   Commentary

This model shows two expanded definitions from Web sites. The first is a definition of *wind* and the accompanying drawing from a Web site of the U.S. Department of Energy, Energy Information Administration (http://www. eia.doe.gov). This Web page discusses the importance of wind as a source of energy and is designed for student readers. The line drawing illustrates the explanation of air movement in the definition.

The second excerpt is from the Federal Bureau of Investigation (FBI) Web site (http://www.fbi.gov) and provides expanded definitions of *criminal enterprise* and *organized crime*.

Discussion

1. Discuss the usefulness of the wind definition and the graphic aid for the intended reader.

2. Discuss the FBI definitions, and analyze the intended readers.

3. What expansion strategies did the writers use for these definitions?

4. In groups, develop an expanded definition for readers who are in the sixth grade for one of the following terms, and then design a Web page focused on that definition:

- Landslide
- Kidnapper
- Burglar
- Paint brush
- Swiss army knife
- Ceiling fan

5. Discuss the Web page you designed in Exercise 4. What changes would you make if asked to prepare a printed fact sheet based on your definition for adult readers? Draft a fact sheet for adult readers for class discussion.

Energy Kid's Page
energy information administration
energy facts fun & games energy history classroom activities

search kid's page go

glossary

Wind Energy -- Energy from Moving Air

Wind is air in motion. It is produced by the uneven heating of the earth's surface by the sun. Since the earth's surface is made of various land and water formations, it absorbs the sun's radiation unevenly. When the sun is shining during the day, the air over landmasses heats more quickly than the air over water. The warm air over the land expands and rises, and the heavier, cooler air over water moves in to take its place, creating local winds. At night, the winds are reversed because the air cools more rapidly over land than over water. Similarly, the large atmospheric winds that circle the earth are created because the surface air near the equator is warmed more by the sun than the air over the North and South Poles. Wind is called a renewable energy source because wind will continually be produced as long as the sun shines on the earth. Today, wind energy is mainly used to generate electricity.

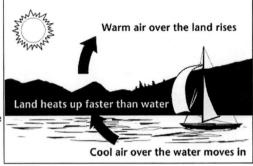

MODEL 8-1   U.S. Department of Energy Web Page

FBI Priorities
About Us
Press Room
What We Investigate
Counterterrorism
Most Wanted
Your Local FBI Office
Reports & Publications
FBI History
For the Family
Employment
How Do I...
Search
Home

Submit A Tip
Apply Today
Links
Contact Us
Site Map
Privacy Policy

Investigative Programs
Organized Crime

Glossary

About Organized Crime

Italian Organized Crime – Labor Racketeering Unit

Eurasian Organized Crime Unit

Asian/African Organized Crime

Case Summaries

Statutes

Glossary of Terms

Criminal Enterprise

The FBI defines a criminal enterprise as a group of individuals with an identified hierarchy, or comparable structure, engaged in significant criminal activity. These organizations often engage in multiple criminal activities and have extensive supporting networks. The terms Organized Crime and Criminal Enterprise are similar and often used synonymously. However, various federal criminal statutes specifically define the elements of an enterprise that need to be proven in order to convict individuals or groups of individuals under those statutes.

The Racketeer Influenced and Corrupt Organizations statute, or Title 18 of the United States Code, Section 1961(4) defines an enterprise as "any individual, partnership, corporation, association, or other legal entity, and any union or group of individuals associated in fact although not a legal entity."

The Continuing Criminal Enterprise statute or Title 21 of the United States Code, Section 848(c)(2) defines a criminal enterprise as any group of six or more people, where one of the six occupies a position of organizer, a supervisory position, or any other position of management with respect to the other five, and which generates substantial income or resources, and is engaged in a continuing series of violations of Subchapters I and II of Chapter 13 of Title 21 of the United States Code.

Organized Crime

The FBI defines organized crime as any group having some manner of a formalized structure and whose primary objective is to obtain money through illegal activities. Such groups maintain their position through the use of actual or threatened violence, corrupt public officials, graft, or extortion, and generally have a significant impact on the people in their locales, region, or the country as a whole.

Federal Bureau of Investigation Web Page

Chapter 8   Exercises

1. Identify the reason the following sentences are not formal sentence definitions. Rewrite each into a correct formal sentence definition.

   a. A curtain is a piece of fabric.

   b. A detective gathers information and clues.

   c. A stapler staples two pieces of paper together.

   d. Acupuncture is an alternative health treatment.

   e. A veterinarian treats animals.

   f. A free throw is a basketball shot.

   g. The Dow Jones average is an index of prices.

   h. A grapefruit is a citrus fruit.

   i. An editorial influences opinions.

   j. Rowing is the movement of rowers.

2. Write a formal sentence definition for each of the following terms. Compare your definitions with those of your classmates. Discuss how many possible definitions each word has.

rock	ring	pit	roll
skin	tie	mint	leaf
glass	pin	fly	crown

3. Write an expanded definition of one of the formal sentence definitions you wrote for Exercise 2.

4. Write a memo to your supervisor at your present job to request a new piece of equipment or a new model of supplies you use. You know that your supervisor is reluctant to spend money on new materials right now, but you think this request is important. Include a formal sentence definition and an expanded definition of the materials you are requesting. Memo format is discussed in Chapter 14.

COLLABORATIVE EXERCISES

5. In groups, write a formal sentence definition for *three* of the following terms. Share your definitions with the class. Next, write an expanded definition of *one* of the terms.

golf club	kayak	rink
hurdle	luge	parallel bars
puck	ice skate	springboard

6. In groups, assume that you are the Student Committee on Campus Equipment. Your job is to identify equipment that will enhance the academic activities of campus life. Select a piece of equipment, and research the product details on the Internet. Write a memo to the appropriate administrator on your campus to request this equipment for use in classes, the library, or another study site. Include an expanded definition of the equipment in your memo. Memo format is discussed in Chapter 14.

7. In groups, assume that you are the student representatives on the University Council. Your job is to monitor the recreational/sports needs of the campus. Select some equipment or additional facility that you think will enhance campus sports or recreational activities. Research details about your choice on the Internet. Write a memo, including expanded definitions of the equipment or facility you are requesting, to the president of the University Council, Dr. Jonathan Gatlin. Explain where Dr. Gatlin can get more information. Memo format is covered in Chapter 14.

INTERNET EXERCISES

8. Search the Internet for information about one of the following terms. Based on your research, develop an expanded definition and a Web page about the term for the general reader. Include links to relevant sources of information.

user interface	laser
fiber optics	water purification
solar energy	food irradiation
cloning	genetically modified food

9. Select a charity or nonprofit organization, and search the Internet for information about the activities of this group. Develop a fact sheet about the group that includes definitions of the group's major purposes. *Or,* if your instructor prefers, develop an information Web page with the definitions and links to relevant sources.

10. Search the Internet for information about the following types of stone, and write expanded definitions for the general reader. Design an information Web page, using your definitions and advising readers on the best uses for these stones in home building.

granite

marble

terrazzo

limestone

soapstone

CHAPTER 9

Description

UNDERSTANDING DESCRIPTION

Technical description provides readers with precise details about the physical features, appearance, or composition of a subject. A technical description may be a complete document in itself, such as a product Web page or a technical manual. Technical descriptions often are included as parts of longer documents, such as reports or operating manuals. The following are some general uses for technical description:

1. *Web pages*—Company Web sites usually include technical descriptions of all available products or services. A consumer planning to buy a camera might visit the Web sites of several camera manufacturers to find product descriptions. These product descriptions generally would include the physical specifications, special features, and consumer appeal of the cameras.

2. *Sales brochures*—Printed product information helps dealers advise customers about a product's specific features. A sales representative who visits a potential customer can leave printed material with that customer to reinforce the sales presentation.

3. *Proposals and other reports*—Readers usually need descriptions of equipment and locations in a report before they can make decisions. A report discussing the environmental impact of a solid-waste landfill at a particular site, for example, probably would include a description of the location to help readers visualize the site and the potential changes.

4. *Manuals*—Descriptions of equipment help operators understand the principles behind running a piece of machinery. Technicians need a record of every part of a machine and its function to effectively assess problems and make repairs. Consumer instruction manuals often include descriptions to help readers locate important parts of the product.

5. *Magazine articles and brochures for general readers*—Articles and brochures about science and technology often include descriptions of mechanisms, geologic sites, or natural phenomena to help readers understand the subject. An article for general readers about a new wing design on military aircraft may include a description of the wings and their position relative to the body of the aircraft to help readers visualize the design. A brochure written for visitors at a science institute describing the development of a solar heat collector may include several descriptions of the mechanism in development so readers can understand how the design changed over time.

Plan and draft a technical description with the same attention to readers and their need for the information that you apply to other writing tasks.

PLANNING DESCRIPTIONS

In deciding whether to include descriptions in your document, consider your readers' knowledge of the subject and why they need your information. A student in an introductory botany class has little expert knowledge about the basic parts of a flower. Such a student needs a description of these parts to understand the variations in structure of different flowers and how seeds are formed. A research report to botanists about a new hybrid flower, however, probably would not include a description of these basic parts because botanists already know them well. The botanists, on the other hand, do need a description of the hybrid flower because they are not familiar with this new variety, and they need the description to understand the research results in the report. Both types of readers, then, need descriptions, but they differ in (1) the subjects they need described, (2) the kinds of information they need, (3) the appropriate language and detail, and (4) the appropriate graphic aids.

Subjects

As you analyze your readers' level of knowledge about a subject and their need for information, consider a variety of subjects that, if described in detail, might help your readers understand and use the information in your document. Here are some general subjects to consider as you decide what to describe for readers:

1. *Mechanism*—any machine or device with moving parts, such as a steam turbine, camera, automobile, or bicycle

2. *Location*—any site or specific geologic area, such as the Braidwood Nuclear Power Plant, the Lake Michigan shoreline at Milwaukee, or the Moon's surface

3. *Organism*—any form or part of plant or animal life, such as an oak tree, a camel, bacteria, a kidney, or a hyacinth bulb

4. *Substance*—any physical matter or material, such as cocaine, lard, gold, or milk

5. *Object*—any implement without moving parts, such as a paper cup, a shoe, a photograph, or a floppy disk

6. *Condition*—the physical state of a mechanism, location, organism, substance, or object at a specific time, such as a plane after an accident, a tumor before radiation, or a forest area after a fire

Whenever your readers are not likely to know exactly what these subjects look like and need to know this to understand the information in your document, provide descriptions.

Kinds of Information

To be useful, a technical description must provide readers with a clear image of the subject. Include precise details about the following whenever such information will help your readers understand your subject better:

- Purpose or function
- Weight, shape, measurements, and materials
- Major and minor parts, their locations, and how they are connected
- Texture, sound, odor, and color
- Model numbers and names
- Operating cycle
- Special conditions for appropriate use, such as time, temperature, and frequency

The writer of this excerpt from an architectural book about the development of greenhouses uses specific details about measurements, materials, and shapes to describe a particular building:

> The wooden tie beams in the Great Conservatory at Chatsworth represent a building innovation in the application of not only cast-iron but also wrought-iron semicircular ribs, in arches and in frameworks. In this building the creation of space and the various forms of iron rib were further developed structurally. This freestanding building, 124 feet wide and 68 feet high, was created by the construction of two glass vaults, built in the ridge-and-furrow system, with a basilicalike cross-section. Rows of cast-iron columns joined by cast-iron girders carried the semicircular wooden ribs of the Paxton gutters to form a high nave with a span of 70 feet. The glass vaults of the aisles, reaching down to the low masonry base, abutted from both sides onto the main iron frame at a height of 40 feet. The aisle had the same profile, but it had quadrant-shaped ribs. The ribs of the surrounding aisles also functioned as buttresses to support the lateral thrust of the main ribs of the nave.[1]

To determine the amount and kinds of details appropriate for a technical description, assess your reader's level of knowledge about the subject and purpose in reading. The reader of this description probably is familiar with architectural terms, such as trusses and vaults. The description focuses on the components of the building's supporting structure and provides measurements—all details relevant to readers who are interested in architecture. If the readers were horticulturists, a description of the interior design and arrangement of plants and flowers probably would be of more interest.

The following description of the effects of lead poisoning in children is from a report to Congress by the U.S. Department of Housing and Urban development. The emphasis is on the typical physical condition of a child suffering from lead poisoning:

> Very severe childhood lead poisoning—involving such symptoms as kidney failure, gastrointestinal problems, comas, convulsions, seizures, and pronounced mental retardation—can occur at blood lead levels as low as 80 µg/dl. At or above 40 µg/dl, children may experience reduced hemoglobin (the oxygen-carrying substance in blood), the accumulation of a potential neurotoxicant known as ALA, and mild anemia. Near 30 µg/dl, studies have found slowed nerve conduction velocity. And between 10–15 and 25 µg/dl, researchers have documented slower reaction time, reductions in intelligence and short-term memory, other neurobehavioral deficits, and adverse effects on heme biosynthesis and vitamin D and calcium metabolism.[2]

In this description, the writer begins by naming potential diseases in general terms and then shifts to specific technical details about blood factors. Notice the informal definition included for the primary (nonexpert) readers.

This description of a location centers on the area around the site of a sunken Spanish galleon:

> The anchor has been found in 25 feet of water on the southern edge of an area known as the "Quicksands." The name didn't mean the area swallowed divers, but that the bottom was covered with a sand composed of loose shell fragments that are constantly shifted by the waves and tides. The result is a series of sinuous "dune" formations on the bottom. Tidal channels cross these dunes, running from northeast to southwest. Around the anchor, the sand is nearly 15 feet deep. Below it is a bed of limestone, the surface of which is rippled and pocked with small hollows and crevices. The sand piled on top of the limestone gradually thins out until it disappears about a quarter of a mile from the anchor. Southeast of the Quicksands, the water depth increases gradually to about 40 feet, and the sand covering the bedrock thins out to a veneer just two and a half inches thick. Small sponges and sea fans grow in this area, which we called the Coral Plateau. Continuing to the southeast, away from the anchor, the Coral Plateau slopes downward into the "Mud Deep." Here, the bedrock dips sharply, possibly following the channel of a river that may have flowed through the area when it was above sea level thousands of years ago. This channel is filled with muddy silt.[3]

In this description, the writer includes a few measurements, but he emphasizes the natural formations of sand and rock present in the area. Notice that the writer organizes spatially—the area around the anchor first and then progressively farther away, moving southeast.

Remember that descriptions may cover either (1) a particular mechanism, location, organism, substance, object, or condition or (2) the general type. The description of the typical child with lead poisoning is a general one. Individual cases may differ, but most children with lead poisoning will fit this description in some way. The descriptions of the conservatory and the sea area, however, are specific to that building and that location. In the same manner, a product description of Model XX must be specific to that model alone and, therefore, different from a description of Model YY. Be sure to clarify for the reader whether you are describing a general type or a specific item.

Details and Language

All readers of technical descriptions need accurate detail. However, some readers need more detail and can understand more highly technical language than others. Here is a brief description from a book about the solar system written for general readers interested in science:

> The Galaxy is flattened by its rotating motion into the shape of a disk, whose thickness is roughly one-fiftieth of its diameter. Most of the stars in the Galaxy are in this small disk, although some are located outside it. A relatively small, spherical cluster of stars, called the nucleus of the Galaxy, bulges out of the disk at the center. The entire structure resembles a double sombrero with the galactic nucleus as the crown and the disk as the brim. The Sun is located in the brim of the sombrero about three-fifths of the way out from the center to the edge. When we look into the sky in the direction of the disk we see so many stars that they are not visible as separate points of light, but blend together into a luminous band stretching across the sky. This band is called the Milky Way.[4]

In this description, the writer compares the galaxy to a sombrero, something most general readers are familiar with, and then locates the various parts of the galaxy on areas of the sombrero, thus creating an image that general readers will find easy to use. The writer also uses an informal tone, drawing readers into the description by saying, "When we look into the sky . . . " This informal tone is appropriate for general readers, but not for expert readers who need less assistance in visualizing the subject.

Here is a description paragraph from the Toshiba Web site. The description is included in a press release from Toshiba America Medical Systems announcing a new ultrasound system:

> The multifunctional Nemio 30 adds continuous wave Doppler and phased array probe capabilities that make it ideal for the cardiology environment, including the private practice cardiologist, as well as hospital radiology departments and diagnostic imaging centers. In addition to the linear and convex probe ports featured in the 10 and 20, Nemio 30 also has a third port for use of a transesophageal probe and a separate port for a pencil probe.[5]

Clearly, the readers of this press release are expected to have expert knowledge of the technical terms, such as "array probe," as well as understand the purpose of the system. The nonexpert reader would not understand this description. Company Web sites usually contain product information designed for potential customers. The language used in the descriptions is selected to appeal to these readers. All technical descriptions must include appropriate language for the readers and the kinds of details they need.

Follow these guidelines for using appropriate language in technical descriptions:

1. Use specific rather than general terms. Notice these examples:

General terms	*Specific terms*
short	1 in. tall
curved	S-shaped
thin	⅛ in. thick
light	cream-colored
nearby	4 in. from the base
fast	400 rpm
large	7 ft high
light	1.5 oz
heavy	16 t
noise	high-pitched shriek

2. Indicate a range in size if the description is of a general type that varies, or give an example:

   Poor: Bowling balls for adults usually weigh 12 lb.

   Better: Bowling balls for adults vary in weight from 10 to 16 lb.

3. Use precise language, but not language too technical for your readers. Do not write "a combination of ferric hydroxide and ferric oxide" if "rust" will do.

4. If you must use highly technical terms or jargon and some of your readers are nonexperts, define those terms:

   The patient's lipoma (fatty tumor) had not grown since the previous examination.

5. Compare the subject to simple, well-known items and situations to help general readers visualize it:

   The difference in size between the Sun and the Earth is similar to the difference between a baseball and a grape seed.

Be specific and accurate, but do not overwhelm your readers with details they cannot use. Consider carefully the language your readers need and the number of details they can use appropriately.

Graphic Aids

Graphic aids are an essential element in descriptions because readers must develop a mental picture of the subject, and words alone may not be adequate to paint that picture successfully. Chapter 6 discusses graphic aids in detail along with format devices to highlight certain facts. These graphic aids usually are effective ways to illustrate details in descriptions:

1. *Photographs*—Photographs supply a realistic view of the subject and its size, color, and structure, but they may not properly display all features or show locations clearly. Photographs are most useful for general readers who want to know the overall appearance of a subject. A guide to house plants for general readers should include photographs, for instance, so readers can differentiate among plants. A research article for botanists, however, probably would include line drawings showing stems, leaves, whorls, and other distinctive features of the plants being discussed.

2. *Line drawings*—Line drawings provide an exterior view showing key features of mechanisms, organisms, objects, or locations and special conditions, as well as how these features are connected. Line drawings are useful when readers need to be familiar with each part or area.

3. *Cutaway and exploded drawings*—Cutaway drawings illustrate interior views, including layers of materials that cannot be seen from photographs or line drawings of the exterior. Exploded drawings emphasize the separate parts of a subject, especially those that might be concealed if the object is in one piece. Often readers need both line drawings of the exterior and interior and exploded drawings to fully understand a subject.

4. *Maps, floor plans, and architects' renderings*—Maps show geographic locations, and floor plans and architects' renderings show placement of objects within a specified area. Features that may be obscured in a photograph from any angle can be marked clearly on these illustrations, giving readers a better sense of the composition of the area than any photograph could.

Use as many graphic aids as you believe your readers need to get a clear picture of your subject. Also, be sure to direct your readers' attention to the graphic aids by references in the text, and always use the same terms in the text and in your graphic aids for specific parts or areas.

STRATEGIES—Complete Comparisons

When comparing two elements, identify them fully in the sentence.

No: The Washburn Clinic is better staffed.

Yes: The Washburn Clinic is better staffed than the Winslow Health Center is.

No: Replacing the spark plugs is easier than the muffler.

Yes: Replacing the spark plugs is easier than replacing the muffler.

WRITING DESCRIPTIONS

After identifying your expected readers and the kinds of information they need, consider how to organize your descriptions.

Organization

In some cases, company policy or institution requirements dictate the organization of a description. An example of a predetermined description format appears in Model 9-5, a portion of a police report. Although police report formats may vary from city to city, most departments require a narrative section describing the incident that led to police intervention. The writer must follow the department's established format because readers—police, attorneys, judges, and social-service workers—handle hundreds of similar reports and rely on seeing all reports in the same format with similar information in the same place.

If you are not locked into a specific description format for the document you are writing, select an organizational pattern that will best serve your readers, usually one of these:

Spatial—The spatial pattern most often is used for descriptions of mechanisms, objects, and locations because it is a logical way to explain how a subject looks. Remember that you must select a specific direction and follow it throughout the description, such as from base to shade for a lamp, from outside to inside for a television set, and from one end to the other for a football field. Writers often also explain the function of a subject and its main parts. This additional information is particularly important if readers are not likely to know the purpose of the subject or it will not be apparent from the description of parts or composition. Thus, a description of a stomach pump may include an explanation of the function of each valve and tube, so readers can understand how the parts work together.

Chronological—The chronological pattern describes features of a subject in the order they were produced or put together. This information may be central to a description of, for example, a Gothic church that took two centuries to build or an anthropological dig where layers of sediments containing primitive tools have been deposited over two million years.

In all cases, consider your readers' knowledge of the subject and the organization that will most help them use the information effectively. A general reader of a description of a Gothic church may want a chronological description that focuses on the historical changes in exterior and interior design. A structural engineer, however, may want a chronological description that focuses on changes in materials and construction techniques.

Sections of a Technical Description

Model 9-1 presents a typical outline for a technical description of an object or mechanism. As this outline illustrates, a technical description has three main sections: an introduction, a description of the parts, and a description of the cycle of operations.

Introduction

The introduction to a technical description orients readers to the subject and gives them general information that will help them understand the details in the body of the description. Follow these guidelines:

1. Write a formal sentence definition of the complete mechanism or object to ensure that your readers understand the subject. If the item is a very common one, however, you may omit the formal definition unless company policy calls for it.

2. Clarify the purpose if it is not obvious from the definition. In some cases, how the object or mechanism functions may be most significant to your readers, so explain this in the introduction.

3. Describe the overall appearance, so readers have a sense of size, shape, and color. For general readers, include extra details that may

I. Introduction

   A. Definition of the object

   B. Purpose of the object

   C. Overall description of the object—size, color, weight, etc.

   D. List of main parts

II. Description of Main Parts

   A. Main part 1

      1. Definition of main part 1

      2. Purpose of main part 1

      3. Details of main part 1—color, shape, measurements, etc.

      4. Connection to main part 2 or list of minor parts of main part 1

      5. Minor part 1 (if relevant)

         (a) Definition of minor part 1

         (b) Purpose of minor part 1

         (c) Details of minor part 1—color, shape, measurements, etc.

      6. Minor part 2 (if relevant)

         ⋮

   B. Main part 2

      ⋮

III. Cycle of Operation

   A. How parts work together

   B. How object operates

   C. Limitations

MODEL 9-1   Model Outline of Technical Description of a Mechanism

be helpful, such as a comparison with an ordinary object. Also include, if possible, a graphic illustration of the object or mechanism.

4. List the main parts in the order you plan to describe them.

Description of Parts

The body of a technical description includes a formal definition and the physical details of each major and minor part presented in a specific organizational pattern, such as spatial or chronological. Include as many details as your readers need to understand what the parts look like and how they are connected. If your readers plan to construct the object or mechanism, they will need to know every bolt and clamp. General readers, on the other hand, usually are interested only in major parts and important minor parts. Graphic illustrations of the individual parts or exploded drawings are helpful to readers in this section.

Cycle of Operation

Technical descriptions for in-house use or for expert readers seldom have conclusions. General readers, however, may need conclusions that explain how the parts work together and what a typical cycle of operation is like. If the description will be included in sales literature, the conclusion often stresses the special features or advantages of the product.

Model 9-2 (page 221) is a technical description of a specific model of a toaster written by a student as a section of a consumer manual that includes a description, operating instructions, and maintenance guidelines. Because this description is of a particular model, the writer describes the parts and their locations in detail. The organization is spatial, moving from the lowest part, the base, to the cord, which enters the housing above the base, and then to the housing. Because the housing is so large, the writer partitions it, describing the outer casing first, the heating wells second, and then the control levers. The writer ends with a brief description of the operating cycle. Notice that this description contains (1) formal sentence definitions of the toaster and each main part, (2) headings to direct readers to specific topics, and (3) specific measurements for each part.

■■■■■ CHAPTER SUMMARY

This chapter discusses writing technical descriptions. Remember:

- Descriptions provide readers with precise details about the physical features, appearance, or composition of a subject.

- In planning a description, writers consider what subjects readers need described, the kinds of information readers need, appropriate detail and language, and helpful graphic aids.

- Subjects for descriptions may include mechanisms, locations, organisms, substances, objects, and conditions.

- Descriptions should include precise details about purpose, measurements, parts, textures, model numbers, operation cycles, and special conditions when appropriate for readers.

- The amount of detail and degree of technical language in a description depends on the readers' purpose and their technical knowledge.

I. Introduction

The Kitchen King toaster, Model 49D, is an electric appliance that browns bread between electric coils inside a metal body. The toaster also heats and browns frozen and packaged foods designed especially to fit in the vertical heating wells of a toaster. The Kitchen King Model 49D is $6\frac{3}{4}$ in. high, $4\frac{1}{2}$ in. wide, $8\frac{3}{4}$ in. long, and weighs 2 lb. There are three main parts: the base, the cord, and the housing.

II. Main Parts

The Base

The base is a flat stainless steel plate, $4\frac{1}{2}$ in. wide and $8\frac{3}{4}$ in. long. It supports the housing of the toaster and is nailed to it at all four corners. The base rests on four round, black, plastic feet, each $\frac{1}{2}$ in. high and $\frac{3}{4}$ in. in diameter. The feet are attached to the base $\frac{1}{2}$ in. inside the four corners. The front and back of the base ($2\frac{1}{2}$ in. wide) are edged in black plastic strips that protrude $\frac{3}{4}$ in. past the base. A removable stainless steel plate, $3\frac{1}{2}$ in. wide and $5\frac{1}{4}$ in. long, called the crumb tray, is centered in the base. One of the shorter sides of the crumb tray is hinged. The opposite side has a catch that allows the hinged plate to swing open for removal of trapped crumbs.

The Cord

The cord is an insulated cable that conducts electric current to the heating mechanism inside the body of the toaster. It is 32 in. long and $\frac{1}{8}$ in. in diameter. Attached to the base of the toaster at the back, it enters the housing through an opening just large enough for the cable. At the other end of the cord is a two-terminal plug, measuring 1 in. long, $\frac{3}{4}$ in. wide, and $\frac{1}{2}$ in. thick. The two prongs are $\frac{1}{2}$ in. apart and are each $\frac{5}{8}$ in. long, $\frac{1}{4}$ in. wide, and $\frac{1}{16}$ in. thick. The plug must be attached to an electrical outlet before the toaster is operable.

Housing

The housing is a stainless steel boxlike cover that contains the electric heating mechanism and in which the toasting process takes

MODEL 9-2 Technical Description: Kitchen King Toaster, Model 49D

place. The housing is $8\frac{3}{4}$ in. long, $6\frac{1}{4}$ in. high, and $4\frac{1}{2}$ in. wide. The front and back are edged at the top with black plastic strips that protrude past the housing $\frac{3}{4}$ in. The housing also contains two heating wells and two control levers.

Heating Wells. The heating wells are vertical compartments, $5\frac{1}{2}$ in. long, $1\frac{1}{4}$ in. wide, $5\frac{3}{4}$ in. deep. The wells are lined with electric coils that brown the bread or toaster food when the mechanism is activated. Inside each heating well is a platform on which the toast rests and which moves up and down the vertical length of the well.

Control Levers. The control levers are square pieces of black plastic attached to springs that regulate the heating and the degree of browning. The levers are $1\frac{1}{2}$ in. square and are on the front of the toaster.

Heat Control Lever. The heat control lever is centered 1 in. from the top of the housing and is attached to the platforms inside both heating wells. When the lever is pushed down, it moves along a vertical slot to the base of the toaster, lowering the platforms inside the heating wells. This movement also triggers the heating coils.

Browning Lever. The browning lever is located in the lower right corner on the front of the toaster. The browning lever moves horizontally along a 2-in. slot. The slot is marked with three positions: dark, medium, and light. This lever controls the length of time the heating mechanism is in operation and, therefore, the degree of brownness that will result.

III. Conclusion

The user must insert the cord into a 120-volt AC electrical outlet to begin the operating cycle. Next, the user inserts one or two pieces of bread or toaster food into the heating wells, moves the browning lever to the desired setting, and lowers the heat control lever until it catches and activates the heating mechanism. The toaster will turn off automatically when the heating cycle is complete, and the platforms in the heating wells will automatically raise the bread or toasted food, so the user can remove it easily.

- Graphic aids are essential in descriptions to help readers develop a mental picture of the subject.

- Most descriptions are organized spatially or chronologically.

SUPPLEMENTAL READINGS IN PART 2

Barefoot, D. K. "Ten Tips on Writing White Papers," *Intercom,* p. 483.
Garhan, A. "ePublishing," *Writer's Digest,* p. 495.
McAdams, M. "It's All in the Links: Readying Publications for the Web," *The Editorial Eye,* p. 512.

ENDNOTES

1. From G. Kohlmaier and B. von Sartory, *Houses of Glass, A Nineteenth-Century Building Type* (Cambridge, MA: The MIT Press, 1991), pp. 87–88.

2. From U.S. Department of Housing and Urban Development, *Comprehensive and Workable Plan for the Abatement of Lead-Based Paint in Privately Owned Housing* (Washington, DC: U.S. Government Printing Office, December 7, 1990), pp. 2–3.

3. From R. D. Mathewson III, *Treasure of the Atocha* (New York: E. P. Dutton, 1986), p. 67.

4. From R. Jastrow, *Red Giants and White Dwarfs* (New York: W. W. Norton, 1979), p. 27.

5. "Toshiba Debuts Nemio™ Ultrasound," *Toshiba Web site.* Retrieved May 8, 2001, from http://www. toshiba.com/tams/press/05082001.html.

MODEL 9-3   Commentary

This model is a Web page from the U.S. Bureau of Reclamation describing the Hoover Dam power plant (http://www.usbr.gov/lc/hooverdam/History/workings/powerplant.html). The page includes links to several other types of information about Hoover Dam.

Discussion

1. Discuss the design of the Web page. What elements are useful for the general reader?

2. In groups, find the definitions included in this description. Discuss how helpful these are for the general reader.

3. In groups, decide what features you would include in a Web page description of the building you are in at the moment. Discuss your ideas with the rest of the class.

4. In groups, compare the descriptions in this model with those in Models 6-23 and 9-4. All these models are aimed at the general reader. What differences of style in the descriptions do you notice? What differences of emphasis do you notice? Which descriptions contain sales appeal?

Return to the Story
of Hoover Dam

The Power Plant

How it all works

The Dam

The Reservoir

The Power Plant

Hoover Dam
Statistics

Colorado River
Basin Map

Jet Flow Gates

There are 17 main turbines in Hoover Powerplant. The original turbines were all replaced through an uprating program between 1986 and 1993. With a rated capacity of 2,991,000 horsepower, and two station-service units rated at 3,500 horsepower each, for a plant total of 2,998,000 horsepower, the plant has a nameplate capacity of 2,074,000 kilowatts. This includes the two station-service units, which are rated at 2,400 kilowatts each.

Hoover Dam provides generation of low-cost hydroelectric power for use in Nevada, Arizona, and California. Hoover Dam alone generates more than 4 billion kilowatt-hours a year - enough to serve 1.3 million people. From 1939 to 1949, Hoover Powerplant was the world's largest hydroelectric installation; with an installed capacity of 2.08 million kilowatts, it is still one of the country's largest.

Hoover Dam's $165 million cost has been repaid, with interest, to the Federal Treasury through the sale of its power. Hoover Dam energy is marketed by the Western Area Power Administration to 15 entities in Arizona, California, and Nevada under contracts which expire in 2017. Most of this power, 56 percent, goes to southern California users; Arizona contractors receive 19 percent, and Nevada users get 25 percent. The revenues from the sale of this power now pay for the dam's operation and maintenance. The power contractors also paid for the uprating of the powerplant's nameplate capacity from 1.3 million to over 2.0 million kilowatts.

Hydroelectric Generators

The Hoover Dam Powerplant is arranged in two wings, 650 feet in length, one on each side of the river. There are nine main generator units on the Arizona side of the river and eight on the Nevada side. The larger of these units each have a generating output of 133 megawatts. There are two in-house generators capable of generating 2,400 kilowatts

MODEL 9-3   U.S. Bureau of Reclamation Web Page

each. The total rated capacity of the plant is 2,074 megawatts.

Hydroelectricity is a clean, renewable (the water is not consumed and can be used for other purposes) source of energy that does not result in air pollution, chemical runoff, or toxic waste disposal.

The primary parts of a generating unit are:

The Exciter

The Rotor

The Stator

The Shaft

The Turbine

The **exciter** is itself a small generator that makes electricity, which is sent to the rotor, charging it with a magnetic field.

The **rotor** is a series of electromagnets, also called poles. The rotor is connected to the shaft, so that the rotor rotates when the shaft rotates.

The **stator** is a coil of copper wire. It is stationary.

The **shaft** connects the exciter and the rotor to the turbine.

Water strikes the **turbine** causing it to spin. Hoover Dam uses Francis tubines.

Electricity is produced as the magnets of the rotors spin past the stationary wiring of the stator. This concept was discovered by scientist Michael Faraday in 1831 when he found that electricity could be created by rotating magnets within copper coils.

Most of the Hoover Dam generators have 40 poles, or electromagnets, and spin at 180 revolutions per minute. Electrical current leaves the generator at 16,500 volts, and is then carried to the transformers where it is 'stepped up' to 230,000 volts for transmission.

Contact: Pagemaster
This page was last
updated in September 2003

Privacy | Disclaimer | Accessibility | FOIA | Information Quality | FAQ
DOI | Recreation.gov | FirstGov

226

MODEL 9-4   Commentary

This model is from the Web site of Orion Telescopes & Binoculars (http://www. telescope.com). The site includes Web pages for each of the company's products. This page shows a color photograph of the StarMax 127 and a description of the special features of the product.

Discussion

1. Review this description of the StarMax 127, and discuss whether the language is completely objective or the writer is using sales appeal as well as facts. Cite specific places in the text to support your opinion.

2. Discuss the potential reader for this Web page. What information is the reader likely to be seeking?

3. In groups, draft a Web page description for a small product that is readily available in your group, such as a cell phone, a wristwatch, or a piece of jewelry. Discuss why the description would be on a company Web site and what audience the page should appeal to. Next, discuss designing the Web page. What graphics would you use? What other kinds of information, besides the technical description, would you include on a Web page for this product?

08/23/2001

A Winning Combination of Power, Performance, and Portability!

StarMax™ 127mm EQ Compact "Mak" Telescope

For the serious astronomer who values performance and quality optics but wants portability and easy set-up, the StarMax 127 EQ is ideal. Credit its light-folding Maksutov-Cassegrain optical design and big 127mm (5") aperture for delivering high-resolution imaging performance in a tube only 14.5" long! And with the included EQ-3 equatorial mount providing more than ample support for high-magnification study, this telescope will turn a stargazing pastime into a passion.

With 55% greater light-gathering capacity than a 4" scope and double that of a 90mm, the hundreds of deep-sky objects plotted on your star atlas now become willing targets. On the other hand, the scope's long 1540mm focal length (f/12.1) permits magnifications upwards of 250x with optional eyepieces, allowing study of surface features on Mars and ring detail on Saturn. Internal tube baffling and multi-coatings on the meniscus lens maximize image contrast and brightness. This scope simply doesn't compromise.

The same can be said for the StarMax 127's construction and accessories. No cheap plastic here as on competitors' offerings, just back-to-basics quality. The seamless aluminum tube is enameled a rich metallic burgundy and fitted with quality aluminum front and rear cells. A built-in, two-bolt 1/4"-20 plate assures rigid coupling to the mount (or to a tripod for terrestrial use). Aluminum focus knob with rubber grip. Aluminum eyepiece holder with two large thumbscrew locks. Quality 25mm (62x) Sirius Plössl eyepiece (1.25") and mirror star diagonal. A 6x26 correct-image achromatic finder scope with aluminum dovetail bracket. Even a padded carrying case.

Optional EQ-3M single and dual-axis electronic drives allow hands-free tracking.

The StarMax 127 EQ will quickly become the most frequently used telescope you'll ever own!

Weighs 37 lbs. total. One-year limited warranty.

MODEL 9-4   Orion Telescope Web Page

MODEL 9-5 Commentary

This portion of a police accident report is the required narrative after routine identifications, license numbers, date, and times have been recorded. The readers of police reports include district attorneys, defense lawyers, police, social workers, judges, insurance investigators, and anyone who may have a legal involvement in the case. For consistency, the sections in all such reports are always in the same order.

Each section provides specific details about locations, conditions, and physical damage. The officer includes estimates of distances and notes the paths of the vehicles so that skid marks can be checked later for vehicle speed and direction. In addition, the officer identifies the ambulance service, hospital, and paramedics and lists other police reports that relate to the same case. This information will help readers who need to see all the documents in the case or who may need to interview medical personnel about injuries.

Discussion

1. Discuss the kinds of information included in each section of this report. How would this information help the district attorney prosecuting the case?

2. Assume that an accident has occurred at an intersection you know well. Write a description of the intersection that would be appropriate for the "Description of Scene" section of this type of police report.

3. In groups, assume that an accident has taken place in your classroom. Your group members are witnesses and must write a description of the classroom for the insurance investigator. Compare your description with the ones written by other groups.

SUPPLEMENTAL COLLISION NARRATIVE	Date of Incident: 1/7/06
Location: Newbury Blvd. and Downer Ave. Shorewood, WI	Citation: MM36-T40
Subject: Coronal Primary Collision Report 06-29	Officer ID: 455
	Time: 19:25

Description of Scene: Newbury Boulevard is an east-west roadway of asphalt construction with two lanes of traffic in either direction and divided by a grassy median 10 ft wide. All lanes are 14 ft wide. A curb approximately 6 in. high is on the periphery of the East 2 and West 2 lanes. This is a residential street with no business district. Downer Avenue is a north-south roadway of asphalt construction with two lanes of traffic in either direction separated by a broken yellow line. All lanes are 12 ft wide. A curb approximately 6 in. high is on the periphery of the North 2 and South 2 lanes. At the intersection with Newbury Boulevard, Downer Avenue is a residential street. A business district exists four blocks south. The speed on both streets is posted at 30 mph. The roadways are unobstructed at the intersection. Both roadways are straight at this point, and the intersection is perpendicular. At the time of the accident, weather was clear, temperature about 12°F, and the pavements were dry, without ice. Ice was present along the curb lanes extending about 4 in. from the curbs into the East 2, West 2 lanes of Newbury Boulevard.

Description of Vehicle: The Ford Taurus was found at rest in a westerly direction on Newbury Boulevard. The midrear of the vehicle was jammed into the street light pole at the curb of lane West 2 of Newbury Boulevard. Pole number is S-456-03. The vehicle was upright, all four wheels inflated, brakes functional, speedometer on zero, and bucket front seats pressed forward against the dashboard. The front of the vehicle showed heavy damage from rollover. The grillwork was pushed back into the radiator, which, in turn, was pushed into the fan and block. The driver's door was open, but the passenger door was wedged closed. The vehicle rear wheels were on the curb, the front wheels in lane West 2 of Newbury. The windshield was shattered, but still within the frame. Oil and radiator coolant were evident on the roadway from the curb to approximately 6 ft into lane West 2 of Newbury, directly in front of the vehicle.

MODEL 9-5 Police Report

Driver: Driver Coronal was standing on the curb upon my arrival. After detecting the distant odor of alcohol on his breath and clothes and noting his red, watery eyes and general unsteadiness, I administered a field sobriety test, which Coronal failed. (See Arrest Report 06-197.)

Passenger: Passenger Throckmorton was in the passenger seat of the vehicle, pinned between the bucket seat and the dashboard. He was in obvious pain and complained of numbness in his right leg. Passenger was removed from the vehicle and stabilized at the scene by paramedics from Station 104 and transported to Columbia Hospital by Lakeside Ambulance Service.

Physical Evidence: A skid mark approximately 432 ft long began at the center line of Downer Avenue and traveled westerly in an arc into lane West 2 of Newbury Boulevard, ending at the vehicle position on the curb. Gouge marks began in the grassy devil's strip between the curb and the sidewalk. The gouge marks ended at the light pole under the rear wheels of the vehicle. A radio antenna was located 29 ft from the curb, lying on the grassy median of Newbury Boulevard. The antenna fits the antenna base stub on the Ford Taurus.

Other reports: Arrest 06-197
Blood Alcohol Test 06-416
Coefficient of Friction Test 06-136
Hospital Report on Throckmorton

Chapter 9 Exercises

1. Write a technical description of one of the following that will appeal to the general reader. Assume that your description will be featured in a brochure that includes photographs of the product.

digital camera	DVD player
cell phone	kayak
barometer	compass
pedometer	wet suit
binoculars	mountain bike

2. Assume you are a technical writer preparing a food encyclopedia that will be used by general readers in nutrition classes. You need to write technical descriptions of the following foods. Because the encyclopedia will be used in nutrition classes, your descriptions should include nutrition information and potential uses. Select one of the following for your description.

applesauce	peanut butter	lemonade
margarine	yogurt	mustard
ice cream	catsup	pizza
tacos	croissant	guacamole

3. Assume you have just taken out an insurance policy on your room in case of destruction of the contents and the interior. The insurance agent has asked you to prepare a full description of the room and its contents to keep on file in case of a claim for damages. Write the description.

4. Select a location on your campus (e.g., football stadium, chemistry building, student center, plaza, library, historical building). Assume that you are working in the development office of your school and the director wants this location presented in a three-panel brochure (front and back) that can be folded to fit into a standard business envelope. The director, Gerald Althorp, plans to mail the brochures to alumni along with an appeal for a generous donation to the school. He believes the brochures will help trigger fond memories. Your brochure should include a description of the location, any special history, the purpose or importance of the location, and special features, such as public events. Consider graphics—photographs, line drawings, maps. If you do not have access to computer software that can create a three-panel brochure, design a two-page flyer (front and back) with the same information. Write a transmittal memo to Althorp to accompany the brochure. Chapter 11 explains transmittal memos.

5. In groups, prepare a description of your campus that could be included in a brochure sent to prospective freshmen. Include the major buildings and special features of the campus as well as the location. You want to consider the sales appeal of this description for your intended reader.

6. In groups, write a description of a movable classroom desk chair with attached writing surface. Begin the description with a formal sentence definition. Compare your group's draft with the drafts of other groups. Discuss how the descriptions differ and appropriate revisions for each.

7. In groups, select a sport that requires at least four pieces of equipment. Prepare a bulletin that describes the essential equipment needed for this sport. Include a checklist of features to look for when a person buys each piece of equipment. The bulletin will be available at sporting goods stores for customers who are newcomers to the sport.

8. In groups, write a description of your classroom, including walls, ceiling, floor, and anything attached to these areas. Do not include the contents of the room. Compare your descriptions with those of other groups, and discuss your choice of organization, language, and details.

INTERNET EXERCISES

9. Read the article "Ten Tips on Writing White Pages" by Barefoot in Part 2. Select three products you use or are familiar with, and find the company Web sites. Print the product information pages, and analyze how closely these pages follow the advice in the article. Decide how appealing you think these pages are for potential customers. Write a short report to your instructor explaining your analysis. *Or,* if your instructor prefers, present your findings to the class. Report format is covered in Chapter 12, and oral presentations are covered in Chapter 16.

10. Design a Web page for an outdoor recreational area in your home state, including links to other relevant topics. *Or,* design a Web page for a particular event in your home state, such as the state fair, a historical reenactment, a classic automobile show, or a gardening expo. Present your Web page design to the class, and explain your reasons for the design.

CHAPTER 10

Instructions, Procedures, and Process Explanations

UNDERSTANDING INSTRUCTIONS, PROCEDURES, AND PROCESS EXPLANATIONS

This chapter discusses three related but distinct strategies for explaining the stages or steps in a specific course of action. Instructions, procedures, and process explanations all inform readers about the correct sequence of steps in an action or how to handle certain materials, but readers and purposes differ.

Instructions

Instructions provide a set of steps that readers can follow to perform a specific action, such as operate a forklift or build a sundeck. Readers of instructions are concerned with performing each step themselves to complete the action successfully.

Procedures

Procedures in business and industry provide guidelines for three possible situations: (1) steps in a system to be followed by one employee, (2) steps in a system to be followed by several employees, and (3) standards and rules for handling specific equipment or work systems. Readers of procedures include (1) those who must perform the actions, (2) those who supervise employees performing the actions, and (3) people, such as upper management, legal staff, and government inspectors, who need to understand the procedures to make decisions and perform their own jobs.

Process Explanations

Process explanations describe the stages of an action or system either in general (how photosynthesis occurs) or in a specific situation (how an experiment was conducted). Readers of a process explanation do not intend to perform the action themselves, but they need to understand it for a variety of reasons.

In some situations, these three methods of presenting information about systems may overlap, but to serve your readers adequately, consider each strategy separately when writing a document. Because so many people rely on instructions and procedures and companies may be liable legally for injuries or damage that results from unclear documents, directions of any sort are among the most important documents a company produces.

Graphic Aids

Instructions, procedures, and process explanations all benefit greatly from graphic aids illustrating significant aspects of the subject or those aspects that readers find difficult to visualize. Readers need such graphic aids as (1) line drawings showing where parts are located, (2) exploded drawings showing how parts fit together, (3) flowcharts illustrating the stages of a process, and (4) drawings or photographs showing how actions should be performed or how a finished product should look.

Do not, however, rely on graphic aids alone to guide readers. Companies have been liable for damages because instructions or procedures did not contain enough written text to adequately guide readers through the steps. Visual perceptions vary, and all readers may not view a drawing or diagram the same way. Written text, therefore, must cover every step and every necessary detail.

STRATEGIES—Telephone Manners

Maintain a professional attitude while handling business over the telephone. Follow these guidelines for an effective business call:

- **Do** identify yourself, your company, and your purpose. ("This is Jason Price with Acme Leasing. I'm calling Ms. Kerr with the estimate she wanted.")
- **Do** speak clearly and pleasantly
- **Do** plan what you are going to discuss and have the materials available. Avoid shuffling papers and searching for the information you wanted to talk about.
- **Do** close with a specific statement. ("I'll fax those figures over to you now.")
- **Do** inform the other person if you are using a speakerphone and identify the other people in the room.
- **Do not** eat or drink while talking on the telephone.
- **Do not** allow other noises (e.g., music, equipment) to overwhelm your conversation.

WRITING INSTRUCTIONS

To write effective instructions, consider (1) your readers and their knowledge of the subject, (2) an organizational pattern that helps readers perform the action, and (3) appropriate details and language. In addition to the initial instructions, readers may need troubleshooting instructions that tell them what to do if, after performing the action, the mechanism fails to work properly or the expected results do not appear.

Readers

As you do before writing anything, consider who your readers will be. People read instructions in one of three ways. Some, but only a few, read instructions all the way through before beginning to follow any of the steps. Others read and perform each step without looking ahead to the next. And still others begin a task without reading any instructions and turn to them only when difficulties arise. Since you have no way to control this third group, assume that you are writing for readers who will read and perform each step without looking ahead. Therefore, all instructions must be in strict chronological order.

Everyone is a reader of instructions at some point, whether on the job or in private. Generally, readers fall into one of two categories:

1. *People on the job who use instructions to perform a work-related task*—Employees have various specialties and levels of technical knowledge, but all employees use instructions at some time to guide them in doing a job. Design your instructions for on-the-job tasks for the specific groups that will use them. Research chemists in the laboratory, crane operators, maintenance workers, and clerks all need instructions that match their needs and capabilities.

2. *Consumers*—Consumers frequently use instructions when they (a) install a new product in their homes, such as a computer or a light fixture; (b) put an object together, such as a toy or a piece of furniture; and (c) perform an activity, such as cook a meal or refinish a table.

Consumers, unlike employees assigned to a particular task, represent a diverse audience in capabilities and situations.

Varied capabilities. A consumer audience usually includes people of different ages, education, knowledge, and skills. Readers of a pamphlet about how to operate a new gasoline lawn mower may be 12 or 80 years old, may have elementary education only or postgraduate degrees, may be skilled amateur gardeners, or may have no experience with a lawn mower. When writing for such a large audience, aim your instructions at the level of those with the least education and experience because they need your help most. Include what may seem like overly obvious directions, such as "Make sure you plug in the television set," or obvious warnings, such as "Do not put your fingers in the turning blades."

Varied situations.   Consider the circumstances under which consumers are likely to use your instructions. A person building a bookcase in the basement probably is relaxed, has time to study the steps, and may appreciate additional information about options and variations. In contrast, someone in a coffee shop trying to follow the instructions on the wall for the Heimlich maneuver to rescue a choking victim probably is nervous and has time only for a quick glance at the instructions.

In all cases, assume that your readers do not know how to perform the action in question and that they need guidance for every step in the sequence.

Analyzing your readers will help you construct effective instructions, but if possible, you should also test the document during the design stage. When American Airlines personnel prepared passenger instructions for evacuating an airplane in an emergency, the company tested four versions. The first version put the unnumbered instructions in one standard paragraph. A usability test showed readers had a 14% comprehension rate. The second version, also in a standard paragraph but with numbered steps, increased reader comprehension to 29%. The third version had numbered steps in separate paragraphs, and reader comprehension rose to 56%. The fourth version added white space between the numbered steps, raising comprehension to 87%.[1] The readers of the airline instructions were not experienced or knowledgeable about the evacuation process. The design of the instructions was a crucial element for increasing understanding.

Organization

Readers have come to expect consistency in all instructions they use—strict chronological order and numbered steps. Model 10-1 presents a typical outline for a set of instructions for either consumers or employees. As the outline illustrates, instructions generally have a descriptive title and three main sections.

Title

Use a specific title that accurately names the action covered in the instructions.

Poor: Snow Removal

Better: Using Your Acme Snow Blower

Poor: Good Practices

Better: Welding on Pipelines

I. Introduction
   A. Purpose of instructions
   B. Audience
   C. List of parts
   D. List of materials/tools/conditions needed
   E. Overview of the chronological steps
   F. Description of the mechanism or object, if needed
   G. Definitions of terms, if needed
   H. Warnings, cautions, notes, if needed
II. Sequential Steps
   A. Step 1
      1. Purpose
      2. Warning or caution, if needed
      3. Instruction in imperative mood
      4. Note on condition or result
   B. Step 2
      ⋮
III. Conclusion
   A. Expected result
   B. When or how to use

MODEL 10-1　Model Outline for Instructions

Introduction

Depending on how much information your readers need to get started, your introduction may be as brief as a sentence or as long as a page or more. If appropriate for your readers, include these elements:

Purpose and Audience.　Explain the purpose of the instructions unless it is obvious. The purpose of a coffeemaker, for instance, is evident from the name of the product. In other cases, a statement of purpose and audience might be helpful to readers:

> This manual provides instructions for mechanics who install Birkins fine-wire spark plugs.

> These instructions are for nurses who must inject dye into a vein through a balloon-tipped catheter.

> This safe practices booklet is for employees who operate cranes, riggers, and hookers.

Lists of Parts, Materials, Tools, and Conditions.　If readers need to gather items to follow the instructions, put lists of those items in the introduction.

Parts are the components needed, for example, to assemble and finish a table, such as legs, screws, and frame. *Materials* are the items needed to perform the task, such as sandpaper, shellac, and glue. *Tools* are the implements needed, such as a screwdriver, pliers, and a small paintbrush. Do not combine these items into one list unless the items are simple and the list is very short. If the lists are long, you may place them after the introduction for emphasis. *Conditions* are special circumstances that are important for completing the task successfully, such as using a dry, well-ventilated room. Be specific in listing items, such as

- One $\frac{3}{4}$-in. videotape

- One Phillips screwdriver

- Three pieces of fine sandpaper

- Five 9-volt batteries

- Maintain a 70° to 80°F temperature

Overview of Steps.   If the instructions are complicated and involve many steps, summarize them for the reader:

> The following sections cover installation, operation, start-up and adjustment, maintenance, and overhaul of the Trendometer DR-33 motors.

Description of the Mechanism.   A complete operating manual often includes a technical description of the equipment so readers can locate specific parts and understand their function. Chapter 9 explains how to prepare technical descriptions.

Definitions.   If you use any terms your readers may not understand, define them. Chapter 8 explains how to prepare formal and informal definitions.

Warnings, Cautions, and Notes.   Readers must be alerted to potential danger or damage before they begin following instructions. If the potential danger or damage pertains to the whole set of instructions, place these in the introduction. When only one particular step is involved, place the warning or caution *before* the step.

Both *Danger* and *Warning* refer to potential injury of a person. *Danger* means that a serious injury or death *will definitely* occur if a hazard is not avoided:

> DANGER: The compressed gas will EXPLODE if the valve is unsealed. DO NOT open the valve.

A *Warning* means that a serious injury or death *may* occur if a hazard is not avoided:

> WARNING: To prevent electrical shock, do not use this unit in water.

Caution refers to possible damage to equipment:

> CAUTION: Failure to latch servicing tray completely may damage the printer.

Notes give readers extra information about choices or conditions:

> Note: A 12-in. cord is included with your automatic slicer. An extension cord may be used if it is 120 volt, 10 amp.

In addition to these written alerts, use an appropriate symbol or graphic aid to give the reader a clear understanding of the hazard or condition. Use capital letters or boldface type to highlight key words.

Sequential Steps

Explain the steps in the exact sequence readers must perform them, and number each step. Explain only one step per number. If two steps must be performed simultaneously, explain the proper sequence:

> While pushing the button, release the lever slowly.

Include the reason for a particular action if you believe readers need it for more efficient use of the instructions:

> Tighten the belt. Note: A taut belt will prevent a shift in balance.

Use headings to separate the steps into categories. Identify the primary stages if appropriate:

- Separating
- Mixing
- Applying

Identify the primary areas if appropriate:

- Top drawer
- Side panel
- Third level

Refer readers to other steps when necessary:

> If the valve sticks, go to step 12.

Then, tell readers whether to go on from step 12 or go back to a previous step:

> When the valve is clear, go back to step 6.

For regular maintenance instructions, indicate the suggested frequency of performing certain steps:

> *Once a month:*   • Check tire pressure
> • Check lights
> • Check hoses
>
> *Once a year:*   • Inspect brakes
> • Change fuel filter
> • Inspect cooling system

Conclusion

If you include a conclusion, tell readers what to expect after following the instructions, and suggest other uses and options if appropriate:

> Your food will be hot, but not as brown as if heated in the oven rather than in the microwave. A few minutes of standing time will complete the cooking cycle and distribute the heat uniformly.

Model 10-2 shows instructions for shutting down a boiler. The step-by-step instructions are numbered, and each step is below a line drawing that illustrates the action or location in the step. Readers are experienced workers who understand the technical terms.

Details and Language

Whether you are writing for consumers or for highly trained technicians, follow these guidelines:

1. Keep parallel structure in lists. Be sure that each item in a list is in the same grammatical form.

   > *Nonparallel:*   This warranty does not cover:
   > a. Brakes
   > b. Battery
   > c. Using too much oil

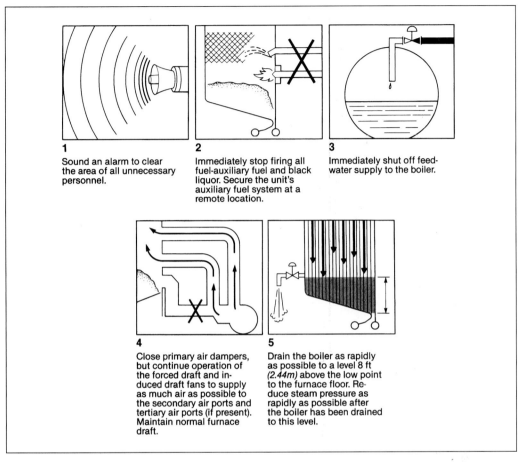

1
Sound an alarm to clear the area of all unnecessary personnel.

2
Immediately stop firing all fuel-auxiliary fuel and black liquor. Secure the unit's auxiliary fuel system at a remote location.

3
Immediately shut off feed-water supply to the boiler.

4
Close primary air dampers, but continue operation of the forced draft and induced draft fans to supply as much air as possible to the secondary air ports and tertiary air ports (if present). Maintain normal furnace draft.

5
Drain the boiler as rapidly as possible to a level 8 ft (2.44m) above the low point to the furnace floor. Reduce steam pressure as rapidly as possible after the boiler has been drained to this level.

MODEL 10-2 Instructions for Emergency Shutdown of an Operating Recovery Boiler

This list is not parallel because the first two items are names of things (nouns) and the last item is an action (verbal phrase). Rewrite so that all items fit the same pattern.

Parallel: This warranty does not cover:
a. Brakes
b. Battery
c. Excess oil consumption

In this revision, all items are things—brakes, battery, and consumption. Lists can confuse readers if the listed items do not all fit the same category. Here is an example:

Poor: Parts Included for Assembly:
End frames
Caster inserts

Casters
Screwdriver
Sleeve screws
Detachable side handles

The screwdriver is not a part included in the assembly package, but the consumer might think it is and waste time looking for it. List the screwdriver in a different category, such as "Tools Needed."

2. Maintain the same terminology for each part throughout the instructions, and be sure the part is labeled similarly in illustrations. Do not refer to "Control Button," "Program Button," and "On/Off Button" when you mean the same control.

3. Use headings to help readers find specific information. Identify sections of the instructions with descriptive headings:

Assembly Kit Contents

Materials Needed

Preparation

Removing the Cover

Checking the Fuel Filter

4. Use the imperative mood for sequential steps. Readers understand instruction steps better when each is a command:

Adjust the lever.

Turn the handle.

Clean the seal.

5. Never use passive voice in instructions. The passive voice does not make clear who is to perform the action in the step.

Poor: The shellac should be applied.
Better: Apply the shellac.

6. State specific details. Use precise language in all sections.

Poor: Turn the lever to the right.
Better: Turn the lever clockwise one full revolution.

Poor: Keep the mixture relatively cool.
Better: Keep the mixture at 50°F or less.

Poor: Place the pan near the tube.
Better: Place the pan 1 in. from the bottom of the tube.

Poor: Attach the wire to the terminal.
Better: Attach the green wire to the AUDIO OUT terminal.

7. Write complete sentences. Long sentences in instructions can be confusing, but do not write fragments or clip out articles and write "telegrams."

> *Poor:*   Repairs excluded.
>
> *Better:*   The warranty does not cover repairs to the electronic engine controls.
>
> *Poor:*   Adjust time filter PS.
>
> *Better:*   Adjust the time filter to the PS position.

8. Do not use *should* and *would* to mean *must* and *is*.

> *Poor:*   The gauge should be on empty.
>
> *Better:*   The gauge must be on empty.
>
> *Poor:*   The seal would be unbroken.
>
> *Better:*   The seal is unbroken.

Both *should* and *would* are less direct than *must* and *is* and may be interpreted as representing possible conditions rather than absolute situations. A reader may interpret the first sentence as saying "The gauge should be on empty, although it might not be." Always state facts definitely.

Narrative Style Instructions

Some writers use a *narrative format* for instructions. A narrative format presents the instructions in sentences in a paragraph as in the following sample from a memo by a hospital administrator:

> Effective June 1, 2006, register all pre-admission out-patient testing as OPT. Forward the registration to the Outpatient Surgery Department, and schedule the actual surgery while the patient is in the registration office if possible. Prepare all testing forms and send them to each required department.

Notice that this paragraph presents the steps in chronological order and in commands. The readers are experienced nurses and registration personnel who already understand the general process and can readily understand this slight change.

Model 10-8 (page 260) shows instructions in narrative formats. Because most readers find the style more difficult to follow than numbered steps, writers usually include graphics to illustrate the steps. Use this style only if your readers are experienced in the general process and there are not many steps. In Model 10-8, the dive is illustrated by a line drawing showing the diver in each stage of the dive.

Troubleshooting Instructions

Troubleshooting instructions tell readers what to do if the mechanism fails to work properly or results do not match expectations. Model 10-3 shows a typical three-part table for troubleshooting instructions. These troubleshooting instructions (1) describe the problem, (2) suggest a cause, and (3) offer a solution. Readers find the problem in the left column and then read across to the suggested solution. Troubleshooting instructions are most helpful when simple adjustments can solve problems.

Problem	Possible Cause	Solution
Snow in picture	Antenna Interference	Check antenna connection Turn off dishwasher, microwave, nearby appliances
Fuzzy picture	Focus pilot	Adjust Sharpness Control clockwise
Multiple images	Antenna Lead-in wire	Check antenna connection Check wire condition
Too much color	Color saturation	Adjust Color Control counterclockwise
Too little color	Color saturation	Adjust Color Control clockwise
Too much one color	Tint level	Adjust Tint Control clockwise

MODEL 10-3   Troubleshooting Guide for a Television Set

WRITING PROCEDURES

Procedures often appear to be similar to instructions, and people sometimes assume the terms mean the same thing. However, the term *procedures* covers several types of guidelines, including (1) a system with sequential steps that one person performs, (2) a system with sequential steps that involves several people interacting with and supporting each other, and (3) the standards and methods for handling equipment or events with or without sequential steps.

Because these documents contain the rules and appropriate methods for the proper completion of tasks, they have multiple readers, including (1) employees who perform the procedures; (2) supervisors who must understand the system and oversee employees working within it; (3) company lawyers concerned about liability protection through the use of appropriate warnings, cautions, and guidelines; (4) government inspectors from such agencies as the

Occupational Safety and Health Administration who check procedures to see if they comply with government standards; and (5) company management who must be aware of and understand all company policies and guidelines. Although the primary readers of procedures are those who follow them, remember that other readers inside and outside the company also use them as part of their work.

Procedures for One Employee

Procedures for a task that one employee will complete are similar to basic instructions. These procedures are organized, like instructions, in numbered steps following the general outline in Model 10-1. Include the same items in the introduction: purpose, lists of parts and materials, warnings and cautions, and any information that will help readers perform the task more efficiently. Most company procedures include a description of the principles behind the system, such as a company policy or government regulation, research results, or technologic advances. When procedures involve equipment, they often include a technical description in the introduction.

Procedures for one employee also share some basic style elements with instructions:

- Descriptive title

- Precise details

- Complete sentences

- Parallel structure in lists

- Consistent terms for parts

- Headings to guide readers

The steps in the procedure are always in chronological order. Steps may be written as commands, in passive voice, or in indicative mood:

To prevent movement of highway trucks and trailers while loading or unloading, set brakes and block wheels. (*Command*)

The brakes should be set and the wheels should be blocked to prevent movement of highway trucks and trailers during loading and unloading. (*Passive voice*)

To prevent movement of highway trucks and trailers while loading or unloading, the driver sets the brakes and blocks wheels. (*Indicative mood*)

The indicative mood can become monotonous quickly if each step begins with the same words, such as "The driver places . . . " or "The driver then sets . . . "

Commands and passive voice are less repetitious. Do not use the past tense for procedures because readers may interpret this to mean that the procedures are no longer in effect.

Procedures for Several Employees

When procedures involve several employees performing separate but related steps in a sequential system, the *playscript organization* is most effective. This organization is similar to a television script that shows the dialogue in the order the actors speak it. Playscript procedures show the steps in the system and indicate which employee performs each step. The usual pattern is to place the job title on the left and the action on the same line on the right:

Operator:	1. Open flow switch to full setting.
	2. Set remote switch to temperature sensor.
Inspector:	3. Check for water flow to both chiller and boiler.
Operator:	4. Change flow switch to half-setting.
	5. Tighten mounting bolt.
Inspector:	6. Check for water flow to both boiler and chiller.
Operator:	7. Shut off chiller flow.
Inspector:	8. Test boiler flow for volume per second.
	9. Complete test records.
Operator:	10. Return flow switch to appropriate setting.

Generally, the steps are numbered sequentially no matter how many employees are involved. One employee may perform several steps in a row before another employee participates in the sequence. The playscript organization ensures that all employees understand how they fit into the system and how they support other employees also participating in the system. Use job titles to identify employee roles because individuals may come and go, but the tasks remain linked to specific jobs.

Include the same information in the introduction to playscript procedures as you do in step-by-step procedures for a single employee, such as warnings, lists of materials, overview of the procedure, and the principles behind the system. The steps in playscript organization usually are commands, although the indicative mood and passive voice may be used.

Procedures for Handling Equipment and Systems

Procedures for handling equipment and systems include guidelines for (1) repairs, (2) installations, (3) maintenance, (4) assembly, (5) safety practices, and (6) systems for conducting business, such as processing a bank loan or fingerprinting suspected criminals. These procedures include lists of tools, definitions, warnings and cautions, statements of purpose, and the basic principles. They differ from other procedures in an important way: The steps are not always sequential. Often the individual steps may be performed in any order, and they usually are grouped by topic rather than by sequence.

Model 10-4 shows a typical organization for company procedures about safe practices on a construction site. The introduction establishes the purpose of the procedures and warns readers about possible injury if these and other company guidelines are not followed. The individual items are grouped under topic headings. Notice that the items are not numbered because they are not meant to be sequential. In this sample, the items are written as commands, but, like other procedures, they may be in the passive voice or indicative mood as well.

PROCEDURES FOR SAFE EXCAVATIONS

This safety bulletin specifies correct procedures for all excavation operations. In addition, all Clinton Engineering standards for materials and methods should be strictly observed. Failure to do so could result in injury to workers and to passersby.

PERSONAL PROTECTION:

- Wear safety hard hat at all times.
- Keep away from overhead digging equipment.
- Keep ladders in trenches at all times.

EXCAVATION PROTECTION:

- Protect all sites by substantial board railing or fence at least 48 in. high or standard horse-style barriers.
- Place warning lights at night.

EXCAVATED MATERIALS:

- Do not place excavated materials within 2 ft of a trench or excavation.
- Use toe boards if excavated materials could fall back into a trench.

DIGGING EQUIPMENT:

- Use mats or heavy planking to support digging equipment on soft ground.
- If a shovel or crane is placed on the excavation bank, install shoring and bracing to prevent a cave-in.

MODEL 10-4   A Typical Company Procedures Bulletin Outlining Safe Practices on a Construction Site

WRITING PROCESS EXPLANATIONS

A *process explanation* is a description of how a series of actions leads to a specific result. Process explanations differ from instructions and procedures in that they are not intended to guide readers in performing actions, only in understanding them. As a result, a process explanation is written as a narrative, without listed steps or commands, describing four possible types of actions:

- Actions that occur in nature, such as how diamonds form, how the liver functions, or how a typhoon develops

- Actions that produce a product, such as how steel, light bulbs, or baseballs are made

- Actions that make up a particular task, such as how gold is mined, how blood is tested for cholesterol levels, or how a highway is paved

- Actions in the past, such as how the Romans built their aqueducts or how a battle was waged.

Readers

A process explanation may be a separate document, such as a science pamphlet or a section in a manual or report. Although readers of process explanations do not intend to perform the steps themselves, they do need the information for specific purposes, and one process explanation will not serve all readers.

A general reader reading an explanation in a newspaper of how police officers test drivers for intoxication wants to know the main stages of the process and how a police officer determines if a driver is, indeed, intoxicated. A student in a criminology class needs more details to understand each major and minor stage of the process and to pass a test about those details. An official from the National Highway Traffic Safety Administration may read the narrative to see how closely it matches the agency's official guidelines, while a judge may want to be sure that a driver's rights are not violated by the process. Process explanations for these readers must serve their specific needs as well as describe the sequential stages of the process. Model 10-5 shows how process explanations of sobriety testing may differ for general readers and for students. Notice that the version for students includes more details about what constitutes imbalance and more explanation about the conditions under which the test should or should not be given.

For all process explanations, analyze the intended reader's technical knowledge of the subject and why the reader needs to know about the process. Then, decide the number of details, which details, and the appropriate technical terms to include in your narrative.

For a general reader:

THE WALK-AND-TURN TEST

The police officer begins the test by asking the suspect to place the left foot on a straight line and the right foot in front of it. The suspect then takes nine heel-to-toe steps down the line, turns around, and takes nine heel-to-toe steps back. The suspect is given one point each for eight possible behaviors showing imbalance, such as stepping off the line and losing balance while turning. A score of two or more indicates the suspect is probably legally intoxicated.

For a student reader:

THE WALK-AND-TURN TEST

The test is administered on a level, dry surface. People who are over 60 years old, more than 50 pounds overweight, or have physical impairments that interfere with walking are not given this test.

The police officer begins the test by asking the suspect to place the left foot on a straight line and the right foot in front of it. The suspect must maintain balance while listening to the officer's directions for the test and must not begin until the officer so indicates. The suspect then takes nine heel-to-toe steps down the line, keeping hands at the sides, eyes on the feet, and counting aloud. After nine steps, the suspect turns and takes nine heel-to-toe steps back in the same manner. The officer scores one point for each of the following behaviors: failing to keep balance while listening to directions, starting before told to, stopping to regain balance, not touching heel to toe, stepping off the line, using arms to balance, losing balance while turning, and taking more or less than nine steps each way. If the suspect falls or cannot perform the test at all, the officer scores nine points. A suspect who receives two or more points is probably legally intoxicated.

Model 10-5   Two Process Explanations of Sobriety Testing

Organization

Model 10-6 presents a model outline for a process explanation. As the outline illustrates, process explanations have three main sections: an introduction, the stages in the process, and a conclusion.

Introduction

The introduction should include enough details about the process so that readers understand the principles underlying it and the conditions under

I. Introduction
   A. Definition
   B. Theory behind process
   C. Purpose
   D. Historical background, if appropriate
   E. Equipment, materials, special natural conditions
   F. Major stages

II. Stages in the Process
   A. Major stage 1
      1. Definition
      2. Purpose
      3. Special materials or conditions
      4. Description of major stage 1
      5. Minor stage 1
         a. Definition
         b. Purpose
         c. Special materials or conditions
         d. Description of minor stage 1
      6. Minor stage 2
         ⋮
   B. Major stage 2
      ⋮

III. Conclusion
   A. Summary of major stages and results
   B. Significance of process

MODEL 10-6   Model Outline for a Process Explanation

which it takes place. Depending on your readers' technical expertise and how they intend to use the process explanation, include these elements:

- *Definition*—If the subject is highly technical or readers are not likely to recognize it, provide a formal definition.

- *Theory behind the process*—Explain the scientific principles behind the process, particularly if you are describing a research process.

- *Purpose*—Explain the purpose unless it is obvious from the title or the readers already know it.

- *Historical background*—Readers may need to know the history of a process and how it has changed to understand its current form.

- *Equipment, materials, and tools*—Explain what types of equipment, materials, and tools are essential for proper completion of the process.

- *Major stages*—Name the major stages of the process so that readers know what to expect.

Stages of the Process

Explain the stages of the process in the exact sequence in which they normally occur. In some cases, one sentence may adequately explain a stage. In other cases, each stage may actually be a separate process that contributes to the whole. In describing each stage, (1) define it, (2) explain how it fits the overall process, (3) note any dangers or special conditions, (4) describe exactly how it occurs, including who or what does the action, and (5) state the results at the end of that particular stage. If a major stage is made up of several minor actions, explain each of these and how they contribute to the major stage. Include the reasons for the actions in each stage:

> The valve is closed before takeoff because. . . .

> The technician uses stainless steel implements because. . . .

Conclusion

The conclusion of a process explanation often explains the expected results of the process, what the results mean, and how this process influences or interacts with others, if it does.

The following short process explanation explains the asexual reproduction of sponges:

> Like many invertebrates with little or no mobility, sponges are able to reproduce both asexually and sexually. Asexual reproduction is achieved by budding or breaking off small pieces capable of developing into complete sponges. The buds break away from the parent sponge and drift away in the current. Exactly where the buds settle is a matter of chance, but if bottom conditions are favorable, the bud can develop into a healthy, whole sponge. Asexual reproduction results in genetic clones.[2]

Details and Language

Process explanations, like instructions, can include these elements:

- Descriptive title

- Precise details

- Complete sentences

- Consistent terms for actions or parts

- Headings in long narratives to guide readers

- Simple comparisons for general readers

In addition, remember these style guidelines:

- Do not use commands. Because readers do not intend to perform the actions, commands are not appropriate.

- Use either passive voice or the indicative mood. Passive voice usually is preferred for process explanations that involve the same person performing each action to eliminate the monotony of repeating "The technician" over and over. However, natural processes or processes involving several people are easier to read if they are in the indicative mood (see Model 10-5).

- If you are writing a narrative about a process you were involved in, such as an incident report or a research report, use of the first person is appropriate:

  I arrested the suspect. . . .
  I called for medical attention. . . .

- Use transition words and phrases to indicate shifts in time, location, or situation in individual stages of the process:

  Shifts in time: then, next, first, second, before
  Shifts in location: above, below, adjacent, top
  Shifts in situation: however, because, in spite of, as a result, therefore

CHAPTER SUMMARY

This chapter discusses writing effective instructions, procedures, and process explanations. Remember:

- Instructions, procedures, and process explanations are related but distinct strategies for explaining the steps in a specific action.

- Instructions provide the steps in an action for readers who intend to perform it, either for a work-related task or for a private-interest task.

- Instructions typically have an introduction that includes purpose; lists of parts, materials, tools, and conditions; overview of the steps; description of a mechanism (if appropriate); and warnings, cautions, and notes.

- The steps in instructions should be in the exact sequence in which a reader will perform them.

- The conclusion in instructions, if included, explains the expected results.

- Instructions should be in the imperative mood and include precise details.

- Troubleshooting instructions explain what to do if problems arise after readers have performed all the steps in an action.

- Procedures are company guidelines for (1) a system with sequential steps that must be completed by one person, (2) a system with sequential steps that involves several people, and (3) handling equipment or events with or without sequential steps.

- Procedures for a task that one employee will perform are similar to instructions with numbered steps.

- Procedures for several employees are best written in playscript organization.

- Procedures for handling equipment and systems often are organized topically rather than sequentially because the steps do not have to be performed in a specific order.

- Process explanations describe how a series of actions leads to a specific result.

- The stages of the process are explained in sequence in process explanations, but because readers do not intend to perform the process, the steps are not written as commands.

SUPPLEMENTAL READINGS IN PART 2

Garhan, A. "ePublishing," *Writer's Digest*, p. 495.

McAdams, M. "It's All in the Links: Readying Publications for the Web," *The Editorial Eye*, p. 512.

ENDNOTES

1. Elaine L. Ostrander, "Usability Evaluations: Rationale, Methods, and Guidelines," *Intercom* 46.6 (June 1999): 18–21.

2. Marty Snyderman, "Sponges: The World's Simplest Multicellular Animals," *Dive Training* 11.3 (March 2001): 56–64.

3. Mike Collins, David Holloway, Brenda Legge, and Diane Carr, *The Ultimate Do-It-Yourself Book* (London: Anness Publishing Ltd., 2004), p. 82.

MODEL 10-7   Commentary

These instructions are part of a Ciba Corning reference manual that includes operating instructions for technicians running a computer-operated testing system. The instructions shown here cover filling water bottles and emptying waste bottles—essential procedures. The technical writer who prepared these instructions uses the internationally recognized caution symbol—an exclamation point inside a triangle. The biohazard symbol tells the reader to read an appendix containing precautions for working with biohazardous materials.

Notice that the first step in the instructions is a major one, followed by the six steps necessary to perform the major step. Also notice that the writer directs the reader to other sections in the manual in Steps 2 and 5.

Discussion

1. Identify the design elements in these instructions. Discuss how effectively the format serves the reader who must follow the instructions.

2. In groups, write a process explanation for "Preparing the ACS Bottles." Assume your reader is a supervisor who has never performed the process but needs to understand it before attending a meeting with the clinical biology technicians who perform the process daily.

3. Write instructions for cleaning a piece of equipment in your field or for cleaning and adjusting a piece of classroom equipment, such as an overly full pencil sharpener or an out-of-focus overhead projector. Identify a specific reader before you begin.

Preparing the ACS Bottles

Use this procedure to fill the ACS water bottle and empty the ACS waste bottle, if required.

If the Extended Operation Module (EOM) is connected to the ACS:180, complete this procedure when you prepare the EOM bottles at the start of each day or shift. Ensure that the EOM is inactive when you prepare the ACS bottles by pressing the EOM button and checking to see that the LED next to the EOM is off. Press the EOM button to activate the EOM when you complete the procedure.

BIOHAZARD: Refer to Appendix B, *Protecting Yourself from Biohazards,* for recommended precautions when working with biohazardous materials.

CAUTION: Do not perform this procedure while the system is assaying samples. The system stops the run and the tests in process are not completed.

NOTE: For optimum assay and system performance, fill the ACS water bottle with fresh deionized water at the start of each day or shift.

1. Remove the ACS water bottle and the ACS waste bottle.

    a. Lift the lid above the water bottle until the tubing clears the mouth of the bottle.

    b. Grasp the handle of the water bottle and lift the bottle up and out, and set it aside.

    c. Lift the waste bottle lid until the tubing clears the mouth of the waste bottle.

    d. If the EOM is connected to the ACS:180, disconnect the ACS waste bottle from the waste tubing by pressing the metal latch on the tubing fitting.

    e. Grasp the handle of the waste bottle and lift the bottle up and out.

    f. Gently lower the lid.

2. Empty the contents of the waste bottle into a container or drain approved for biohazardous waste.

    If required, clean the waste bottle as described in Section 2, *Cleaning the ACS Water and the Waste Bottles,* in your *ACS:180 Maintenance and Troubleshooting Manual.*

3. Fill the water bottle with fresh deionized water.

MODEL 10-7   ACS Instructions

(Continued)

Preparing the ACS Bottles

4. Install the bottles on the system, as shown in Figure 6 – 2.
    a. Lift the waste bottle lid and return the waste bottle to its location.
    b. If the EOM is connected to the ACS:180, connect the waste tubing to the waste bottle by pressing the tubing into the fitting on the bottle, if required.
    c. Lower the waste bottle lid.
    d. Install the water bottle, insert the tubing into the bottle, and lower the water bottle lid.

5. Proceed to Replenishing Supplies, Page 6 – 13.

Figure 6 – 2. Installing the ACS Water and Waste Bottles

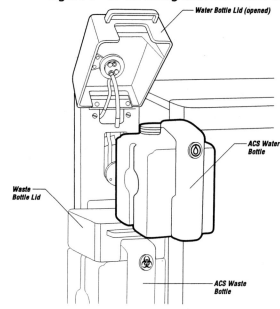

Water Bottle Lid (opened)

ACS Water Bottle

Waste Bottle Lid

ACS Waste Bottle

MODEL 10-8   Commentary

This model shows one page of a brochure that covers effective diving techniques published by the National Spa and Pool Institute. The instructions are in narrative format. Notice that the instructions focus on body position and the importance of a shallow dive. The illustration shows the diver in the correct position throughout the dive.

Discussion

1. Discuss the combination of text and illustrations and how useful it is to a reader seeking diving tips.

2. In groups, revise these diving instructions into a set of numbered steps. Discuss the need for different illustrations with numbered steps.

3. In groups, revise these diving instructions into a process explanation. Compare your draft with those of others in the class.

4. In groups, draft narrative instructions with an illustration for a specific action in another sport, such as dribbling in basketball.

Diving is a sport that almost everyone can enjoy, either as a participant or as a spectator. But, as with every other sport, injuries can spoil the fun for everyone. So to get the most pleasure from diving and to avoid serious injuries, don't take needless risks. Learn some basic rules for safe diving.

Think Ahead.

Once you've started your dive, you don't have time to think. Know the depth of the water. Plan your dive path. Never dive where you don't know the water depth or where there may be hidden obstructions.

Steer Up.

When you dive down, you must be ready to steer up. As you enter the water, your arms must be extended over your head, hands flat and aiming up. Hold your head up and arch your back. This way, your whole body helps you steer up, away from the bottom.

Plan a shallow dive. Immediately steering up. Don't try the straight vertical-entry dives you see in competition. These dives take a long time to slow down and must be done only after careful training and in pools designed for competitive diving.

Head and Hands Up.

Your extended arms and hands not only help you to steer up to the surface, they can also protect your head. If a diver's head hits bottom, major injury to neck and spine can result. So always remember, head and hands up!

Control Your Dive.

Sometimes divers lose control through improper use of hands and arms. Practice holding your arms extended, hands flat and tipped up. Like learning to swim or ride a bicycle, you have to learn to make the right moves automatically. Carefully rehearse the proper diving techniques before you dive.

Plan your dive

Back arched

Arms extended

Head and hands up
Hands tipped up

Hold diving form

"Steer up for a safe dive."

MODEL 10-8   Safe Diving Instructions

MODEL 10-9   Commentary

This model is a Web page from the U.S. Environmental Protection Agency (http://www.epa.gov/epaoswer/non-hw/reduce/catbook/Tip2.htm). The Web page is part of a guide for consumers on how to reduce solid waste. It contains instructions for consumers regarding steps they can take at home to reduce the toxic nature of solid waste.

Discussion

1. Discuss the design of this Web page and its readability for the average consumer. How easy is it for consumers to find the specific instructions on the page?

2. Discuss whether these instructions should be presented as one of the following: (a) chronological steps, (b) procedures in a nonsequential list, or (c) narrative instruction format.

3. In groups, redesign this Web page to make it a more effective set of instructions for the typical consumer. Consider visuals that you might add. Compare your Web design with those of other groups. Discuss the reasons for the design elements your group chose. Decide as a class what two specific design elements you think are essential for a Web page like this one.

Too Much Trash

A Basic Solution

Making it Work

The 4 Principles

The 12 Tips

Conclusion

Tip 2: Adopt Practices That Reduce Waste Toxicity

In addition to reducing the amount of materials in the solid waste stream, reducing waste toxicity is another important component of source reduction. Some jobs around the home may require the use of products containing hazardous components. Nevertheless, toxicity reduction can be achieved by following some simple guidelines.

Take actions that use nonhazardous or less hazardous components to accomplish the task at hand. Instead of using pesticides, for example, plant marigolds in your garden to ward off certain pests. In some cases, you may be using less toxic chemicals to do a job, and in others, you may use some physical method, such as sandpaper, scouring pads, or just a little more elbow grease, to achieve the same results.

Learn about alternatives to household items containing hazardous substances. In some cases, products that you have around the house can be used to do the same job as products with hazardous components. (See Source Reduction Alternatives Around the Home, or check with local libraries or bookstores for guidebooks on nonhazardous household practices).

If you do need to use products with hazardous components, use only the amounts needed. Leftover materials can be shared with neighbors or donated to a business, charity, or government agency, or, in the case of used motor oil, recycled at a participating service station. Never put leftover products with hazardous components in food or beverage containers.

For products containing hazardous components, read and follow all directions on product labels. Make sure the containers are always labeled properly and stored safely away from children and pets. When you are finished with containers that are partially full, follow local community policy on household hazardous waste disposal (See Household Hazardous Waste Collection). If at any time you have any questions about potentially hazardous ingredients in products and their impacts on human health, do not hesitate to call your local poison control center.

MODEL 10-9   U.S. Environmental Protection Agency Web Page

MODEL 10-10   Commentary

This Carbon Monoxide Fact Sheet is available on the National Safety Council Web site (http://nsc.org). The fact sheet begins with a definition of carbon monoxide and descriptions of the effects of the gas. The fact sheet includes three individual sets of procedures.

Discussion

1. Discuss the format and organization of this fact sheet. How helpful are the headings for the reader looking for general information?

2. Discuss why the writer probably decided to include a definition of carbon monoxide.

3. In groups, review the sets of procedures in the fact sheet. Discuss whether some procedures might be revised into a numbered list.

4. In groups, develop a fact sheet for a specific campus activity.

Carbon Monoxide

What Is It?

Carbon monoxide (CO) is an odorless, colorless gas that interferes with the delivery of oxygen in the blood to the rest of the body. It is produced by the incomplete combustion of fuels.

What Are the Major Sources of CO? Carbon monoxide is produced as a result of incomplete burning of carbon-containing fuels including coal, wood, charcoal, natural gas, and fuel oil. It can be emitted by combustion sources such as unvented kerosene and gas space heaters, furnaces, woodstoves, gas stoves, fireplaces and water heaters, automobile exhaust from attached garages, and tobacco smoke. Problems can arise as a result of improper installation, maintenance, or inadequate ventilation.

What Are the Health Effects? Carbon monoxide interferes with the distribution of oxygen in the blood to the rest of the body. Depending on the amount inhaled, this gas can impede coordination, worsen cardiovascular conditions, and produce fatigue, headache, weakness, confusion, disorientation, nausea, and dizziness. Very high levels can cause death.

The symptoms are sometimes confused with the flu or food poisoning. Fetuses, infants, elderly, and people with heart and respiratory illnesses are particularly at high risk for the adverse health effects of carbon monoxide.

An estimated 1,000 people die each year as a result of carbon monoxide poisoning and thousands of others end up in hospital emergency rooms.

What Can Be Done to Prevent CO Poisoning?

- Ensure that appliances are properly adjusted and working to manufacturers' instructions and local building codes.
- Obtain annual inspections for heating system, chimneys, and flues and have them cleaned by a qualified technician.
- Open flues when fireplaces are in use.
- Use proper fuel in kerosene space heaters.
- Do not use ovens and gas ranges to heat your home.
- Do not burn charcoal inside a home, cabin, recreational vehicle, or camper.
- Make sure stoves and heaters are vented to the outside and that exhaust systems do not leak.
- Do not use unvented gas or kerosene space heaters in enclosed spaces.
- Never leave a car or lawn mower engine running in a shed or garage, or in any enclosed space.
- Make sure your furnace has adequate intake of outside air.

What If I Have Carbon Monoxide Poisoning?

Don't ignore symptoms, especially if more than one person is feeling them. If you think you are suffering from carbon monoxide (CO) poisoning, you should

MODEL 10-10   Carbon Monoxide Fact Sheet Web Page

- Get fresh air immediately. Open doors and windows. Turn off combustion appliances and leave the house.
- Go to an emergency room. Be sure to tell the physician that you suspect CO poisoning.
- Be prepared to answer the following questions: Is anyone else in your household complaining of similar symptoms? Did everyone's symptoms appear about the same time? Are you using any fuel-burning appliances in the home? Has anyone inspected your appliances lately? Are you certain they are working properly?

What About Carbon Monoxide Detectors?

Carbon monoxide (CO) detectors can be used as a backup *but not as a replacement* for proper use and maintenance of your fuel-burning appliances. CO detector technology is still being developed and the detectors are not generally considered to be as reliable as the smoke detectors found in homes today. You should not choose a CO detector solely on the basis of cost; do some research on the different features available.

Carbon monoxide detectors should meet Underwriters Laboratories Inc. standards, have a long-term warranty, and be easily self-tested and reset to ensure proper functioning. For maximum effectiveness during sleeping hours, carbon monoxide detectors should be placed close to sleeping areas.

If your CO detector goes off, you should

- Make sure it is the CO detector and not the smoke alarm.
- Check to see if any member of your household is experiencing symptoms.
- If they are, get them out of the house immediately and seek medical attention.
- If no one is feeling symptoms, ventilate the home with fresh air and turn off all potential sources of CO.
- Have a qualified technician inspect your fuel-burning appliances and chimneys to make sure they are operating correctly.

Related Links

NSC Environmental Health Center Indoor Air Quality Program
EPA Automobiles and Carbon Monoxide
Hamel Volunteer Fire Department
Wayne State University
See other Fact Sheets.

Use Policy | Fact Sheet Menu | Online Resources | NSC Home | Site Map | Comments

National Safety Council
A Membership Organization Dedicated to Protecting Life and Promoting Health
1121 Spring Lake Drive, Itasca, IL 60143-3201
Tel: (630) 285-1121; Fax: (630) 285-1315

June 22, 2000

MODEL 10-11   Commentary

This model is from an industrial guide. The process explanation covers a specific type of boiler and how it functions. The text is divided into numbered stages. The line drawing of the boiler includes numbers keyed to these stages and arrows that indicate the movements of various elements. Thus, the drawing also is a flowchart showing how the process works within the boiler. Readers are experienced workers who understand the language used.

Discussion

1. Discuss the effectiveness of separating the process explanation into numbered stages.

2. Discuss how the text guides readers through the flowchart. How does the flowchart aid readers?

3. In groups, draft a process explanation for one of the following. Create a flowchart to accompany your process explanation.

    a. Checking a book out of your school library

    b. Ordering and getting food in a fast-food restaurant

    c. Making a photocopy on a copy machine in your school library

    d. Putting air in a tire

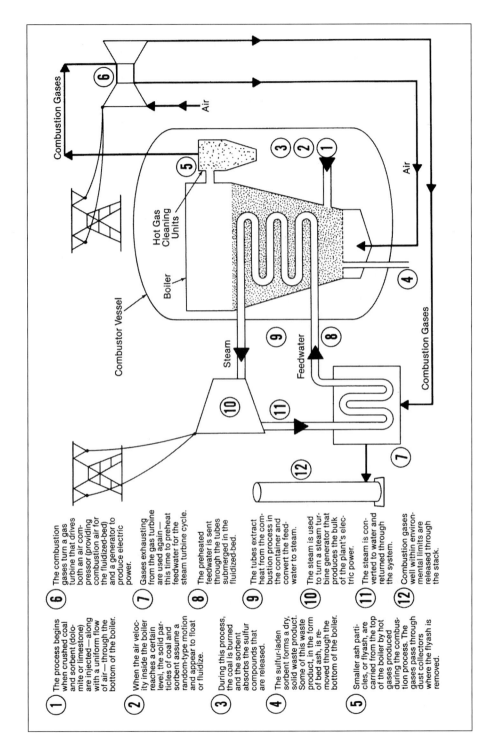

1. The process begins when crushed coal and sorbent (dolomite or limestone) are injected—along with a uniform flow of air—through the bottom of the boiler.

2. When the air velocity inside the boiler reaches a certain level, the solid particles of coal and sorbent assume a random-type motion and appear to float or fluidize.

3. During this process, the coal is burned and the sorbent absorbs the sulfur compounds that are released.

4. The sulfur-laden sorbent forms a dry, solid waste product. Some of this waste product, in the form of bed ash, is removed through the bottom of the boiler.

5. Smaller ash particles, or flyash, are carried from the top of the boiler by hot gases produced during the combustion process. The gases pass through dust collectors where the flyash is removed.

6. The combustion gases turn a gas turbine that drives both an air compressor (providing combustion air for the fluidized-bed) and a generator to produce electric power.

7. Gases exhausting from the gas turbine are used again—this time to preheat feedwater for the steam turbine cycle.

8. The preheated feedwater is sent through the tubes submerged in the fluidized-bed.

9. The tubes extract heat from the combustion process in the container and convert the feedwater to steam.

10. The steam is used to turn a steam turbine generator that produces the bulk of the plant's electric power.

11. The steam is converted to water and returned through the system.

12. Combustion gases well within environmental limits are released through the stack.

MODEL 10-11   Process Explanation and Flowchart

MODEL 10-12   Commentary

This process explanation of how a jet engine produces power appeared in *CODE ONE*, a magazine read by pilots and technicians involved in aircraft production. The process explanation begins with a brief review of the overall process. The explanation continues with a partition of the major pieces of the engine and an identification of their purpose. The graphic is a cutaway drawing of the engine.

Discussion

1. Discuss the organization and details of this process explanation. How well do you think the reader with a technical background can understand the process?

2. Discuss the cutaway drawing. What other types of graphics might be helpful with this process explanation?

3. In groups, select a household appliance, such as a toaster, and write a process explanation. Create an illustration to accompany the process explanation.

4. In groups, draft a process explanation for a general activity, such as using a specific type of public transportation or making a simple repair, such as putting grout in a bathroom crack.

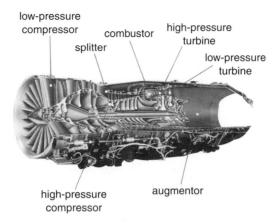

JSF119-611 Primer

A jet engine produces power by compressing air, adding fuel, and sustaining a continuous combustion. The hot gas of combustion expands rapidly out of the back end of the engine, which produces forward thrust. The primary parts of a military jet engine are the compressor, combustor, turbine, and augmentor. The compressor forms the front part of the engine. The first three sets of blades of the JSF119-611 form the three stages of the low-pressure compressor (also called a fan). The next six sets of blades form the six stages of the high-pressure compressor. These nine compressor stages draw air in, pressurize it, and deliver it to the combustor, where it is mixed with fuel and ignited. A portion of the compressed air bypasses the combustor as well. This bypass air is used to cool hot portions of the engine and to provide airflow for the augmentor. The Lockheed Martin STOVL variant uses bypass air to power the roll ducts in the wings.

Extremely hot and rapidly expanding gases produced in the combustor enter the turbine section, which consists of three stages of alternating stationary and rotating blades. The first stage of the turbine (the high-pressure turbine) is connected by a shaft to the high-pressure compressor in the front of the engine. The back two stages of the turbine (the low-pressure turbine) are connected by another independent shaft to the first three stages of the compressor (the fan). The turbine stages essentially absorb enough energy from the hot expanding gases to keep the compressor stages rotating at an optimum speed.

The hot gases exit the turbine and enter the augmentor (also called the afterburner), where they are joined with bypass air. When the engine is in afterburner, additional fuel is injected into the augmentor. This secondary combustion produces a significant amount of additional thrust.

The JSF119-611 shares a common engine core with the F119 that powers the F-22 Raptor. The core consists of the high-pressure compressor, the combustor, and the high-pressure turbine.

MODEL 10-12   Jet Engine Process

Chapter 10 Exercises

1. Write a set of instructions for one of the following activities. Identify your reader and purpose before you begin. Include appropriate warnings and cautions, and develop at least one graphic aid, illustrating a specific point in your instructions.

- Driving from your campus (A) to two points of interest (B and then to C) in your community and then driving back to campus (A)
- Performing a specific health test, such as taking blood pressure, testing blood sugar levels, using an inhaler, giving yourself a shot
- Performing a specific exercise for a specific purpose, such as calf stretches for plantar fasciitis, weight exercises for strengthening back muscles
- Performing a specific laboratory test or research activity in your major
- Performing an indoor household chore
- Performing an outdoor chore

2. Using the process explanation of the "Walk-and-Turn Test" in Model 10-5, write a set of instructions for a new police officer who is administering the test for the first time.

3. The following guidelines are from a safety booklet at a company that specializes in cleaning up after natural disasters, such as floods. Write a formal set of procedures, using all these guidelines. These are not sequential steps.

- When working in hot environments, have plenty of water available.
- Workers should ensure that all ladders and scaffolds are properly secured before use.
- Do not touch downed power lines or any object in water that is in contact with downed lines.
- Wear a NIOSH-approved dust respirator if working with moldy building materials or vegetable matter (e.g., hay, grain, or compost).
- Fuel-powered generators should be used only outdoors.
- Be sure all fire exits are clear of debris.
- Sticks or poles should be used to check for pits, holes, and protruding objects before entering a flooded area.
- Working in the sun requires sunscreen and frequent rest breaks.
- Workers should wear light-colored, loose-fitting clothing.
- Always wear watertight boots with steel toes and insole, long pants, and safety goggles during cleanup.
- Caution is needed when moving ladders near overhead power lines.

- Workers should use life vests when engaged in activities that involve exposure to deep water.
- Take frequent breaks when lifting heavy, waterlogged objects.
- Hands should be washed often, especially before eating, drinking, or applying cosmetics.
- There should be a plan for contacting medical personnel in case of an accident.
- Report any downed power lines, gas leaks, or snakes to proper authorities.
- When using a generator, be sure the main circuit is off and locked out before starting it.
- Workers should exercise extreme caution when handling containers holding unknown substances or substances known to be toxic.
- Clean water should be available for eyewash and other first-aid treatments.
- Washouts, trenches, excavations, and gullies must be inspected for support and stability before workers enter the area.
- Trenches must be supported with a trench box or sloped at no less than a 45° angle for packed soil or a 34° angle for granular soil.

4. Convert the following process explanation to a set of instructions for the do-it-yourself homeowner:

> To inspect floorboard space, the worker must lift existing boards. The worker may find that some have been cut and lifted earlier to provide access to pipes and ducts. These boards can be levered up with a chisel. The worker must not use a screwdriver because that will damage the floorboard. If boards have not been previously cut, the worker must lever up the floorboard with a chisel after first checking that the boards are not tongued-and-grooved. Any tongue-and-groove placement must be cut with a floorboard saw or a circular saw. After raising the boards, the worker looks for signs of woodworm or dry rot. Damaged wood must be removed next. The worker then replaces any floorboards by using a plane to shave the boards to fit the gaps. Floorboards are then secured with two floorboard nails at each joist, about 1 in. from the edge of the board and exactly in the middle of the joist.[3]

COLLABORATIVE EXERCISES

5. In groups, write a set of playscript procedures for a process on your campus, such as registering for classes, checking out library materials, participating in student activities.

6. In groups, select one of the following sports techniques: baseball—bunting, football—placekicking, golf—putting, tennis—serve and volley, basketball—free throw, and rock climbing—rappelling. Write a process explanation for the technique. Compare your results with those of other groups.

7. The following is a process explanation of the technique of "self-arrest" used in rock climbing to stop a fall. In groups, rewrite this process into a set of instructions for a rock climber.

> The climber who begins to slip must start the process of self-arrest immediately. The climber must roll onto his/her stomach and firmly grip the ice axe. One hand is on the axe head, and one hand grips the shaft that should run diagonally across the climber's chest so the head rests on the front of the shoulder. If the climber is wearing crampons, he/she jams them into the snow, at the same time ramming the head of the ice axe into the snow. If the climber is not wearing crampons, pressing the knees and boots into the snow helps slow the fall. Climbers typically practice these techniques for several hours before each climb.

INTERNET EXERCISES

8. Search for Web sites with information about coffee. Find out the following: (a) how coffee is grown, (b) how coffee is harvested, (c) how coffee is roasted, (d) varieties of coffee, and (e) tips for brewing coffee. A gourmet coffee store wants to give its customers a fact sheet with information about good coffee—from the bean to the cupful. Based on the information you find, prepare one of the following for the customers: (a) a 1–2-page information fact sheet or (b) a three-panel brochure (front and back) with coffee information.

9. Search the Internet for information about mold and how to prevent it. Prepare a consumer fact sheet that includes a process explanation about how mold forms and instructions for combating it.

10. Search the Internet for information about acid rain. Prepare a process explanation for general readers.

CHAPTER 11

Formal Report Elements

SELECTING FORMAL REPORT ELEMENTS

Most in-house reports, whatever their purpose, are in memo format, as illustrated in Chapter 12. Long reports, however, whether internal or external, more often include the formal report elements discussed in this chapter.

Management or company policy usually dictates when formal report elements are appropriate. In some companies, certain report types, such as proposals or feasibility studies, always include formal report elements. In addition, long reports addressed to multiple readers often require formal report elements, such as glossaries and appendixes, to effectively serve all reader purposes.

The formal report is distinguished from the informal report by the inclusion of some or all of the special elements described in this chapter.

WRITING FRONT MATTER

Front matter includes all the elements that precede the text of a report. Front-matter elements help a reader (1) locate specific information and (2) become familiar with the general content and organization of the report.

Title Page

A *title page* records the report title, writer, reader, and date. It usually is the first page of a long formal report. In some companies, the format is standard. If it is not, include these items:

- Title of the report, centered in the top third of the page
- Name, title, and company of the primary reader or readers, centered in the middle of the page
- Name, title, and company of the writer, centered in the bottom third of the page
- Date of the report, centered directly below the writer's name

The title should accurately reflect the contents of the report. Use key words that identify the subject quickly and inform readers about the purpose of the report.

Title: Parking

Revision: Feasibility Study of Expanding Parking Facilities

Title: Office Equipment

Revision: Proposal for Computer Purchases

Transmittal Letter or Memo

A *transmittal letter* or *memo* sends the report to the reader. In addition to establishing the title and purpose of the report, the transmittal letter provides a place for the writer to add comments about procedures, recommendations, or other matters that do not fit easily into the report itself. Here, too, the writer may offer to do further work or credit others who assisted with the report. Place the transmittal letter either immediately after the title page or immediately before it, as company custom dictates. In some companies, writers use a transmittal memo, but the content and placement are the same as for a transmittal letter. If the transmittal letter simply sends the report, it often is before the title page. However, if the transmittal letter contains supplementary information about the report or recommends more study or action, some writers place it immediately after the title page, where it functions as a part of the report. In your transmittal letter:

- State the report title, and indicate that the report is attached.
- Establish the purpose of the report.
- Explain why, when, and by whom the report was authorized.
- Summarize briefly the main subject of the report.
- Point out especially relevant facts or details.
- Explain any unusual features or organization.
- Acknowledge those who offered valuable assistance in gathering information, preparing appendixes, and so on.
- Mention any planned future reports.
- Thank readers for the opportunity to prepare the report, or offer to do more study on the subject.
- Recommend further action, if needed.

Table of Contents

The *table of contents* alerts the reader to (1) pages that contain specific topics, (2) the overall organization and content of the report, and (3) specific and supplemental materials, such as appendixes. Place the table of contents directly after the transmittal letter or title page, if that is after the transmittal letter, and before all other elements. The title page is not listed in the table of contents, but it is counted as page i. All front matter for a report is numbered in small Roman numerals. The first page of the report proper is page 1 in Arabic numbers, and all pages after that have Arabic numbers. In your table of contents:

- List all major headings with the same wording used in the report.

- List subsections, indented under major headings, if the subheadings contain topics that readers are likely to need.

- List all formal report elements, such as the abstract and appendixes, except for the title page.

- Include the titles of appendixes, for example:

   APPENDIX A: PROJECTED COSTS

- Do not underline headings in the table of contents even if they are underlined in the text.

List of Figures

Any graphic aid, such as a bar graph, map, or flowchart that is not a table with numbers or words in columns, is called a *figure*. The list of figures follows the table of contents. List each figure by both number and title, and indicate page numbers:

List of Tables

The list of tables appears directly after the list of figures. List each table by number and title, and indicate page numbers:

If your report contains only two or three figures and tables, you may combine them into one "List of Illustrations." In this case, list figures first and then tables.

Abstract and Executive Summary

An *abstract* is a synopsis of the most important points in a report and provides readers with a preview of the full contents. An abstract, which can be either descriptive or informative, usually is one paragraph of no more than 200 words. An *executive summary* is a longer synopsis of one to two pages that provides a more comprehensive overview than the abstract does. An executive summary covers a report's main points, conclusions, recommendations, and the impact of the subject on company planning. In some cases, readers may rely completely on such a synopsis, as, for instance, when a nonexpert reader must read a report written for experts. In other cases, readers use these synopses to orient themselves to the main topics in a report before reading it completely.

The abstract or executive summary in a formal report usually follows the list of tables, if there is one. In some companies, the style is to place an abstract on the title page or immediately after the title page. Whether you decide to include the longer executive summary or the shorter abstract depends on company custom and the expectations of your readers. In some companies, executives prefer to see both an abstract and an executive summary with long reports. If a report is very long, an executive summary allows a fuller synopsis and provides readers with a better understanding of the report contents and importance than an abstract will. Even though they are all synopses, the two types of abstracts and the executive summary provide readers with different emphases.

Descriptive Abstract

A *descriptive abstract* names the topics covered in a report without revealing details about these topics. Here is a descriptive abstract for a proposal to modify and redecorate a hotel restaurant:

> This proposal recommends a complete redesign of the Bronze Room in the Ambassador Hotel to increase our appeal to hotel guests and local customers. A description of the suggested changes, as well as costs, suggested contractors, and management reorganization, is included.

Because a descriptive abstract does not include details, readers must read the full report to learn about specifics, such as cost and planned decor. Descriptive abstracts have become less popular in recent years because they do not provide enough information for busy readers who do not want to read full reports.

Informative Abstract

An *informative abstract,* the one frequently used for formal reports and technical articles, describes the major subjects in a report and summarizes the conclusions and recommendations. This informative abstract is for the same report covered by the preceding descriptive abstract:

> This proposal recommends converting the Bronze Room of the Ambassador Hotel into a nineteenth-century supper club. This style has been successful in hotels in comparable markets, and research shows that it creates an upscale atmosphere for the entire hotel. The construction and decorating will take about 10 weeks and cost $245,000. The conversion is expected to halt a 5-year decline in local customer traffic and occupancy rates. This redesign of the Bronze Room should put the hotel in a favorable position to compete for convention groups interested in history and the arts.

This informative abstract includes more details and gives a more complete synopsis of the report contents than the descriptive abstract does.

Executive Summary

An *executive summary* includes (1) background of the situation, (2) major topics, (3) significant details, (4) major conclusions or results, (5) recommendations, and (6) a discussion of how the subject can affect the company. Writers often use headings in executive summaries to guide readers to information. Here is an executive summary of the hotel proposal:

> This proposal recommends complete redecoration of the Bronze Room from its present style into a late nineteenth-century supper club.
>
> **Problem**
>
> The Ambassador Hotel has experienced declining occupancy rates and declining local customer traffic for the past 5 years. Three new hotels, all part of well-known chains, have opened within one mile of the Ambassador. The Ambassador's image of elegant sophistication has been eroded by the competition of larger, more modern hotels.
>
> **Comparable Markets**
>
> Research indicates that the image of the main dining room in a hotel has a direct impact on marketing the hotel. A late nineteenth-century style for a hotel main dining room has been successful in St. Louis, Milwaukee, and Cleveland. These markets are comparable to ours. A survey by Hathaway Associates indicated that local customers would be interested in a dining atmosphere that reflects old-time sophistication and elegance.

Beryl Whitman, president of Whitman Design Studio, has prepared a proposal for specific design changes and suggestions for a new name for the Bronze Room.

Advantages

- A late nineteenth-century style will create an upscale sophisticated atmosphere that contrasts with the ultramodern look of the competing hotels.
- Our distinction will come from a perceived return to gracious and efficient hotel service of the past.
- Other hotel facilities will be undisturbed.
- Renovation will take 10 weeks and can be completed by November 1, in time for the holiday season.
- Projected cost of $245,000 can be regained within 18 months if we attract convention groups interested in history, such as the North Central Historians Association.
- We can expect an immediate 20% increase in local customers.

The recommendation is to redesign the Bronze Room with a target completion date of November 1.

Abstracts and executive summaries, important elements in formal reports, are difficult to write because they require summarizing in a few words what a report covers in many pages. Remember that an informative abstract or executive summary should stand alone for readers who do not intend to read the full report immediately. In preparing abstracts and executive summaries, follow these guidelines:

1. Write the abstract or executive summary after you complete the report.

2. Identify which topics are essential to a synopsis of the report by checking major headings and subheadings.

3. Rewrite the original sentences into a coherent summary. Simply linking sentences taken out of the original report will not produce a smooth style.

4. Write full sentences, and include articles *a, an,* and *the.*

5. Avoid overly technical language or complicated statistics, which should be in the data sections or an appendix.

6. Do not refer readers to tables or other sections of the original report.

7. If conclusions are tentative, indicate this clearly.

8. Do not add information or opinions not in the original report.

9. Edit the final draft for clarity and coherence.

WRITING BACK MATTER

Back matter includes supplemental elements that some readers need to understand the report information or that provide additional specialized information for some readers.

References

A *reference list* records the sources of information in the report and follows the final section of the report body. Preparation of a reference list is explained later in this chapter under "Documenting Sources."

Glossary/List of Symbols

A *glossary* defines technical terms, such as *volumetric efficiency,* and a *list of symbols* defines scientific symbols, such as *Au.* Include a glossary or list of symbols in a long report if your readers are not familiar with the terms and symbols you use in the report. Also include informal definitions of key terms or symbols in the text to aid readers who do not want to flip pages back and forth looking for definitions with every sentence. If you do use a glossary or list of symbols, say so in the introduction of the report to alert readers. The glossary or list of symbols usually follows the references. In some companies, however, custom may dictate that the two lists follow the table of contents. In either case, remember these guidelines:

- Arrange the glossary or list of symbols alphabetically.

- Do not number the terms or symbols.

- Include any terms or symbols that you are using in a nonstandard or limited way.

- List the terms or symbols on the left side of the page, and put the definition on the right side on the same line:

  bugseed   an annual herb in northern temperate regions
  Pt   platinum

Appendixes

Appendixes contain supplemental information that is too detailed and technical to fit well into the body of the report or information that some readers need and others do not. Appendixes can include documents, interviews, statistical results, case histories, lists of pertinent items, specifications, or lists of legal references. The recent trend in formal reports has been to place highly technical or statistical information in appendixes for those readers who are interested in such material. Remember these guidelines:

- Label appendixes with letters, such as "Appendix A" and "Appendix B," if you have more than one.

- Provide a title for each appendix, such as "Appendix A. Questionnaire Sample."

- Indicate in the body of the report that an appendix provides supplemental information on a particular topic, such as "See Appendix C for cost figures."

STRATEGIES—Protecting Your Computer

Computer viruses in the form of email attachments are a serious threat to your computer and stored files. Unfortunately, new viruses are sent out each day. To protect yourself, follow these tips:

- **Do not** forward an email warning about a virus to others. Often, viruses are sent with the subject line "Warning" to entice the recipients to open the attachment

- **Do not** open attachments if you do not know the sender. Watch out for garbled messages or an attachment without an accompanying message.

- **Do not** open attachments from unknown sources because they have a catchy title, such as "You won!" or "Here it is!"

- **Do not** open attachments that have extensions ".exe," ".vbs," or ".com." When opened, these files can take over your system.

- **Do** install virus protection software.

DOCUMENTING SOURCES

Documenting sources refers to the practice of citing original sources of information used in formal reports, journal articles, books, or any document that includes evidence from published works. Cite your information sources for the following reasons:

- Readers can locate the original sources and read them if they want.

- You are not personally responsible for every fact in the document.

- You will avoid charges of plagiarism. *Plagiarism* is the unacknowledged use of information discovered and reported by others or the use of their exact words, copied verbatim. See Chapter 2 for a longer discussion of plagiarism.

In writing a report that relies somewhat on material from other sources, remember to document information when you are doing either of the following:

- Using a direct quotation from another source

- Paraphrasing information from another source

If, however, the information you are presenting generally is known and readily available in general reference sources, such as dictionaries and encyclopedias, you need not document it. You would not have to document a statement that water is made up of two parts hydrogen and one part oxygen or that it boils at 212°F and freezes at 32°F. In addition, if your readers are experts in a particular field, you need not document basic facts or theories that all such specialists would know.

The documentation system frequently used in the natural sciences, social sciences, and technical fields is the *American Psychological Association (APA)* system, also called the *author-date* system. Another system often used in the sciences or technical fields is the *number-reference* system. Although all documentation systems are designed to help readers find original sources, the systems do vary slightly.

The following is a list of style guides used in particular associations or fields:

ACS Style Guide: A Manual for Authors and Editors

American Medical Association Manual of Style

American National Standard for the Preparation of Scientific Papers for Written or Oral Presentations

The Chicago Manual of Style

A Manual for Authors of Mathematical Papers

MLA Handbook for Writers of Research Papers

Publication Manual of the American Psychological Association

Scientific Style and Format: The CBE Manual for Authors, Editors, and Publishers

Suggestions to Authors of the Reports of the United States Geological Survey

APA System: Citations in the Text

In the APA system, when you refer to a source in the text, you must state the author and the date of publication. This paragraph is from a student report on lumber supply and illustrates in-text citation:

> A decade ago, Europeans took the lead in trying to save the world's lumber supply and an international movement resulted in the Forest Stewardship Council, the international organization that sets standards for forest certification (Newcombe, 2003). During the 1990s, the "combined import value of logs, veneer, and plywood dropped almost 40%" (Krickstein & Rafter, 2001, p. 67). Safin, Coetzer, and Fleming (2003) report that major markets are unlikely to recover quickly from that drop. Williams (2002a, 2002b) in studies of world markets, notes that Brazilian pine exports have fallen, but the Brazilian mahogany exports have risen, showing that not all timber markets have moved in the same direction. Another study (Howard et al., 2004) reports that technology has made it possible to get the same amount of product out of 82 trees today as out of 100 trees in the 1980s. Howard et al. also states that technology advances are saving more trees than environmental efforts are.

Notice these conventions of the APA citation system:

1. If an author's name begins a sentence, place the date of the work in parentheses immediately after.

2. If an author is not referred to directly in a sentence, place both the author's last name and the year of publication, separated by a comma, in parentheses.

3. If there are two authors, cite both names every time you refer to the source in your text. For multiple authors, cite all the names up to six. If there are more than six authors, cite only the first author and follow with et al. For subsequent references to sources with more than two authors, cite only the first author and et al.:

   Jones, Dean, Hutton, and Angeli (2001) found evidence. . . . (first citation)
   Jones et al. (2001) also reported. . . . (subsequent citation)

4. If multiple authors are cited in parentheses, separate names with commas, and use an ampersand (&) between the last two names.

5. If an author is referred to more than once in a single paragraph, do not repeat the year in parentheses as long as the second reference clearly refers to the same author and date as in the first reference.

6. If you are quoting from a source, state the page numbers immediately after the quotation:

   "interactive forums" (Jackson, 2002, p. 46).
   Jackson (2002) mentioned "interactive forums" (p. 46).

   If you are quoting from an Internet source without page numbers, cite the section and paragraph number if possible:

   "interactive forums" (Jackson, 2002, Usability, para. 2).

7. If you are citing two or more works by the same author, state all publication dates:

   Young (1999, 2001) disagrees. . . .

8. If you are citing two works by the same author, published in the same year, distinguish them by *a, b, c,* and so on:

   Lucas (2001a, 2001b) refers to. . . .

9. If you are citing several works by different authors in the same parentheses, list the works alphabetically by first author:

   Several studies tested for side effects but found no significant results (Bowman & Johnson, 1990; Mullins, 1979; Roberts & Allen, 1975; Townsend, 2001).

10. Use last names only unless two authors have the same surname, and then include initials to avoid confusion:

    W. S. Caldwell (1993) and R. D. Caldwell (2001) reported varied effects. . . . Heightened effects were noted (R. D. Caldwell, 2001; W. S. Caldwell, 1993).

11. Use an *ellipsis* (three spaced periods) to indicate omissions from direct quotations:

    Bagwell (2002) commented, "This handbook will not fulfill most needs of a statistician, but . . . model formulas are excellent."

12. If the omission in a quotation comes at the end of the sentence, use four periods to close the quotation:

    According to Martin (2002), "Researchers should inquire further into effects of repeated exposure. . . ."

13. Use brackets to enclose any information you insert into a quotation:

    Krueger (1997) cited "continued criticism from the NCWW [National Commission on Working Women] regarding salary differences between the sexes" (p. 12).

14. If you are citing personal communication, such as email, telephone conversations, letters, or messages on an electronic bulletin board, state the initials and last name of the sender and the date:

K. J. Kohls said the sales reports were unavailable (personal communication, January 7, 2002).

The sales reports were unavailable (K. J. Kohls, personal communication, January 7, 2002).

15. If the work has no author, use the first two or three main words in the title in place of the author. If an agency or association is the author, use the full name in the first citation with the abbreviation in brackets. Use the abbreviation in the citations that follow the first one:

Online fraud activity is increasing every month ("Cyberspace Fraud," 2005).

The study reported significant differences (National Safety Council [NSC], 2004).

Michigan had a higher accident rate than Wisconsin did (NSC, 2004).

APA System: List of References

This list of references for a report on the job or college paper includes each source cited in the document. Follow these conventions for business reports and college papers:

1. Do not number the list.

2. Indent the second line of the reference.

3. List items alphabetically according to the last name of the first author.

4. If there is no author, alphabetize by the title, excluding *a, an,* and *the*.

5. Alphabetize letter by letter.

Bach, J. K.
Bachman, D. F.
DeJong, R. T.
DuVerme, S. C.
MacArthur, K. O.
Martin, M. R.
McDouglas, T. P.
Realm of the Incas
Sebastian, J. K.
St. John, D. R.
Szartzar, P. Y.

6. Place single-author works ahead of multiple-author works if the first author is the same:

    Fromming, W. R.
    Fromming, W. R., Brown, P. K., & Smith, S. J.

7. Alphabetize by the last name of the second author if the first author is the same in several references:

    Coles, T. L., James, R. E., & Wilson, R. P.
    Coles, T. L., Wilson, R. P., & Allen, D. R.

8. List several works by one author according to the year of publication. Repeat the author's name in each reference. If two works have the same publication date, distinguish them by using *a, b, c,* and so on, and alphabetize by title:

    Deland, M. W. (2001).
    Deland, M. W. (2004a). Major differences. . . .
    Deland, M. W. (2004b). Separate testing. . . .

Remember that each reference should include author, year of publication, title, and publication data. Here are some sample references for typical situations:

1. *Journal article—one author:*

    Taylor, H. (2004). Vanishing wildlife. *Modern Science, 36,* 356–367.
    Note:
    - Use initials, not first names, for authors.
    - Capitalize only the first word in a title, except for proper names.
    - Underline or italicize journal names and volume numbers.

2. *Journal article—two authors:*

    Hawkins, R. P., & Pingree, S. (1981). Uniform messages and habitual viewing: Unnecessary assumptions in social reality effects. *Human Communication Research, 7,* 291–301.
    Note:
    - Use a comma and ampersand between author names.
    - Capitalize the first word after a colon in an article title.
    - Capitalize all important words in journal names.

3. *Journal article—more than two authors:*

    Geis, F. L., Brown, V., Jennings, J., & Corrado-Taylor, D. (1984). Sex vs. status in sex-associated stereotypes. *Sex Roles, 11,* 771–785.
    Note:
    - Use an ampersand between the names of the last two authors.
    - List all authors even if et al. was used in text references.

- If an article has more than six authors, list only the first six names and follow with a comma and et al.

4. *Journal article—issues separately paginated:*

Battison, J. (1988). Using effective antenna height to determine coverage. *The LPTV Report, 3*(1), 11.

Note: If each issue begins on page 1, place the issue number in parentheses directly after the volume number without a space between them.

5. *Magazine article:*

Lamana, D. (2001, August). Tossed at sea. *Cinescape, 1,* 18–23.

Note:

- Include the volume number.
- Include day, month, and year for weekly publications.

6. *Newspaper article—no author:*

Proposed diversion of Great Lakes. (2005, December 7). *Cleveland Plain Dealer,* pp. C14–15.

Note:

- If an article has no author, alphabetize by the first word of the title, excluding *a, an,* and *the.*
- If a newspaper has several sections, include the section identification as well as page numbers.
- Use pp. before page numbers for newspapers.
- Capitalize all proper names in a title.

7. *Newspaper article—author:*

Sydney, J. H. (2004, August 12). Microcomputer graphics for water pollution control data. *The Chicago Tribune,* pp. D6, 12.

Note: If an article appears on nonconsecutive pages, cite all page numbers, separated by commas.

8. *Article in an edited collection—one editor:*

Allen, R. C. (1987). Reader-oriented criticism and television. In R. C. Allen (Ed.), *Channels of discourse* (pp. 74–112). Chapel Hill, NC: Univ. of North Carolina Press.

Note:

- Capitalize only the first word in a book title, except for proper names.
- Underline or italicize book titles.
- Use initials and the last name of the editor in standard order.
- Use pp. with page numbers of the article in parentheses after the book title.
- Use the Postal Service ZIP code abbreviations for states in the publication information.

9. *Article in an edited collection—two or more editors:*

Zappen, J. P. (1983). A rhetoric for research in sciences and technologies. In P. V. Anderson, R. J. Brockmann, & C. R. Miller (Eds.), *New essays in technical and scientific communications: Research theory, practice* (pp. 123–138). Farmingdale, NY: Baywood.

Note:

• List all editors' names in standard order.
• Use *Eds.* in parentheses if there is more than one editor.

10. *One book—one author:*

Sagan, C. (1980). *Cosmos.* New York: Random House.

Note: Do not include the state or country if the city of publication is well known for publishing, such as New York, Chicago, London, or Tokyo.

11. *Book—more than one author:*

Johanson, D. C., & Edey, M. A. (1981). *Lucy: The beginnings of humankind.* New York: Simon and Schuster.

Note: Capitalize the first word after a colon in a book title.

12. *Edited book:*

Hunt, J. D., & Willis, P. (Eds.). (1990). *The genius of the place: The English landscape garden 1620–1820.* Cambridge, MA: MIT Press.

Note:

• Place a period after Eds. in parentheses.
• Include the state if the city is not well known or could be confused with another city by the same name (e.g., Cambridge, England).

13. *Book edition after first edition:*

Reep, D. C. (2006). *Technical writing: Principles, strategies, and readings* (6th ed.). New York: Longman.

Note: Do not place a period between the title and the edition in parentheses.

14. *Article in a proceedings:*

Pickett, S. J. (2002). Editing for an international audience. *Proceedings of the Midwest Conference on Multicultural Communication, 12,* 72–86.

Note: If the proceedings are published annually with a volume number, treat them the same as a journal.

15. *Unpublished conference paper:*

Dieken, D. S. (2000, May). *The legal aspects of writing job descriptions.* Paper presented at the meeting of the Southwest Association for Business Communication, San Antonio, TX.

Note: Include the month of the meeting if available.

16. *Report in a document deposit service:*

Cooper, J. F. (2001). *Clinical practice in nursing education* (Report No. CNC-95-6). Nashville, TN: Council for Nursing Curriculum. (ERIC Document Reproduction Service No. ED 262 116)

Note: Include document number at the end of reference in parentheses.

17. *Report—corporate author, author as publisher:*

American Association of Junior Colleges. (2001). *Extending campus resources: Guide to selecting clinical facilities for health technology programs* (Rep. No. 67). Washington, DC: Author.

Note:

- If the report has a number, insert it between the title of the report and the city of publication, in parentheses, followed by a period.
- If the report was written by a department staff, use that as the author, such as "Staff of Accounting Unit." Then give the corporation or association in full as the publisher.

18. *Dissertation—obtained on microfilm:*

Jones, D. J. (2000). Programming as theory: Microprocessing and microprogramming. *Dissertation Abstracts International, 61,* 4785B-4786B. (UMI No. 00–08, 134).

Note: Include the microfilm number in parentheses at the end of the reference.

19. *Dissertation—obtained from a university:*

Heintz, P. D. (1985). Television and psychology: Testing for frequency effects on children (Doctoral Dissertation, The University of Akron, 1985). *Dissertation Abstracts International, 45,* 4644A.

Note: When using the printed copy of a dissertation, include the degree-granting university and the year of dissertation in parentheses after the title.

20. *Reviews of books, films, and television programs:*

Hanson, V. D. (2001, July 29). Marching through Georgia [Review of the book *Sherman*]. *The New York Times Book Review,* p. 11.

Kenny, G. (2001, July). *Moulin Rouge's* moveable feast [Review of the film *Moulin Rouge*]. *Premiere, 14,* 82–83.

Flaherty, M. (2001, August 17). [Review of television program *Russian trinity*]. *Entertainment Weekly, 59.*

Note:

- Identify the items reviewed as a book, film, or television program in brackets.
- If the review has no title, retain the brackets around the text identifying the book, film, or television program.

21. *Films and television programs:*

Oz, F. (Director). (2001). *The score* [Motion picture]. United States: Paramount.

Koppie, B. (Producer). (2005, August 17). *My generation* [Television broadcast]. New York: American Broadcasting Company.

Franklin, B. (Producer). (2002). *Swimming with predators* [Television series]. New York: Animal Planet.

Hunter, T. (Writer), & Gentry, R. (Director). (2005). Home from the hills [Television series episode]. In A. Wilkes (Producer), *Westward bound.* New York: Fox.

Note:

- For a film, cite the primary creative contributor, usually the director.
- Identify the title as a motion picture in brackets before the period.
- Cite the country of origin and the production studio.
- For a small independent film or industrial film, provide the name and address of the distributor or owner in parentheses after the brackets.
- For an individual television program, cite the producer, director, and writers, if available, along with the date of broadcast.
- For a television series, cite the producer and the network that broadcast the series, along with the year the series appeared.
- For a single episode of a series, put the writer and the director first; put the producer or production company in the same position as an editor of a collection.
- If the names of television producers, directors, and writers are not available, cite the title of the program or series first as you would an article without an author.

22. *Computer software or program:*

Lippard, G. L. (2003). Torsion analysis and design of steel beams [Computer software]. Astoria, NY: Structural Design Software, Inc. (AX-P34-57)

Note:

- Identify the title as a computer program or software in brackets following the title without any punctuation between the brackets and the title.
- Cite the city and producer of the software or program.
- If the item has an identification number, place it in parentheses following the name of the software producer.
- Do not list references for standard software (e.g., Microsoft Word, Excel).

23. *Information on CD-ROM:*

Mitchell, R. N. (2001, July 12). World mining trends [CD-ROM]. *Foreign economic projections.* U.S. Department of Commerce.

Brown, P. K. (2002). Investment banking in Norway [CD-ROM]. *Business Today, 23*(5), 16–18. Abstract from: ABI/INFORM: 12560.00

Note:

- Cite CD-ROM in brackets immediately after the title, followed by a period.
- Give source and retrieval number if possible.
- Do not put a period after a retrieval number because it may appear to be part of the electronic address.

24. *Data file:*

U.S. Department of Labor, Occupational Safety & Health Administration. (2001). *Fatal workplace injuries* [Data file]. Retrieved May 6, 2001, from http://www.osha-slc.gov/OshDoc/data_FatalFacts

Note:

- Identify the source as a data file in brackets after the title and before the period.
- State the retrieval date with the full Internet address last.

25. *Internet home page:*

National Safety Council. Available from http://www.nsc.org

Note: If the reference is to a source rather than to specific material, cite the Internet address where the source is available.

26. *Newspaper—electronic version:*

Rogers, S. (2005, June 6). Women in banking. *USA Today.* Retrieved June 6, 2005, from http://www.usatoday.com

New rules for tennis rankings. (2000, October 3). *New York Times.* Retrieved October 3, 2000, from http://www.nytimes.com

Note:

- Cite the author first if available. If no author is given, begin with the title of the material, followed by a period.
- Put the date of the item in parentheses directly after the author if available or after the title.
- Spell out the name of the newspaper, and underline or italicize it.
- Give the date you found the information and the Internet address for the home page.

27. *Internet journal or newsletter:*

Lopez, G. A. (2004 March). Restoring classic yachts. *The Modern Yacht Newsletter, 5*(3). Retrieved April 8, 2004, from http://www.awb.org/modyacht/32004.html

Note:

- Cite the author first if available. If no author is listed, begin with the article title followed by a period.

- Use the full journal or newsletter title, and cite the volume and issue numbers if available.
- State the full Internet address and the retrieval date.

28. *Individual document on the Internet:*

Cenni, D. (2002, February 12). Neanderthal tools. Retrieved March 22, 2002, from http://www.anthr.org/tools

Note:

- Cite the author and date of the material if available. If there is no date, put n.d. in parentheses after the author's name.
- If there is no author, begin with the title of the document.
- Provide an Internet address that links directly to the material.

29. *Email communication:*

Smith, J. (smith@aol.com). (2005, September 16). Suggestions for conference. Email to D. C. Reep (dreep@uakron.edu).

Note:

- Although the current APA guide advises against including email in a reference list, most writers now are including citations for email. Consult your instructor.
- Cite the sender's name first, followed by his or her email address in parentheses, followed by a period.
- Cite the date the email was sent in parentheses, followed by a period.
- State the subject if there is one.
- Cite the receiver's name and give his or her email address in parentheses.

30. *Message Posted to a Discussion Group*

Mike22. (2004, May 16). Special-purpose chisels. Message posted to http://groups.yahoo.com/builders/4316

Note:

- State the writer's name if available. If a discussion name is used, cite that name.
- State the date the message was posted.
- State the subject line used by the writer.
- If the message has an identification number and it is not already in the address, state that at the end of the Internet address.

Number-Reference System

In the number-reference system, the references are written as shown above, but the reference list is numbered. The list may be organized in one of two ways:

1. List the items alphabetically by last name of author.

2. List the items in the order they are cited in the text.

The in-text citations use only the number of the reference in parentheses. Following is the same paragraph shown in APA style earlier to illustrate in-text citations. Here, the number-reference system is used; the reference list is in the order the references appear in the text.

> A decade ago, Europeans took the lead in trying to save the world's lumber supply and the international movement resulted in the Forest Stewardship Council, the international organization that sets the standards for forest certification (1). During the 1990s, the "combined import value of logs, veneer, and plywood dropped almost 40%" (2:67). Reference 3 reports that major markets were unlikely to recover quickly from that drop. Studies of world markets (4,5) note that Brazilian pine exports have fallen, but the Brazilian mahogany exports have risen, showing that not all timber markets have moved in the same direction. Another study (6) reports that technology has made it possible to get the same amount of product out of 82 trees today as out of 100 trees in the 1980s. The same study also states that technology advances are saving more trees than environmental efforts are.

Notice these conventions for the number-reference system:

1. Page numbers are included in the parentheses, separated from the reference number by a colon.

2. Sentences may begin with "Reference 6 states. . . . " However, for readability, rewrite as often as possible to include the reference later in the sentence.

For some readers, the APA system is easier to use because dates and the names of authors appear in the text. The number-reference system requires readers to flip back and forth to the reference list to find dates and authors. Whichever system you use, be consistent throughout your report.

CHAPTER SUMMARY

This chapter explains how to prepare front and back matter for formal reports and how to document sources of information. Remember:

*   Front matter consists of elements that help readers locate specific information and become familiar with report organization and content before reading the text. Included are the title page, transmittal letter or memo, table of contents, list of figures, list of tables, abstract, and executive summary.

*   Back matter consists of elements that some readers need to understand the report information or that provide additional information for readers. Included are a glossary, list of symbols, and appendixes.

- Documenting sources is the practice of citing the original sources of information used in formal reports and other documents.

- Sources of information should be documented whenever a direct quotation is used or whenever information is paraphrased and used.

- The APA system of documentation is used frequently in the natural sciences, social sciences, and technical fields for citations in the text and for references.

- The number-reference system, similar to APA style, also is used in the sciences and technical fields.

SUPPLEMENTAL READINGS IN PART 2

Caher, J. M. "Technical Documentation and Legal Liability," *Journal of Technical Writing and Communication,* p. 486.

MODEL 11-1   Commentary

The following samples of front matter are from the same formal report—a feasibility study prepared by an outside consultant. The writer has evaluated three possible solutions to the company's need for more space and presents his findings along with his recommendation.

- The title page reflects the contents and purpose of the report with a descriptive title.

- The transmittal letter presents the report to the client, identifies the subject and purpose, and includes the recommendation. The writer also credits two company managers who assisted him. In a goodwill closing, the consultant also comments favorably on his working relationship with the client and offers to consult further.

- The table of contents shows the main sections of the report. Notice that the writer uses all capitals for main headings but capitals and lowercase letters for subheadings. The pagination is in small Roman numerals for front matter and in Arabic numbers for the report body and back matter.

- The list of illustrations includes both figures and tables. Notice that tables are listed second even though they appear in the report ahead of the figures.

- The executive summary for this report identifies the purpose, briefly describes the problem, reviews the main elements for each alternative solution, and states the recommendation. This report probably will have several secondary readers before company officers make a decision on relocation. The executive summary provides more detail for these readers than an abstract would. Notice that the page includes the report title and the words *executive summary*. The Roman numeral pagination is centered at the bottom of the page.

Discussion

1. Compare this transmittal letter with those that appear in Model 12-5 and Model 13-6. Discuss which elements in each are meant to help the reader use the report.

2. Discuss the visual aspects of the table of contents. What features help guide the reader to specific information?

3. Based on the information in the executive summary, draft an informative abstract for this report. Discuss which information you can cut out, and compare your draft with those of your classmates.

FEASIBILITY STUDY OF OFFICE EXPANSION
FOR UNITED COMPUTER TECHNOLOGIES, INC.

Prepared for
Joanne R. Galloway
Senior Vice President
United Computer Technologies, Inc.

By
William D. Santiago
Senior Partner
PRT Management Consultants, Inc.

March 3, 2006

MODEL 11-1   Title Page

PRT MANAGEMENT CONSULTANTS, INC.
10 City Center Square
Sherwood Hills, Ontario N5A 6X8

(519) 555-3000   Fax (519) 555-3002

March 3, 2006

Ms. Joanne R. Galloway
Senior Vice President
United Computer Technologies, Inc.
616 Erie Street
Sherwood Hills, Ontario N5A 2R4

Dear Ms. Galloway:

Enclosed is the feasibility study PRT has prepared at your request. The report analyzes the options United Computer Technologies has in expanding the present physical space.

As we discussed in our February 2 meeting, our firm investigated three reasonable options—redesigning your current facilities, renovating the Wallhaven Complex and moving your offices there, or leasing space in the Huron Circle Building. Our staff researched these possibilities and consulted with the two managers you suggested, Daniel Kaffee in Design Technology and Paul Weinberger in Marketing Systems. We also obtained data from the central office of Retail Facilities, Inc., in Toronto.

Although each option has merit, our report concludes that leasing space in the Huron Circle building is the best choice for your company at this time. The attached report reviews all three options.

Working with you and your staff has been a pleasure. If you have any questions after reading the report or would like to discuss our conclusions, please call me at 555-3000.

Sincerely,

William D. Santiago

William D. Santiago
Senior Partner

WDS:sd

Transmittal Letter

TABLE OF CONTENTS

Table of Contents

LIST OF ILLUSTRATIONS

Figures

Tables

List of Illustrations

EXECUTIVE SUMMARY

Feasibility Study of Office Expansion
for United Computer Technologies, Inc.

This report evaluates three options for expanding the office space of United Computer Technologies, Inc. At present, the company has 640 full-time employees and 46 regular part-time employees in office space designed to accommodate 600 people. Also, the company's client traffic has increased 14% in the past year, creating a need for at least three more conference rooms.

Options were evaluated on the basis of cost, time constraints, office layout, location, and appropriateness for the client base.

- **Redesign of the current offices**—Redesign to accommodate present needs would be extensive, requiring wall removal and other major construction. Cost is estimated at $350,000. Clients would have to visit amid noise and debris for the 4 to 5 months of construction.

- **Renovating and moving to the Wallhaven Complex**—The Wallhaven Complex would require some renovation, such as painting, redecorating interior offices, and creating conference rooms. Cost is estimated at $200,000 with a time estimate of 6 weeks until the company could move in. A major drawback is the location near a busy landfill. The area might deter clients.

- **Leasing offices in the Huron Office Building**—This building has adequate space and its location is easy to reach from a major freeway. Some redesign would be needed here, but the building is only 12 years old, so no major construction is necessary. Cost is estimated at $60,000.

Based on the estimated costs and the desire of United Computer Technologies to expand as soon as possible, the recommendation is to lease offices in the Huron Office Building.

v

Executive Summary

MODEL 11-2   Commentary

This model contains samples of back matter.

- The reference list is from a student research report on ancient Celtic weapons. The writer uses the APA documentation style. Notice that the header at upper right includes the writer's name and the page number.

- The glossary is from a long technical report to multiple readers. The writer knows that at least half the readers are not experts in the technical subject and includes the glossary to help readers understand the information.

- The sample appendix is from a company reference manual and lists the specifications for a technical procedure. Notice that a brief description of the particular features or capacity of each component is included.

Discussion

1. Discuss the visual design of the glossary and the appendix. What features help readers who need this information?

2. As a self-test, identify the type of source (e.g., article, book chapter) in each item of the reference list. Compare your results with those of a classmate. If you disagree about an item, compare it to the documentation samples in this chapter, and identify the type of source.

REFERENCE LIST

Albers, G. T. (2002, March 3). Celtic chariots at Telemon. Historic Trails, 12, 23–26, 87.

The Battersea shield. (2001, June 23). Boston Globe, p. B8.

Buford, J. G., & Longstreet, J. R. (1994). Celtic excavations in 1990. Journal of the Ancient World, 42(3), 25–41.

Celtic swords. (2005, March 17). USA Today. Retrieved March 17, 2005, from http://www.usatoday.com

Cunliffe, B. R. (2003). Pits, preconceptions and propitiation in the British Iron Age. Oxford Journal of Archaeology, 11, 69–83.

Dukes, J. T. (Director). (2003). Iron Age hillforts [Motion picture]. (Available from History Cinema, 62 South Main, Chicago, IL 60610)

Hancock, G. H., Lee, R. T., & Berenger, T. N. (2001). The Celtic longboat. Proceedings of the Northeast Association of Historians, 16, 43–49.

Merriman, N. (1987). Value and motivation in prehistory: The evidence for Celtic spirit. In I. Hodder (Ed.), The archaeology of Celtic meanings (pp. 111–116). Cambridge, England: Cambridge Univ. Press.

Niblett, R. T. (1992). A chieftain's burial from St. Albans. Antiquity, 66, 917–929.

Pleiner, R. (2004). The Celtic sword. Oxford: Oxford Univ. Press.

Rigby, V. (1986). The Iron Age: Continuity or invasion? In P. Longworth & C. J. Cherry (Eds.), Archaeology in Britain since 1945 (pp. 52–72). London: British Museum.

Shapiro, H. D. (n.d.). Celtic long swords. Retrieved July 29, 2004, from http://museums.ncl.ac.uk/reticulum/NORTHERNFRONTIER/The WretchedBritons/WARFARE/weapons.htm

Watkins, G. (Producer). (2005, April 12). Arthur's weapons [Television broadcast]. New York: Arts and Entertainment.

MODEL 11-2   Reference List

GLOSSARY

Accelerometer ....An electrical device that measures vibration and converts the signal to electrical output.

Air/Fuel Mixture ....A ratio of the amount of air mixed with fuel before it is burned in the combustion chamber.

Ammeter.....................................An electric meter that measures current.

Amplitude ....The maximum rise or fall of a voltage signal from 0 volts.

Battery-Hot ....Circuit fed directly from the starter relay terminal. Voltage is available whenever the battery is charged.

Blown ....A melted fuse filament caused by overload.

Capacitor....................................A device for holding or storing an electric charge.

DVOM
   (Digital Volt-Ohmmeter) ....A meter that measures voltage and resistance and displays them on a liquid crystal display.

ECA
   (Electronic Control Assembly) ...See Processor

Fuse...A device containing soft metal that melts and breaks the circuit when it is overloaded.

Induced Current ....The current generated in a conductor as a magnetic field moves across the conductor.

Oscillograph ....A device for recording the waveforms of changing currents or voltages.

APPENDIX D: ACS:180 SPECIFICATIONS

This section summarizes the ACS:180 specifications. The ACS:180 design meets the requirements of the following agencies:

- UL
- CSA
- IEC
- JIS
- FCC Class A
- VDE Class A

Component Specifications

Cuvette	disposable reaction vessel with a 1.0-mL maximum fill volume
Cuvette Loading Bin	holds 250 cuvettes and automatically feeds the cuvettes into the track
Cuvette Preheater	regulated at a temperature of 37.0°C (+/−0.5°C) to heat the cuvettes
Cuvette Track	aluminum track, thermostatically controlled by four thermal electric devices (TEDs) to maintain the present temperature of 37.0°C (+/−0.5°C)
Cuvette Waste Bin	autoclavable collection bin fitted with a disposable, biohazard liner that holds up to 250 used cuvettes
Heating Bath	static fluid bath that contains a specially conditioned circulating fluid; fluid temperature is regulated at 37.5°C (+/−1.0°C)
Waste Bottle	removable container that holds approximately 2.8 liters (capacity of 200 tests) of waste fluid
Water Bottle	removable container that holds approximately 2.6 liters (capacity for 200 tests) of deionized water at temperatures between 15.0° and 37.0°C

Appendix

Chapter 11 Exercises

1. Assume that you used the following sources for a report. Prepare a reference list in APA style, following the guidelines in this chapter.

a. A book by Sam Duncan titled The Ultimate Home Improvement Handbook, published in 2003 by SelfHelp Press in Cleveland, Ohio.

b. An edited book called The Cutting Edge. The sixth edition was published in 2001 by Chester and Lang Publishers in London, England. The editors are Sean Sullivan and Lila Serna. You are citing chapter 8, pages 243–259, titled "repairing doors." The author of the chapter is Matthew Newbury.

c. A television program called "clearing sinks and drains" that appeared on the House & Home TV Network on June 26, 2004, broadcast from New York. The program was part of a series called "maintenance and repairs," produced by Luis Costa. The episode did not have a writer and director listed in the credits.

d. An article from volume 4, number 6 of an Internet newsletter called "building for tomorrow." The article by Michael Redbird is titled "wood finishing" and is dated June 6, 2005. You retrieved it from http://www.buildtomorrow.com/woodfinishing/6605 on July 23, 2005.

e. An article from a newspaper Web site dated April 7, 2003. The article was written by Jake Polsky and titled "door locks." The paper was USA Today, and you retrieved the article from http://www.usatoday.com/life/door.htm on April 8, 2003.

f. An article from the newspaper San Francisco Chronicle, page D7, October 12, 2004. There was no author. The title of the article was "home security."

g. A company Web page that you found on November 6, 2004. The company is Randall Paints at www.randallpaints.com.

h. An article by Steve Winchell, Susan Chang, Sonia Deal, and Carole Szabb titled "saving historical homes." It appeared in the spring 2004 issue of the journal of home décor on pages 435–438, Vol. 12, Number 2.

i. An edited book titled Wood Design for the Ages: A Review of 2000 Years, published in 2000 by the University of Texas Press in Austin, Texas. The editors are George Hengst and John Horsa.

2. If you are writing a long report for one of your classes, write an appropriate transmittal memo to your instructor.

3. The following short paragraphs of product information are intended for the Web site of a manufacturer of rehabilitation equipment. The Web design requires that these paragraphs be no longer than 40 words. Revise these paragraphs to retain the information and meet the word limitation. Compare your results with those of your classmates.

    a. The Hercules Semi-Recumbent Cycle is an ideal cycle for patients who are too weak to use a standard cycle. The adjustable pedal cranks can create a smooth knee movement that avoids a painful jolt to reconstructed ligaments. The patient can pedal forward or backward, and the Constant Power Mode allows the patient to exercise at a specific workload that exercises different muscle groups. The adjustable contoured seat supports the lower back and accommodates patients with heights from 4 ft. 6 in. to 6 ft. 7 in.

    b. The Hercules Gait Trainer is designed for stroke victims and patients with orthopedic problems. It promotes normal walking patterns based on the patient's leg length. When a patient uses it regularly, he or she can increase cardiovascular capacity and endurance while providing a safe environment for rehabilitation. The Gait Trainer comes with video and audio biofeedback as well as arm and crutch supports, oxygen tank holders, and geriatric or pediatric handrails. It is tested to handle up to 350 lb.

COLLABORATIVE EXERCISES

4. In groups, read "Technical Documentation and Legal Liability" by Caher in Part 2. Draft an abstract of the article. Compare your results with those of other groups.

5. In groups, read "Taking Your Presentation Abroad" by Weiss in Part 2. Draft an abstract of the article, and compare your results with those of other groups. Next, revise your abstract, and cut 20 words out of it. If possible, revise your abstract a second time, and cut out 10 more words.

INTERNET EXERCISES

6. Find an online journal in your major or a topic for another course. Print an article, and write an abstract for it. If the online journal includes an abstract with the article, revise the original abstract to cut out one-third of the words. Submit the original article and your abstract or your revision of the original abstract to your instructor with a transmittal memo.

7. Select a topic from your major and search the Internet for information about it. Develop a glossary of 5–8 key terms relevant to the topic. Exchange glossaries with a classmate in another field, and review the entries for clarity.

8. Using your school library's on-line catalog, search for 3–6 articles on a topic relevant to your major or to a class you are taking. Draft a reference list of the articles in the correct APA style.

CHAPTER 12

Short and Long Reports

UNDERSTANDING REPORTS

Next to correspondence, reports are the most frequently written documents on the job. Reports usually have several purposes, most commonly the following:

- To inform readers about company activities, problems, and plans so readers are up to date on the current status and can make decisions—for example, a progress report on the construction of an office building or a report outlining mining costs.

- To record events for future reference in decision making—for example, a report about events that occurred during an inspection trip or a report on the agreements made at a conference.

- To recommend specific actions—for example, a report analyzing two production systems and recommending adoption of one of them or a report suggesting a change in a procedure.

- To justify and persuade readers about the need for action in controversial situations—for example, a report arguing the need to sell off certain company holdings or analyzing company operations that are hazardous to the environment and urging corrective action.

Reports may vary in length from one page to several hundred, and they may be informal memos, formal bound manuscripts, rigidly defined form reports (e.g., an accident report at a particular company), or documents for which neither reader nor writer has any preconceived notion of format and organization. Reports may have only one reader or, more frequently, multiple readers with very different purposes.

Whether a report is short or long depends on how much information the reader needs for the specific purpose in the specific situation, not on the subject or the format of the report. Busy people on the job do not want to read reports any longer than necessary to meet their needs. Long reports usually are necessary in the following situations:

- Scientists reporting results of experiments and investigations

- Consultants evaluating company operations and recommending changes

- Company analysts predicting future trends in the industry

- Writers reporting on complex industry or company developments

For all reports, analyze reader and purpose before writing, as discussed in Chapter 3, and organize your information, as outlined in Chapter 4. In addition, use the guidelines for planning, gathering information, evaluating sources, taking notes, interpreting data, and drafting discussed later in this chapter.

Reports for specific purposes are discussed in Chapter 13. The rest of this chapter focuses on the general structure of short and long reports and on ways to develop long reports that deal with complex subjects and require information from many sources.

DEVELOPING SHORT REPORTS

Most short reports within an organization are written as memos. Short reports written to people outside the organization may be formal reports, with the elements described in Chapter 11, or letters, as shown in Models 4-6 and 12-4. Because short reports provide so much of the information needed to conduct daily business and because readers on the job are always pressed for time, most short reports, whatever their format, are organized in the opening-summary, or front-loaded, pattern. The delayed-summary pattern is used primarily when readers are expected to be hostile to the report's subject or conclusions.

Opening-Summary Organization

The *opening-summary organization* is often called *front loaded* because this pattern supplies the reader with the most important information—the conclusions or results—before the specific details that lead to these results or conclusions. The pattern has two or three main sections, depending on the writer's intent to emphasize certain points.

Opening Summary

The opening is a summary of the essential points covered in the report. Through this summary, the reader has an immediate overall grasp of the situation and understands the main point of the report. Include these elements:

- The subject or purpose of the report
- Special circumstances, such as deadlines or cost constraints
- Special sources of information, such as an expert in the subject
- Main issues central to the subject
- Conclusions, results, recommendations

Readers who are not directly involved in the subject often rely solely on the opening summary to keep them informed about the situation. Readers who are directly involved prefer the opening-summary pattern because it previews the report for them. This way, when they read the data sections, they already know how the information is related to the report conclusions. Here is an opening from a short report sent by a supervisor of a company testing laboratory to his boss:

> The D120 laboratory conducted a full temperature range test on the 972-K alternator. The test specifications came from the vendor and the Society of Automotive Engineers and are designed to check durability while the alternator is functioning normally. Results indicate that the alternator is acceptable, based on temperature ranges, response times, and pressure loads.

In this opening summary, the supervisor identifies the test he is reporting, cites the standards on which the test is based, and reports the general result—the alternator is acceptable. The reader who needs to know only whether the alternator passed the test does not have to read further.

Data Sections

The middle sections of a short report provide the specific facts relevant to the conclusions or recommendations announced in the opening. The data sections for the report about the alternator would provide figures and specific details about the three tested elements—temperature ranges, response times, and pressure loads—for readers who need this information.

Closing

Reports written in the opening-summary pattern often do not have closings because the reader knows the conclusions from the beginning and reads the data sections for details, making any further conclusions unnecessary. However, in some cases you may want to reemphasize results or recommendations by repeating them in a brief closing. Depending on the needs of your readers, your closing should:

- Repeat significant results/recommendations/conclusions
- Stress the importance of the matter
- Suggest future actions
- Offer assistance or ask for a decision

Here is the closing paragraph from the short report written by the supervisor in the testing laboratory:

> In conclusion, the 972-K alternator fits our needs and does not require a high initial cost. The quality meets our standards as well

as those set by the Society of Automotive Engineers. Because we need to increase our production to meet end-of-the-year demands, we should make a purchase decision by November 10. May I order the 972-K for a production run starting November 20?

In this closing, the supervisor repeats the test results—that the alternator meets company standards. He also reminds his reader about production deadlines and asks for a decision about purchasing the alternators. Because of a pressing deadline, the supervisor believes he needs a closing that urges the reader to reach a quick decision.

Model 12-1 shows a short report in the opening-summary pattern. The writer was told to review safety procedures at a shipyard. The opening summary reminds the reader of the report's purpose, notes injury statistics, and states that improvements could be made to prevent eyesight and hearing injuries. The middle sections review the safety equipment in use and the potential for injuries. The writer then provides a detailed recommendation section.

Delayed-Summary Organization

The *delayed-summary organization,* as its name suggests, does not reveal the main point or result in the opening. This organizational pattern slows down rather than speeds up the reader's understanding of the situation and the details. Use the delayed-summary pattern when you believe that your reader is not likely to agree with your conclusions readily and that reading these conclusions at the start of the report will trigger resistance to the information rather than acceptance. In some cases, too, you may know that your reader usually prefers to read the data before the conclusions or recommendations or that your reader needs to understand the data to consider your conclusions or recommendations. The delayed-summary organization has three main sections.

Introductory Paragraph

The introductory paragraph provides the same information as the opening summary except that it does not include conclusions, results, or recommendations.

Data Sections

The data sections cover details relative to the main subject, just as the data sections in the opening-summary pattern do.

NORTHERN STAR SHIPYARD

To: Michael Greyeagle
         Vice President—Operations

From: Rick Hernandez
         Yard Supervisor   *RH*

Date: May 23, 2005

Subject: Eyesight and Hearing Safety

As you directed, I checked our safety procedures and equipment for yard workers to be sure we are up to date. Overall, we appear to be in compliance with recognized good practices, but we might be able to improve. Our reported accidents were up slightly in 2004 (1.2%), and we want to improve this year, especially in eyesight and hearing.

Safety Problems

In shipyards like ours, there are two recurring problems no matter what job the worker is actually doing. One is eye injuries. The second is hearing loss.

Sight Dangers. We had 14 injuries to eyes from flying particles in the last fiscal year. These were not serious, but we should increase efforts to prevent eye injuries. Our processes in building and repair generate a great deal of dust and metal particles that get into the eyes of the workers. Since the natural tendency is to rub the eyes, workers end up with corneal abrasions that require medical treatment.

Last year we also had two welder's flash injuries that resulted from exposure to the intense light from a welding arc. Although this injury is temporary, repeated exposure can create cataracts or retina damage. We have supplies of welding helmets, goggles, and the shaded lenses to help prevent welder's flash. However, when I walked the yard last week, I saw two workers wearing street sunglasses instead of regulation eye protection. I ordered them to change to goggles, but I am sure this kind of inattention to safety regulations occurs occasionally.

Hearing Dangers. Workers who are not using the pneumatic tools may think they are in no danger from noise, but our yard sound level is so high that all workers are exposed to dangerous noise levels. Although

MODEL 12-1   A Short Report Organized in an Opening-Summary Pattern

M. Greyeagle   -2-   May 23, 2005

we have containers of earplugs throughout the yard, some workers
told me that they don't like to wear them because they are afraid
they will not hear warning sounds or conversation from other
workers.

Recommendations

Based on my observations and talks with individual workers, I
believe we need to increase the safety levels for our yard workers.

Eyesight Recommendations. Although we have all the approved eye
protection equipment, I think we need to add the following:

- Eye-wash stations set up throughout the yard will allow a quick
  eye flush whenever workers need one.

- On board ships, portable eye-wash stations can allow workers
  to rinse quickly without leaving the ship.

- Wipes and glass cleaning solutions at various locations in the
  yard will stop workers trying to clear goggles, etc., with their
  hands and distributing more debris near their eyes.

Hearing Recommmendations. We are currentiy providing the
disposable foam earpiugs. We could provide more variety in protection
equipment, such as the following:

- Some workers might prefer ear muffs, and muffs coulds be worn
  in additions to the ear plugs in a very noisy locations.

- Premolded flexible ear plugs have to be individually fitted, but
  they could be more comfortable for some workers.

- Audiograms should be added to our annual fitness test for
  workers. An initials audiogram when the workers is hired can
  be compared with later audiograms, so we can spot hearing
  changes quickly.

Finally, I believe all work stations supervisors need a training session
on eyesight and hearing safty, so they will act promptly whenever
they see a problem. I will be glad to discuss these safety issues with
you further.

Closing Summary

Because the introductory paragraph does not include results, conclusions, or recommendations, a closing that summarizes the main points, results, general observations, and recommendations is essential. The closing also may include offers of assistance, reminders about deadlines or other constraints, and requests for meetings or decisions.

The short report in Model 12-2 shows the delayed-summary pattern. The writer is asking for expensive equipment, and he probably suspects that the reader will be resistant to his recommendation. The writer uses the delayed-summary organization, so he can present his rationale, supported by data, before his request. In his closing, the writer stresses the need for the new equipment and suggests a meeting to discuss the matter. In this situation, the opening-summary organization could trigger immediate rejection by the reader, who might not even read the data. Also, depending on the relationship between writer and reader, the opening-summary might seem to be inappropriately demanding.

STRATEGIES—Emphasis

Emphasize a key term by placing it at the beginning or at the end of a sentence.

No: In the past decade, an analytical technique called **X-ray fluorescence** has become important in the cement industry.

Yes: **X-ray fluorescence,** an analytical technique, has become important in the cement industry.

Yes: In the past decade, the cement industry has accepted an analytical technique called **X-ray fluorescence.**

DEVELOPING LONG REPORTS

Long reports (more than five pages) may be written as memos or letters, but frequently, they are written as formal documents with such elements as appendixes and a table of contents, as explained in Chapter 11. Long reports tend to deal with complex subjects that involve large amounts of data, such as an analysis of how well a company's 12 branch offices are performing with recommendations for shutting down or merging some offices. In addition to dealing with complicated information, a long report nearly always has multiple readers

TO: T. R. Lougani   January 16, 2006
             Vice President, Support Services

FROM: C. S. Chen ^{CSC}
             Asst. Director of Radiology

SUBJECT: **Ultrasound Equipment**

Since ultrasound is a noninvasive procedure and presents no radiation
risk to the patient, it has become increasingly popular with most
physicians. In our department, ultrasound use over the past five years
has increased 32%. In 2004, we performed 1479 procedures, and last
year we performed 2186 procedures. We expect this increased demand
to continue. To prepare for the demand on our equipment (estimated
2297 this year), we need to consider ways to increase our capability.

Updating Current Equipment.   One possibility is to update the current
ultrasound unit with new software. Costs for updating the present
equipment are as follows:

Current unit update	$31,000
Maintenance contract	8,500
(first year)	
Total	$39,500

The maintenance contracts will increase slightly in each succeeding year.
The maintenance charge is rather high because, as the equipment ages,
more replacement parts will be needed. The basic hardware is now
seven years old. Although updating the present equipment will improve
the performance of the equipment, it will only marginally increase the
number of procedures that can be performed over a 24-hour period. We
cannot reasonably expect to make up the cost of updating our equipment
because our volume will not increase significantly.

Purchasing New Equipment.   Purchasing new ultrasound equipment
will provide Columbia Hospital with the ability to keep up with the
expected growth in requests for the procedure. A new unit with
increased capability will allow us to expand by 15% in the first year. The
unit alone will enable this expansion; we will not need new staff or floor
space. Increased capability will also increase our patient referrals. At an
average cost of $240 per examination, an additional $41,000 will result

MODEL 12-2   A Short Report Organized in a Delayed-Summary Pattern

in the first year, and the full cost of equipment should be recovered within the first two years. Patients may also use other services at the hospital once they come here for ultrasound, resulting in further income. Cost of new equipment is as follows:

New unit	$90,000
Maintenance contract	0
Total	$90,000

Conclusion.   Updating old equipment is merely a holding technique, even though such a decision would save money initially. To increase our capability and provide the latest technology for our patients, I recommend that we purchase a new unit to coincide with the opening of the new community clinic on our ground floor, March 1. I would appreciate the opportunity to discuss this matter further. If you wish to see sales and service literature, I can supply it.

who are concerned with different aspects of the subject and have different purposes for using the information in the report.

Planning Long Reports

Consider these questions when you begin planning a long report:

- *What is the central issue?* In clarifying the main subject of a long report, consider the questions your readers need answered: (1) Which choices are best? (2) What is the status of the situation, and how does it affect the company's future? (3) What changes must we make and why? (4) What results will various actions produce? (5) What are the solutions to certain problems? (6) How well or how badly does something work? Not all your readers are seeking answers to the same questions, so in planning a long report you should consider all possible aspects that might concern your readers.

- *Who are your readers, and what are their different purposes?* Because long reports tend to have multiple readers, such reports usually cover several aspects of a subject. You must consider your readers' purposes and how to effectively provide enough information for each purpose. Remember that long reports generally need to include more background information than short reports because (1) multiple readers rarely have an identical understanding of the subject and (2) long reports may remain in company files for years and serve as an information source to future employees dealing with a similar situation. Chapter 3 explains specific strategies for identifying readers' purposes.

- *How much information do I need to include in this report?* The relevant data for all readers and all purposes sometimes constitute masses of information. Determine what kinds of facts you need before you begin to gather information so that you are sure to cover all areas.

Gathering Information

When you gather information for a long report, consult secondary and primary sources for relevant facts.

Secondary Sources

Secondary sources of information consist of documents or materials already prepared in print, on tape, or on film. For a long project, consult secondary sources first because they are readily available and easy to use. Also, the information in secondary sources reflects the work of others on the same subject and may help direct your plans for further research or trigger ideas about types of information your readers need to know.

Unless you have a specific author or title of an article in mind when you use these secondary sources, you must search for information by subject. For effective searches, check under other key topics as well as the main subject. If you are looking for information about diamonds, check under the key word *diamond,* as well as under such topics as mining, precious gems, minerals, South Africa, and carbon. Major secondary sources are as follows:

Library Catalog. Most libraries now use an on-line catalog that lists all books, films, tapes, disks, periodicals, and any materials on microfilm or microfiche in the library. The catalog information will tell you the call number of each item and where it is located. The catalog also usually indicates whether the item is currently available or is checked out and due back on a certain date.

Periodical Indexes. Published articles about science, technology, and business topics are listed in periodical indexes. Your library may have these available in print, on-line, or on CD-ROM. Your librarian can help you get started using an index that is likely to cover the topic you are researching. The most useful indexes for people in professional areas include *ABI/Inform, Engineering Index, Applied Science and Technology Index, Business Periodicals Index,* and *General Science Index.* Most indexes cover several fields, but no one index covers all the relevant journals in a particular field.

Newspaper Indexes. The two newspaper indexes available in most college libraries are *The New York Times Index* and *The Wall Street Journal Index.* Some periodical indexes include these newspapers and others, such as *The Financial Times of Canada.*

Abstract Indexes. Abstract indexes provide summaries of published articles about science, technology, and business. Read the abstract to decide whether the article contains the kinds of information you need. If it does, look for the article in the journal listed. Specialized abstract indexes include *Biological Abstracts, Chemical Abstracts, Computer and Control Abstracts, Electrical and Electronic Abstracts, Geological Abstracts, Mathematical Reviews, Microbiology Abstracts, Nuclear Science Abstracts,* and *Water Resources Abstracts.*

Company Documents. Company Web sites offer a variety of information, such as product descriptions, financial information, annual reports, products currently in development, names of divisions and officers, and contact information. Many company Web sites include a system whereby visitors can order printed material on various topics. Some companies offer to answer email questions.

Government Documents.   The federal government maintains Web sites for all departments, such as the U.S. Department of Commerce. From these sites, you can retrieve information sheets, statistics, technical reports, and newsletters. These are updated regularly, so your information will be the latest available. Your library may have the *Monthly Catalog of United States Government Documents,* which lists unclassified reports and brochures available to the public.

Encyclopedias, Directories, and Almanacs.   College research requires specialized encyclopedias and directories, such as *Meteorology Source Book, Encyclopedia of Energy,* and *McGraw-Hill Encyclopedia of Science and Technology.* Almanacs give specific dates and facts about past events.

Internet.   Currently, there are more than a billion Web sites, and finding specific information can be difficult. To use the Internet effectively, you must use a search engine to sort through available Web sites. Search engines are relatively easy to use, but no two really work the same way. Generally, the following strategies are useful:

- Put your key words or phrases in quotation marks ("educator's handbook").

- Link key words with AND ("technical AND writing"). Some engines provide a drop-down menu for combining words.

- Eliminate a term by using NOT ("horses NOT pintos"). Some engines have a drop-down menu to eliminate or streamline search phrases.

- Use OR to find either word in your key phrase ("pools OR spas"). Some engines automatically search for each word in a phrase individually.

- Search for an unusual term alone first before adding any general terms ("Excalibur").

- Search for synonyms or subcategories for your main concept to enlarge the response ("dark ages" as well as "middle ages").

- Shorten the URL to the home page if you get a broken link.

Click on a link that looks as though it contains the information you want. The following search engines, each indexing millions of Web pages, are most helpful for research in science, technology, and business:

Alta Vista (http://www.altavista.com)

Northern Light (http://www.northernlight.com)

Yahoo (http://www.yahoo.com)

Lycos Pro (http://www.lycospro.com)

Google (http://www.google.com)

Excite (http://www.excite.com)

Webcrawler (http://www.webcrawler.com)

Dogpile (http://www.dogpile.com)

Primary Sources

Primary sources of information involve research strategies to gather unpublished or unrecorded facts. Scientific research reports, for example, usually are based on original experiments, a primary source. A marketing research report may be based on a consumer survey, also a primary source. Secondary sources can provide background, but they cannot constitute a full research project. Only primary sources can provide new information to influence decision making and scientific advances. The major primary sources of information follow:

Personal Knowledge.   When you are assigned to a report project on the job, usually you are already deeply involved in the subject. Therefore, much of the information needed for the report may be in your head. Do not rely on remembering everything you need, however—make notes. Writing down what you know about a subject will clarify in your own mind whether you know enough about a particular aspect of the subject or need to gather more facts.

Observation.   Gathering information through personal observation is time-consuming, but it can be essential for some subjects. Experienced writers collect information from other sources before conducting observations so that they know which aspects they want to focus on in observation. Scientists use observation to check, for example, how bacteria are growing under certain controlled conditions. Social scientists may use observation to assess how people interact or respond to certain stimuli. Technology experts may use observation to check on how well machinery performs after certain adjustments. Do not interfere with or assist in any process you are observing. Observation is useful only when the observer remains on the sidelines, watching but not participating.

Interviews.   Interviews with experts in a subject can yield valuable information for long reports. Writers usually gather information from other sources first and then prepare themselves for an interview by listing the questions they need answered. Know exactly what you need from the person you are interviewing, so you can cover pertinent information without wasting time. Use the guidelines for conducting interviews in Chapter 3.

Tests. Tests can be useful in yielding information about new theories, systems, or equipment. A scientist, for example, must test a new drug to see its effects. A market researcher may test consumer reaction to new products.

Surveys. Surveys collect responses from individuals who represent groups of people. Results can be analyzed in various ways, based on demographic data and responses to specific questions. A social scientist who wants to find out about voting patterns for people in a specific geographic area may select a sample of the population that represents all voting age groups, men and women, income groups, and any other categories thought to be significant. Internet surveys may seem an easy way to gather responses from a large number of people because of the speed and ease of sending and receiving messages. The disadvantages are that respondents are not anonymous, the sample is limited to Internet users, and computer problems could destroy data. Designing an effective questionnaire is complicated because questions must be tested for reliability. If you are not trained in survey research, consult survey research handbooks and ask the advice of experts before attempting even an informal survey on the job.

Evaluating Sources

Part of the research process is evaluating the credibility of your sources before you use the information they provide. Researchers have long-established criteria for evaluating print sources. Articles and books have been reviewed by other experts before being published. Consider the following when you want to use a print source:

- Does the author have a reputation for research in this subject? Is the author connected to a university or a research institute? If you are not sure the author of a source is an expert, check with your instructor or a librarian before using that source.

- Is the book or article recent enough for your purpose? Labor statistics in an article published in 1994 will not be relevant to your current study unless you want the old statistics as a comparison.

- Is the article in an established professional journal? Is the journal sponsored by a professional association or a university?

- Is the book published by an established university or trade press? Do not rely on books published by vanity presses (the author pays the publisher) or self-published books (the author publishes the book and pays all costs).

- Does the book or article have a bibliography of other reliable sources?

Relying on information from the Internet might be especially risky. The Internet can provide both primary and secondary sources. A document written by a scientist who conducted an environmental study for the U.S. Department of the Interior is a primary source. A consumer Web site that discusses the test is a secondary source. Be cautious with the Internet. Anyone can put anything on the Internet without any review process. John Doe's ideas about the environment expressed in a chat room or on his personal Web page are not a legitimate source of information for a report. Consider the following when you evaluate online sources of information:

- Does the author have a scholarly or professional reputation in this subject? Does the site give background information on the author and the research?

- Is the Web site maintained by a professional organization, a university, a government agency, or a research foundation? These Web sites are likely to be the most reliable. Company Web sites are designed to promote their products or services. Personal Web sites or social-issue Web sites are likely to be biased and will present information to support the bias.

- How regularly is the Web site updated?

- Does the Web site provide links to other reliable sources?

- Do the visuals on the Web site provide information, or are they just for entertainment?

Ask yourself how important the Web site is to your research. Are you using it only because it is available? Can you get more information from better sources by using the library?

Taking Notes

Taking useful notes from the multiple sources you consult for a long report is necessary if you are to write a well-developed report. No one can remember all the facts gathered from multiple sources, so whenever you find relevant information, take notes.

Although putting notes in a spiral notebook may seem more convenient than shuffling note cards while you are gathering information, notes in a notebook will be more difficult to use when you start to write. Because such notes reflect the order in which you found the information and not necessarily the order relevant to the final report organization, the information will seem to be "fixed" and nearly impossible to rearrange easily, thus interfering with effective organization for the reader. You would have to flip pages back and forth, wasting time searching for particular facts as you try to organize and write.

A more efficient way to record facts is to use note cards because such cards are easy to rearrange in any order and are durable enough to stand up to frequent handling. Some people prefer to put notes on computers. If you do that, record the same information that would be on a note card. Keep one note per page. Make a backup file every time you add notes. Follow these guidelines for using notes on cards or on a computer to keep track of the information you gather for a long report:

- Fill out a separate note card for each source of information, or create a separate computer file. For printed material, record complete bibliographic data, including the library call number. For film or television material, record the producer, director, title, studio or network, and the date. For interviews, record the name, date, subject discussed, and any other identifying material. For electronic sources, record author, date, title, database, date you accessed the material, and electronic address. Keep these cards or computer file as your bibliography.

- For information notes, record the source name, page number for written material, and date of access for electronic material. For convenience in sorting, also record a key word identifying the topic, such as "Travel Costs."

- Put only one item of information from one source on each card. If you mingle notes on cards, you will end up with the same unorganized information you would have had in a spiral notebook.

- If you write down the exact words from the original source, use quotation marks.

- Condense and paraphrase information for notes, but do not change the meaning. If the original source states, "Consumer reports appear to indicate . . . ," do not condense as "Consumer reports indicate . . . " The original sentence implied uncertainty with the word *appear*. Keep the meaning of the original clear in your notes.

- Record exact figures and dates, and double-check before leaving the source.

- Record enough facts so you can recall the full meaning of the original. One or two key words usually are not enough to help you remember detailed information.

Interpreting Data

Readers need the facts you present in a long report, but they also need help in interpreting those facts, particularly as to how the information affects their decision making and company projects. When you report information, tell readers what it means. Explain why the information is relevant to the central issue and how it supports, alters, or dismisses previous decisions or calls for new directions. Your objective in interpreting data for readers is to help them use the facts efficiently. Whatever your personal bias about the subject, be as objective as you can, and remember that several interpretations of any set of facts may be possible. Alert your readers to all major possibilities. Declining sales trends in a company's southwest sales region may indicate that the company's marketing efforts are inadequate in the region, but the trends also may indicate that consumers are dissatisfied with the products or that the products are not suitable for the geographic region because of other factors. Give the readers all the options when you interpret facts. After you have gathered information for your report and before you write, ask yourself these questions:

- Which facts are most significant for which of my readers?
- How do these facts answer my readers' questions?
- What decisions do these facts support?
- What changes do these facts indicate are needed?
- What option seems the best, based on these facts?
- What trends that affect the company do these facts reveal?
- What solutions to problems do these facts support?
- What actions may be necessary because of these facts?
- What conditions do these facts reveal?
- What changes will be useful to the company, based on these facts?
- How does one set of facts affect another set?
- When facts have several interpretations, which interpretations are the most logical and most useful for the situation?
- Is any information needed to further clarify my understanding of these facts?

Presenting data is one step in helping readers, and interpreting their meaning is the second step. If you report to your readers that a certain drainage system has two-sided flow channels, tell them whether this fact makes the system appropriate or inappropriate for the company project. Your report should include the answers to questions relevant to your subject and to your readers' purposes.

Drafting Long Reports

Long reports usually include the formal report elements discussed in Chapter 11. Like short reports, long reports may be organized in either the opening-summary or delayed-summary pattern, depending on your analysis of your readers' probable attitudes toward the subject. Whatever the pattern, long reports generally have three major sections.

Introductory Section

The introductory section in a long report usually contains several types of information. If this information is lengthy, it may appear as subsections of the introduction or even as independent sections. Depending on your subject and your readers, these elements may be appropriate in your introduction:

Purpose.   Define the purpose of your report—for example, to analyze results of a marketing test of two colognes in Chicago. Include any secondary purposes—for example, to recommend marketing strategies for the colognes or to define the target consumer groups for each cologne. For a report on a scientific research project, tell your readers what questions the research was intended to answer. Also, clarify in this section why you are writing the report. Did someone assign the writing task to you or is the report subject your usual responsibility? Be sure to say if one person told you to write the report, and address it to a third person.

Methods.   Explain the methods you used to obtain information for your report. In a research report, the methods section is essential for readers who may want to duplicate the research and for readers who need to understand the methods to understand the results. Include in this section specific types of tests used, who or what was tested, the length of time involved, and the test conditions. If appropriate, also describe how you conducted observations, interviews, or surveys to gather information and under what circumstances and with which people. You need not explain going to the library or searching through company files for information.

Background.   If the report is about a subject with a long history, summarize the background for your readers. The background may include past research on the same subject, previous decisions, or historical trends and developments. Some readers may know nothing about the history of the subject; others may know only one aspect of the situation. To use the report information effectively, readers need to know about previous discussions and decisions.

Report Limitations.   If the information in the report is limited by what was available from specific sources or by certain time frames or conditions, clarify this for your readers, so they understand the scope of the information and do not assume that you neglected some subject areas.

Report Contents.   For long reports, include a section telling readers the specific topics covered in the report. Some readers will be interested only in certain information and can look for that at once. Other readers will want to know what is ahead before they begin reading the data. Indicate also why these topics are included in the report and are presented in the way they are to help your readers anticipate information.

Recommendations/Results/Conclusions.   If you are using the opening-summary pattern of organization, include a section in which you provide an overview of the report conclusions, research results, or recommendations for action.

Data Sections

The data sections include the facts you have gathered and your interpretations of what they mean relative to your readers. Present the facts in ways best suited to helping your readers use the information. The patterns of organization explained in Chapter 4 will help you present information effectively. Chapter 13 explains data sections in specific types of reports. For a report of a research study, the data sections cover all the results obtained from the research.

Concluding Section

Long reports generally have a concluding section even if there is an opening summary because some readers prefer to rely on introductory and concluding sections. If the report is in the opening-summary pattern, the concluding section should summarize the facts presented in the report and the major conclusions stemming from these facts. Also, explain recommendations based on the conclusions. Some readers prefer that recommendations always appear in a separate section, and if you have a number of recommendations, a separate section will help readers find and remember them.

If the report is in the delayed-summary pattern, the concluding section is important because your readers need to understand how the facts lead to specific conclusions. Present conclusions first, and then present any recommendations that stem from these conclusions. Include any suggestions for future studies of the subject or future consideration of the subject. Do not, however, as you may do in short reports, include specific requests for meetings or remind readers about deadlines. The multiple readers of a long report and the usual formal presentation of long reports make these remarks inappropriate. You may include such matters in the transmittal letter that accompanies your long report, as explained in Chapter 11.

CHAPTER SUMMARY

This chapter discusses the usual organization of short and long reports and provides guidelines for report planning, information gathering, source evaluation, note taking, data interpretation, and drafting. Remember:

- Reports generally have one or more of these purposes: to inform, to record, to recommend, to justify, and to persuade.

- Most short internal reports are written as memos, and most short external reports are written as letters or as formal documents.

- Most short reports are organized in the opening-summary pattern.

- Long reports may be in the opening-summary pattern or the delayed-summary pattern and usually are written as formal documents.

- Planning reports includes identifying the central issue, analyzing readers and purpose, and deciding how much information is necessary for your readers.

- Writers gather information for reports from both primary and secondary sources.

- Notes are best collected on note cards, one item per card.

- For effective reports, writers must interpret the data for their readers.

- Long reports usually have three major divisions: an introduction, information sections, and a concluding section.

SUPPLEMENTAL READINGS IN PART 2

Allen, L., and Voss, D. "Ethics in Technical Communication," *Ethics in Technical Communication,* p. 475.

Garhan, A. "ePublishing," *Writer's Digest,* p. 495.

Hart, G. J. S. "PowerPoint Presentations: A Speaker's Guide," *Intercom,* p. 501.

MODEL 12-3   Commentary

This short report in memo format is organized in the opening-summary pattern. The writer researched the dangers and physical discomfort involved in long-term use of video display terminals (VDTs). The report suggests specific changes that will help employees adjust to VDTs and avoid discomfort.

Discussion

1. Discuss the writer's choice of headings and how helpful they are to the reader.

2. In groups, discuss what kinds of issues would have to be covered in a report on the following topics:

- A comparison of two locations for a new gas station
- A comparison of two software programs for accounting purposes
- A review of the dangers facing construction workers
- A comparison of two models of SUVs for a coach of a softball team

3. In groups, draft a memo, as the director of human resources, to department supervisors telling them to arrange a presentation by Cathy Powell, the company nurse. You want Powell to explain exercises that help avoid wrist and hand pain.

4. In groups, draft a memo, as a department supervisor, to Cathy Powell asking her to make a presentation to your department workers about wrist and hand pain.

5. In groups, draft a memo, as a department supervisor, to your department workers, and assure them that working at VDTs is not dangerous.

To: Michael O'Toole   October 12, 2005
     Director of Human Resources

From: Leila Bracco *LB*
       Division II Supervisor

Re: Problems with Video Display Terminals

As you know, employees who spend most of their day word processing or entering data have been complaining about safety concerns, eye fatigue, and general physical discomfort in using video display terminals (VDTs). Following your suggestion, I researched safety and comfort issues. Since Electric Power, Inc., has a work situation similar to ours, I consulted with Beth Reinhold, Assistant Director of Human Resources, and I obtained an OSHA fact sheet on safety issues (#00-24, January 1, 2003). Overall, I believe we can assure our employees there are no safety concerns in working with VDTs. I also have suggestions for simple modifications to help ease fatigue.

Employee Safety Concerns

The central concern of employees is possible exposure to radiation. I found that the National Institute for Occupational Safety and Health (NIOSH) and the U.S. Army Environmental Hygiene Agency have measured radiation emitted by VDTs. Tests show that levels for all types of radiation are below those allowed in current standards. Employees, therefore, appear to be in no danger from excess radiation.

Although OSHA has no data that any birth defects have resulted from working with VDTs, I suggest we assign employees to other duties during pregnancy to minimize their concerns. We need to encourage employees to report pregnancy at once.

Employee Discomfort

Employees have a variety of complaints about discomfort, including (1) eye fatigue and headaches; (2) pain or stiffness in back, shoulders, or neck; and (3) hand and wrist pain. All of these can be minimized with some adjustments or relatively inexpensive new equipment.

Eye Fatigue. Eye fatigue and headaches can result from glare off the screen and poor lighting. I suggest three changes:

- Consult each employee for preferences in placing the VDT to avoid window glare or reflections. There is no work-related

MODEL 12-3   A Short Report Organized in an Opening-Summary Pattern

reason to have all computers located on the left side of the station as they are now.

- Supplement the current lighting with individual "task" lighting as selected by each employee. We may find that some additional desk lighting will reduce eye strain.

- Purchase viewers to magnify the screen. A new Bausch & Lomb PC Magni-Viewer magnifies the screen text 175% and is easy to attach to a monitor. Cost is about $250 each, and the viewers can be easily moved from computer to computer. Electric Power has purchased these, and employees are quite pleased.

Pain or Stiffness. Pain and stiffness can usually be corrected by adjusting the physical environment of the employee. Work stations should allow flexibility in movement. I suggest we have the vendor for our office furniture make the following adjustments to each employee's specifications:

- Seat and backrest of the employee's chair should support comfortable posture.

- Chair height should allow the entire sole of the foot to rest on the floor.

- Computer screen and document holder should be the same distance from the eye and close together to avoid strain as the employee looks back and forth.

Wrist and Hand Pain. This pain comes from repetitive movement. I suggest the following:

- Regular five-minute breaks should be scheduled so employees can perform simple hand exercises. Our company nurse, Cathy Powell, tells me she has a set of recommended exercises that takes only 3 to 4 minutes to perform.

- Keyboards should be movable and at a height that allows the hands to be in a reasonably straight line with the forearm.

Conclusion

I suggest we make these simple changes in the work environment by December 1. We should assure employees at once that they are in no danger from excess radiation. I would be happy to make a brief report on this topic at the October 20 meeting of the section supervisors.

MODEL 12-4   Commentary

This letter report was written in the delayed-summary organization. The president of the student senate chose to write her report as a letter because she felt uncomfortable sending a memo to the director of student services. Although both reader and writer are connected to the same organization, they are not both employees, and the writer preferred the slightly more formal letter format for her report. She begins by thanking the reader for his past assistance, then explains the survey she conducted, describes the results, and concludes by recommending the renovation (possibly costly) of a room in the student center.

Discussion

1. Discuss why the writer chose to present her report in the delayed-summary pattern. What kinds of difficulties might the opening-summary pattern have created in this situation?

2. Discuss the writer's opening paragraph. What impression is she trying to create?

3. How effectively does the writer present her results? How well do the details support her conclusions? Discuss whether she should have included cost figures for renovating the Hilltop.

4. In groups or alone, as your instructor prefers, draft a letter to the director of student services on your campus, and ask for renovations of a specific room. Explain why the renovations are needed, and discuss your proposed changes in detail.

South Central University
University Park, Georgia 30638

Office of Student Government
(912) 555–1100

April 21, 2005

Mr. Patrick R. Sheehan
Director, Student Services
South Central University
University Park, GA 30638

Dear Mr. Sheehan:

I appreciate your assistance in coordinating student organization events
this year and your efforts to expand services in Jacobs Student Center.
During our meeting in February, you suggested that I survey students
regarding the services in the Student Center and develop a five-year
agenda for issues and projects. I took your advice and conducted a
survey of students on several topics, and I can now report the results
and highlight the area of most interest.

Survey

With the assistance of my fellow officers, Carlos Vega, vice president,
and Marylee Knox, secretary, I surveyed 100 students in the Jacobs
Student Center, April 2–6, 2005. We interviewed only those students
who had at least two more semesters to complete at South Central. The
students surveyed were as follows: males 47% and females 53%,
freshmen 29%, sophomores 32%, juniors 39%.

Each student filled out a written questionnaire about number of hours
spent weekly in the Student Center, eating and recreation time in the
Center, study time in the Center, and preferences in rooms. An open-
ended question asked students to suggest changes in the room they used
most often. We interviewed students at the main entrance to Jacobs
Student Center.

Results

Overall, students spend an average of 12 hours a week in the Center.
Study time averaged 5 hours, and eating/recreation time averaged 7

Mr. Patrick R. Sheehan   -2-   April 21, 2005

hours. There were wide variations in these figures from individual students. The student who spent the least amount of time reported only 1 hour, and the student who spent the most time reported 20 hours. Most students report both study time and eating/recreation time in the Center. Student responses included every public room in the Center—the first-floor food court, bowling alley, billiard room, Hilltop Cafeteria, lounge areas on every floor, and video game room. We were surprised to discover that the most frequently used room was the Hilltop Cafeteria. Students (73%) reported using the Hilltop Cafeteria for both study and eating/recreation at least 50% of the hours they were in the Student Center. Because 79% of the suggestions for changes focused on the Hilltop Cafeteria, I will concentrate on those in this report.

Proposed Changes for Hilltop Cafeteria

Students said they used Hilltop Cafeteria for studying, eating, and meeting with friends, and they had specific criticisms about the facility. As Hilltop also serves as a banquet hall and dance floor for special events, I believe the student concerns merit attention. Comments were primarily about the furniture, lighting, decor, temperature, and atmosphere.

Furniture.   Students reported that many of the tables and chairs are in disrepair. Recently, a table collapsed while students were eating lunch, and chair backs often snap if leaned on. In addition, the tables and chairs are scarred, with chipped tabletops and cracks in the chair seats.

Lighting.   The current lighting is both unattractive and too dim for efficient study. Part of the problem may be the wide spacing between light fixtures. Students requested more lights and higher power.

Decor.   The dull green wall color was cited as making the room look "institutional." Also, the walls near the entrances are dirty from fingerprints, scrapes from chairs, and people brushing against them. Fresh paint in a soft color or muted white was suggested. Because the Hilltop has no windows, there is extensive wall space. Several students suggested wall decorations, such as photographs showing the university's history or student paintings.

Temperature.   Students reported that Hilltop was frequently too warm for comfort. Stacks of cafeteria trays left over from lunch produce an

Mr. Patrick R. Sheehan   -3-   April 21, 2005

unpleasant aroma, making the room unappealing. One suggestion was to install several ceiling fans to circulate air more effectively. Also, the trays could be removed promptly at the end of the main lunch period.

Atmosphere.   Students mentioned that the food court is usually noisy with heavy traffic, and the lounge areas with couches are more suited for socializing than for studying. They prefer the Hilltop if they plan to eat and study. Several suggestions focused on providing classical music in the background, played at a low level. Rearrangement of tables and chairs could create study areas distinct from the eating areas, while still allowing students the option of having soft drinks and snacks while studying.

Conclusions

Based on these student responses to our survey, I believe the first priority for the Office of Student Services should be improvements in the Jacobs Student Center, specifically renovation of the Hilltop Cafeteria. Renovation would include painting the room, installing improved lighting, installing ceiling fans, purchasing new tables and chairs, and installing a music system. I have not obtained any cost figures for these suggested improvements. Student suggestions for improving the Hilltop were so numerous I believe this issue should be discussed. I would be happy to meet with you and provide copies of the questionnaires used in our survey, so you can see the results in greater detail. The officers of the Student Senate are interested in working with you in evaluating student interests and needs on the South Central campus.

Sincerely,

Judith L. Piechowski

Judith L. Piechowski
President
Student Senate

JP:mk

Model 12-5   Commentary

This model is a long report from a summer intern at the American Safety Organization. She was asked to prepare a report on the current status of air bag safety. Because the report would be distributed to multiple readers, the writer prepared a formal report.

The report is in the opening-summary pattern because the writer knows her supervisor prefers that organization. The abstract appears on the title page because that format is standard at this organization. Although this report is an internal document, the writer includes a transmittal letter rather than a memo.

Discussion

1. Discuss the headings in this report. How helpful are they to a reader who is looking for specific information?

2. Discuss why the intern probably chose to write a transmittal letter rather than a memo. Does the letter add to the formality of the report?

3. In groups, draft a memo to employees explaining the major safety points in this report and urging readers to buckle up when they are in a car. Discuss your draft with the class.

REPORT ON THE CURRENT STATUS
OF AIR BAG USE AND SAFETY

Prepared for
Stefanos Cole
Associate Director
American Safety Organization

Submitted by
Nicole Canyon
Intern
Safety Statistics Unit

August 22, 2005

Abstract

Based on statistics from government and safety organizations, air bags, in combination with seat belts, reduce head injuries by 85%. In 2004, 69.2% of light vehicles had air bags. With the advanced frontal air bags being phased in completely with the 2007 models, the United States should improve the survival rate for drivers and passengers in vehicle accidents. No technology can substitute for drivers and passengers properly using seat belts and putting children in the rear seats.

MODEL 12-5   Formal Report

AMERICAN SAFETY ORGANIZATION
1000 Plaza Center
San Antonio, Texas 78254-6465
210-555-2200

August 22, 2005

Mr. Stefanos Cole
Associate Director
American Safety Organization
1000 Plaza Center
San Antonio, TX 78254-6465

Dear Mr. Cole:

Here is the report you asked for concerning current air bag design
and use. I compiled the latest statistics and design information
from the Web sites of the U.S. Department of Transportation, the
National Safety Council, and the Insurance Institute for Highway
Safety. As you directed, I have summarized the key points.

All the statistics indicate that air bags increase driver and
passenger safety, and when they are used in conjunction with
seat belts, the risk of fatalities in accidents is greatly reduced.

The major stumbling block to safety comes from vehicle occupants
neglecting to use seat belts. Strong seat belt laws in some states
have increased the adult use of seat belts, and when adults use
seat belts, they are likely to restrain the children in the vehicle.

I would be glad to show you the original statistics I have collected
for this report. Please let me know if you would like more
information.

Sincerely,

Nicole Canyon

Nicole Canyon
Intern, Safety Statistics Unit

REPORT ON THE CURRENT STATUS OF
AIR BAG USE AND SAFETY

INTRODUCTION

This report reviews the current status of air bag safety and air bag development in the United States since 1990. Statistics and information came from the Web sites of government agencies and safety organizations, such as the National Center for Statistics and Analysis, the National Highway Traffic Safety Administration, the National Safety Council, and the Insurance Institute for Highway Safety. The data point to the conclusion that air bag use, particularly in combination with seat belt use, saves thousands of lives of both drivers and passengers on the nation's highways every year.

AIR BAG SAFETY

The topic of air bag safety concerns both lives saved because of air bags and fatalities caused by air bags.

Lives Saved

Safety agencies calculate "lives saved" using a mathematical analysis of the real-world fatality record of vehicles with air bags compared to the fatality record of vehicles without air bags. This method is called double-pair comparison, and it is widely regarded as an acceptable method of calculating statistical results for safety issues.

Figures for 1998 show that 43.6% of the 204 million vehicles (automobiles and light trucks) on U.S. roads were equipped with at least driver air bags. In 2004, 69.2% of the more than 220 million vehicles (automobiles and light trucks) on the road were equipped with driver air bags. Over 130 million (59.5%) of those vehicles also had passenger air bags. Another 1 million vehicles are sold each month equipped with air bags. These air bags are credited with saving at least 2,000 lives in 2002, and the National

Highway Traffic Safety Administration estimates that 15,158 people since 1990 are alive today because of air bags in their vehicles.

Beginning with the 1998 model year, driver air bags deployed with reduced force, but statistics indicate that they still provided full protection in crashes. Driver death rates were estimated to be 6% lower in 1998 and 1999 model vehicles than in 1997 models that had full-force air bags.

The National Highway Traffic Safety Administration estimates that the combination of an air bag and a lap/shoulder belt reduces the risk of serious head injury to drivers by 85%. Seat belts alone reduce the head injury rate by 60%. For 2003, the most recent figures available, 44% of unrestrained automobile occupants were ejected either partially or totally during crashes. The combination of seat belts and air bags provide the most protection, but either one alone will provide a significant level of protection. The National Center for Statistics and Analysis reports that ejection from vehicles during crashes is a significant cause of fatalities. In 2003, 65% of fatalities involving SUVs were unrestrained occupants who were ejected from the vehicles either totally or partially.

Seat belt use in the United States lags behind that in most industrialized Western countries. Based on state surveys, the U.S. has a 69% rate of use for seat belts. Canada, Australia, the United Kingdom, Germany, and Sweden all have seat belt usage rates of 90% or better. Because the United States rate for seat belt use is only 69%, air bags are a significant force in lowering death rates in traffic accidents.

Air Bag Fatalities

Air bags have caused some fatalities. Since 1990, 246 deaths were determined to have been caused by air bags. The deaths included 85 drivers (34.5%), 11 adult passengers (4.4%), 127 children

(51.6%), and 23 infants (9.3%). Following are the breakdowns of deaths because of air bags:

- Of the 85 drivers killed, 54 were unbelted, and 24 were belted. The rest were either misusing their seat belts or the usage is unknown.
- Of the 11 adult passengers killed, 7 were unbelted; 4 were belted.
- Of the 127 children killed, 96 were unrestrained; 26 children were improperly restrained; and 5 were restrained properly. Twenty-two of the unrestrained children were seated in a front passenger's lap, and three were unrestrained and riding on the lap of the drivers.
- Of the 23 infants killed by air bags, 21 were restrained in rear-facing infant seats. Others were not restrained properly.

In the adult deaths, basic nearness to the air bag unit was the problem. Adults, including short women and older people, can ride safely with air bags if they are belted and at least 10 inches from the air bag cover. Children younger than 12 should always be in the back seat and in proper restraints. Children are more likely to lean forward unexpectedly or wiggle out of their seat belts. Also, the rear seats have always been the safest place to ride even when the vehicles are not equipped with air bags.

AIR BAG DESIGNS

Air bag designs have evolved over the years.

Air Bag Basics

Air bags are considered to be supplemental restraints to support seat belts and work best in combination with seat belts. Air bags are designed to deploy in moderate to severe crashes, and they reduce the possibility that the occupant's upper body or head will strike the inside of the vehicle during a crash. In a moderate to severe crash, inflators fill the air bag with harmless gas. The air bag module consists of the Electronic Control Unit, Crash Sensors, and the ON-OFF Switch.

Electronic Control Unit (ECU). The ECU controls the air bag system. It receives signals from various sensors and decides if and when the air bags should be deployed. The unit is typically in the middle of the vehicle, so it is well protected during a crash.

Crash Sensors. The purpose of the crash sensors is to measure how quickly the vehicle slows down in a frontal crash. The sensors also measure how quickly the vehicle is crushed during a side-impact crash. Sensors are located in the front of the vehicle for frontal air bags and in the door or the pillar between the front and rear doors for side-impact crashes. Severe braking alone will not set off the sensors.

Air Bag ON-OFF Switch. Most vehicles without rear seats or with very small rear seats have a passenger air bag ON-OFF switch as standard equipment. The purpose of the switch is to allow a child age 12 or under to ride in the right front passenger position. If the vehicle has enough space in the rear to provide a seat for a child under age 12, an ON-OFF switch is prohibited. Consumers who want such a switch installed must apply for it.

When a crash begins, a signal goes from the ECU to the inflator within the air bag module. An igniter in the inflator starts a chemical reaction that produces a harmless gas that inflates the air bag in less than 1/20th of a second. Side-impact air bags inflate even more quickly because the side of the vehicle has less protection than the front of the vehicle.

Frontal Air Bags

Frontal air bags do not eliminate the need for seat belts. Occupants who are unbelted or out of position as a crash begins can be injured by a frontal air bag. The bags are not much protection in crashes that involve rollovers, side-impact, or rear-end crashes.

Side-Impact Air Bags

Side-impact air bags (SABs) protect the head and the chest during moderate to severe side-impact crashes. The SABs can deploy from the door, the seat, or overhead, depending upon the mounting. Seat- and door-mounted SABs both provide chest protection. Some SABs are combos; that is, they extend upward from the side of the seatback and protect the head as well. "Curtain" air bags and "tubular" air bags are inflatable and specifically designed to protect the head. The curtain air bags stay inflated to prevent occupants from being ejected during a rollover crash.

At the moment, very few vehicles sold in the United States have seat- or door-mounted SABs in the rear-seat positions where children are likely to be. Also, SABs are not currently regulated by the Federal government. A panel of automotive experts known as the Technical Working Group has developed voluntary SAB testing procedures to minimize potential risk to out-of-position occupants. All vehicle manufacturers have agreed to use these tests when designing SAB systems.

Advanced Frontal Air Bags

Advanced frontal air bags—the third-generation design—were phased in with selected new model 2004 vehicles. All light vehicles will have advanced frontal air bags by the model year 2007. The key difference with the advanced system is that the frontal air bags calculate how to forcefully inflate.

Depending on the specific design, the advanced frontal system automatically determines what level of power is appropriate for inflation. The level of power is based on the sensors that detect (1) occupant size, (2) seat position, and (3) crash severity. The driver and front passenger bags operate independently. Experts believe that the advanced frontal air bags will reduce the risk of fatality when air bags deploy. They were designed primarily to reduce the risk to small adults and children. However, children under 12 should still be in the back seat even if the vehicle has the advanced system.

Air Bag Safety - 6

Vehicles with advanced frontal air bags are required to have
warning labels on the sun visors that direct occupants to keep
children in the back seat. Most important, vehicles will have
an indicator light that says "PASS AIR BAG OFF." This light
informs the driver that the frontal air bag has been turned off
automatically because there is no passenger. The off position also
can occur if a child or a small adult is in the passenger seat.
Drivers will need to study the manual when they get the advanced
frontal air bags to adjust the system. In special circumstances,
such as a driver or passenger with a special medical condition or a
very short driver, vehicle owners may get an OFF switch for this
air bag system.

CONCLUSION

Overall, in spite of a relatively small number of fatalities, air bags
have saved thousands of lives. With the advanced frontal system,
some of the risk from air bags to small adults should be eliminated.
If the occupant of a vehicle is wearing a seat belt in the proper
position, most air bag injuries are only minor cuts or bruises. The
combination of seat belts and air bags significantly reduces the risk
to drivers and passengers in vehicle crashes.

Chapter 12 Exercises

1. The executive director of Forest Glen Resort asks you to research the latest dive computers used by scuba divers. He wants you to recommend at least three models to be available in the resort dive shop where guests rent and buy equipment.

2. You are the assistant to Gabriela Mangeli, communications director at Northern Mutual Insurance Company. The annual sales meeting is scheduled for July, and you are expecting 450 agents, at least half of whom will bring their spouses with them. Ms. Mangeli is creating a list of potential activities for spouses. She wants you to recommend a museum or historical site that would interest out-of-town visitors. Research a local museum or historical site (e.g., battlefield, church, cemetery, home of a famous person), and write a report to Ms. Mangeli that recommends the museum or site be included in the information packet for the people attending the meeting. Attach a brochure that you got from the museum or historical site to your report. Use the opening-summary organization pattern.

3. Assume you are employed in your field. The president of the company you work for, Timothy Phillips, plans to attend a high-level meeting of industry leaders. You have been assigned to write a report in which you identify and describe a recent development in your field. Consider topics such as new technology, new government regulations, the impact of social or environmental problems, foreign competition, safety concerns, and public demands on your field. You know that Mr. Phillips wants current information and is particularly interested in knowing how this subject will affect the future growth of your field.

Write a formal report with the appropriate formal report elements described in Chapter 11. Consider including appropriate tables and figures. To be up to date, use only sources from the last three years in your report. Your instructor may want you to prepare a 5–10-minute oral presentation of your subject for the class. If so, review Chapter 16 and read the oral presentation articles in Part 2.

4. In groups, develop a collaborative plan to write a formal report, using the strategies discussed in Chapter 2. As a group, research information on a topic with widespread impact, such as the condition of the U.S. coastal wetlands, the dangers of commercial fishing, the problem of drugs and alcohol in the workplace, the worldwide impact of AIDS, or the world supply of energy resources. Assign each group member to research a specific aspect of the topic. Share your information in the group, and together, develop appropriate tables and figures for some of the information. Work collaboratively to write the report for submission to your instructor. *Or,* your instructor may prefer that you write the final formal report individually, using the information that the group collected.

INTERNET EXERCISES

5. The increasing spam on cell phones with text messaging is worrying the CEO at Wireless Tomorrow, Inc. Customers are complaining about being flooded with unsolicited ads for everything from discount drugs to cruises. One of your phone plans charges by the incoming message, and those customers are even more upset. Your task is to research the latest software for blocking email spam and write a report that will help your CEO decide what steps to take to satisfy customers.

6. Assume you work for a builder who is starting a large development of at least 75 homes. Research the term *universal design,* and write a report in which you explain the term and make specific recommendations for the design of bathrooms and kitchens in the new homes.

7. You are a summer intern at Jones Financial Services, Inc. Your supervisor, Steven Brooks, has a growing list of clients who are young professionals, newly married, and starting families. He wants to create a list of Web sites with financial information that would be of interest to these clients. Your task is to identify five appropriate Web sites for this list. In your report to Mr. Brooks, explain why you think these five Web sites would be useful to his clients.

CHAPTER 13

Types of Reports

UNDERSTANDING CONVENTIONAL
REPORT TYPES

Over the years, in industry, government, and business, writers have developed conventional organizational patterns for the most frequently written types of reports. Just as tables of contents and glossaries have recognizable formats, so too have such common reports as progress reports and trip reports. Some companies, in fact, have strict guidelines about content, organization, and format for report types that are used frequently. Conventional organizational patterns for specific types of reports also represent what readers have found most useful over the years. For every report, you should analyze reader, purpose, and situation and plan your document to serve your readers' need for useful information. You will also want to be knowledgeable about the conventional structures writers use for common report types.

This chapter explains the purposes and organizational patterns of six of the most common reports. Depending on reader, purpose, situation, and company custom, these reports may be written as informal memos, letters, or formal documents that include all the elements discussed in Chapter 11.

WRITING A FEASIBILITY STUDY

A *feasibility study* provides information to decision makers about the practicality and potential success of several alternative solutions to a problem.

Purpose

Executives often ask for feasibility studies before they consider a proposal for a project because they want a thorough analysis of the situation and all the alternatives. The writer of a feasibility study identifies all reasonable options and prepares a report that evaluates them according to features important to the situation, such as cost, reliability, time constraints, and company or organization goals. Readers expect a feasibility study to provide the information necessary for them to make an informed choice among alternatives. The alternatives may represent choices among products or actions, such as the choice among four types of heating systems, or a choice between one action and doing nothing, such as the decision whether to merge with another company. When you write a feasibility study, provide a full analysis of every alternative, even if one seems clearly more appropriate than the others.

Organization

Model 13-1 shows a feasibility study written by an executive to the president of the company. The company management has already discussed relocation to a new city and the need for this study. In her report, the writer presents information she has gathered about two possible sites, evaluates them both as to how well they match the company's requirements, and recommends one of them. Feasibility studies usually include four sections.

Introduction

The introduction of a feasibility study provides an overview of the situation. Readers may rely heavily on the introduction to orient them to the situation before they read the detailed analyses of the situation and the possible alternatives. Follow these guidelines:

- Describe the situation or problem.
- Establish the need for decision making.
- Identify those who participated in the study or the outside companies that provided information.
- Identify the alternatives the report will consider, and explain why you selected these alternatives, if you did.
- Explain any previous study of the situation or preliminary testing of alternatives.
- Explain any constraints on the study or on the selection of alternatives, such as time, cost, size, or capacity.
- Define terms or concepts essential to the study.
- Identify the key factors by which you evaluated the alternatives.

In Model 13-1 the writer opens with a short statement that includes the report purpose and her recommendation of which location is best.

Comparison of Alternatives

The comparison section focuses equally on presenting information about the alternatives and analyzing that information in terms of advantages and disadvantages for the company. Organize your comparison by topic or by complete subject. For feasibility reports, readers often prefer to read a comparison by key topics because they regard some topics as more important than others and they can study the details more easily if they do not have to move back and forth between major sections. Whether comparing by topic or by complete subject, discuss the alternatives in the same order under each topic, or discuss the topics in the same order under each alternative. Follow these guidelines:

- Describe the main features of each alternative.
- Rank the key topics for comparison by using either descending or ascending order of importance.

- Discuss the advantages of each alternative in terms of each key topic.

- Point out the significance of any differences among alternatives.

The writer in Model 13-1 compares the two possible sites by topic, using subheadings to help readers find specific types of information. The subheadings represent the established criteria for evaluating the locations.

Conclusions

The conclusions section summarizes the most important advantages and disadvantages of each alternative. If you recommend one alternative, do so in the conclusions section or in a separate section if you have several recommendations. State conclusions first because they are the basis for any recommendations. If you believe some advantages or disadvantages are not important, explain why. Follow these guidelines:

- Separate conclusions adequately, so readers can digest one at a time.

- Explain the relative importance to the company of specific advantages or disadvantages.

- Include conclusions for each key factor presented in the comparison.

Recommendations

The recommendations section, if separate from the conclusions, focuses entirely on the choice of alternative. Your recommendations should follow logically from your conclusions. Any deviations will confuse readers and cast doubt on the thoroughness of your analysis. Follow these guidelines:

- Describe your recommendations fully.

- Provide enough details about implementing the recommendations so that your reader can visualize how they will be an effective solution to the problem.

- Indicate a possible schedule for implementation.

In Model 13-1, the writer combines her conclusions and recommendation. She reviews the company's decision to relocate, and she next explains that both sites are acceptable. Because Randolph Center involves less cost and will be ready for occupancy ahead of the company's deadline, she recommends that site.

WRITING AN INCIDENT REPORT

An *incident report* provides information about accidents, equipment breakdowns, or any disruptive occurrence.

Purpose

An incident report is an important record of an event because government agencies, insurance companies, and equipment manufacturers may use the report in legal actions if injury or damage has occurred. In addition, managers need such reports to help them determine how to prevent future accidents or disruptions. An incident report thoroughly describes the event, analyzes the probable causes, and recommends actions that will prevent repetition of such events. If you are responsible for an incident report, gather as many facts about the situation as possible, and carefully distinguish between fact and speculation.

Organization

Model 13-2 is an incident report about a large chemical fire. Although the writer was not present at the onset of the fire, she gathers information from witnesses and the fire chief.

Most large companies have standard forms for reporting incidents during working hours. If there is no standard form, an incident report should include the following sections:

Description of the Incident

The description of the incident includes all available details about events before, during, and immediately after the incident in chronological order. Since all interested parties will read this section of the report and may use it in legal action, be complete, use nonjudgmental language, and make note of any details that are not available. Follow these guidelines:

- State the exact times and dates of each stage of the incident.
- Describe the incident chronologically.
- Name the parties involved.
- State the exact location of the accident or the equipment that malfunctioned. If several locations are relevant, name each at the appropriate place in the chronological description.
- Identify the equipment (by model number, if possible) involved in a breakdown.
- Identify any continuing conditions resulting from the incident, such as the inability to use a piece of equipment.

- Identify anyone who received medical treatment after an accident.

- Name hospitals, ambulance services, and doctors who attended accident victims.

- Name any witnesses to the incident.

- Report witness statements with direct quotations or by paraphrasing statements and citing the source.

- Ask witnesses to sign and date their statements.

- Explain any follow-up actions taken, such as repairs or changes in scheduling.

- Note any details not yet available, such as the full extent of injuries.

The writer of Model 13-2 provides a chronological description of the events, including dates, times, and places. She also identifies the employee who called in the fire and the fire chief she interviewed. Because there were so many minor injuries, none requiring hospitalization, she reports only the numbers of people involved.

Causes

This section includes both direct and indirect causes for the incident. Be careful to indicate when you are speculating about causes without absolute proof. Follow these guidelines:

- Identify each separate cause leading to the incident under discussion.

- Analyze separately how each stage in the incident led to the next stage.

- If you are merely speculating about causes, use words such as *appears, probably,* and *seems.*

- Point out the clear relationship between the causes and the effects of the incident.

In analyzing the cause of the fire, the writer of Model 13-2 reports what the fire investigator told her about improper chemical storage.

Recommendations

This section offers specific suggestions keyed to the causes of the incident. Focus your recommendations on prevention measures rather than on punishing those connected with the incident. Follow these guidelines:

- Include a recommendation for each major cause of the incident.

- List recommendations if there are more than one or two.

- Describe each recommendation fully.

- Relate each recommendation to the specific cause it is designed to prevent.

- Suggest further investigation if warranted.

The recommendation in Model 13-2 focuses on a training session for all personnel who are involved with storing the chemicals.

WRITING AN INVESTIGATIVE REPORT

An *investigative report* analyzes data and seeks answers to why something happens, how it happens, or what would happen under certain conditions.

Purpose

An investigative report summarizes the relevant data, analyzes the meaning of the data, and assesses the potential impact that the results will have on the company or on specific research questions. The sources of the data for investigative reports can include field studies, surveys, observation, and tests of products, people, opinions, or events both inside and outside the laboratory. Investigations can include a variety of circumstances. An inspector at the site of an airplane crash, a laboratory technician testing blood samples, a chemist comparing paints, a mining engineer checking the effects of various sealants, and a researcher studying the effect of air pollution on children are all conducting investigations, and all will probably write reports on what they find. Remember that readers of an investigative report need to know not only what the data are but also what the data mean in relation to what has been found earlier and what may occur in the future.

Organization

Model 13-3 is an investigative report based on survey data collected by the Bureau of Labor Statistics. In such reports, the writer usually examines a question or problem, discusses the importance of the topic, explains the method of collecting information, and then analyzes what the results of the investigation indicate about the question. The conclusions may summarize a key point or indicate the need for further investigation or the need to take further action. Model 13-3 reports the results of a survey to investigate the rates of employment in U.S. families. Scientific investigative reports often appear in journals or on Web sites of research centers and government agencies.

The reports generally include an abstract. Most investigative reports include the following sections:

Introduction

The introduction of an investigative report describes the problem or research question that is the focus of the report. Since investigative reports may be used as a form of evidence in future decision making, include a detailed description of the situation and explain why the investigation is needed. Readers will be interested in how the situation affects company operations or how the research will answer their questions. Follow these guidelines:

- Describe specifically the research question or problem that is the focus of the investigation. Avoid vague generalities, such as "to check on operations" or "to gather employee reactions." State the purpose in terms of the specific questions the investigation will answer.

- Provide background information on how long the situation has existed, previous studies of the subject by others or by you, or previous decisions concerning it. If a laboratory test is the focus of the report, provide information about previous tests involving the same subject.

- Explain your reasons if you are duplicating earlier investigations to see if results remain consistent.

- Point out any limits of the investigation in terms of time, cost, facilities, or personnel.

The report in Model 13-3 explains the survey and overall data collection by the Bureau of Labor Statistics. The introduction also defines the key terms used in the survey to help readers understand the survey results.

Method

This section should describe in detail how you gathered the information for your investigative report. Include the details your readers will be interested in. For original scientific research, your testing method is a significant factor in achieving meaningful results, and readers will want to know about every step of the research procedure. For field investigations, summarize your method of gathering information, but include enough specific details to show how, where, and under what conditions you collected data. Follow these guidelines, where appropriate, for your research method:

- Explain how and why you chose the test materials or specific tests.

- Describe any limits in the test materials, such as quantities, textures, types, or sizes, or in test conditions, such as length of time or options.

- List test procedures sequentially, so readers understand how the experiment progressed.

- Identify by category the people interviewed, such as employees, witnesses, and suppliers. If you quote an expert, identify the person by name.

- If appropriate, include demographic information, such as age, sex, race, physical qualities, and so on, for the people used in your study.

- Explain where and when observation took place.

- Identify what you observed and why.

- Mention any special circumstances and what impact they had on results.

- Explain the questions people were asked or the tests they took. If you write the questions yourself, attach the full questionnaire as an appendix. If you use previously validated tests, cite the source of the tests.

The method section in Model 13-3 identifies the source of the statistics and the scope of the survey. The section also explains possible sampling errors and why the data for 2003 are not comparable to earlier years.

Results

This section presents the information gathered during the investigation and interprets the data for readers. Do not rely on the readers to see the same significance in statistics or other information that you do. Explain which facts are most significant for your investigation. Follow these guidelines:

- Group the data into subtopics, covering one aspect at a time.

- Explain in detail the full results from your investigation.

- Cite specific figures, test results, and statistical formulas, or use quotations from interviewees.

- Differentiate between fact and opinion when necessary. Use such words as *appears* and *probably* if the results indicate but do not prove a conclusion.

- If you are reporting statistical results, indicate (1) means or standard deviations, (2) statistical significance, (3) probability, (4) degrees of freedom, and (5) value.

- Point out highlights. Alert readers to important patterns, similarities to or differences from previous research, cause and effect, and expected and unexpected results.

The results section in Model 13-3 presents the data in three major units: (1) by race and ethnicity, (2) by marital status, and (3) by ages of children. The tables in the appendix provide figures for selected comparisons, and the writer directs the reader to the appendix. The results section also includes some information not shown on the tables.

Discussion

This section interprets the results and their implications for readers. Explain how the results may affect company decisions or what changes may be needed because of your investigation. Show how the investigation answered your original research questions about the problem being investigated. If your results do not satisfactorily answer the original questions, clearly state this. Follow these guidelines where appropriate for your type of investigation:

- Explain what the overall results mean in relation to the problem or research question.

- Analyze how specific results answer specific questions or how they do not answer questions.

- Discuss the impact of the results on company plans or on research areas.

- Suggest specific topics for further study if needed.

- Recommend actions based on the results. Detailed recommendations, if included, usually appear in a separate recommendations section.

The discussion section in Model 13-3 points out probable causes of these high employment rates. The writer also mentions the high rates for working mothers, a trend that developed in the last half of the twentieth century.

Conclusion

When a research investigation is highly technical, the conclusion summarizes the problem or research question, the overall results, and what they mean in relation to future decisions or research. Nontechnical readers usually rely on the conclusion for an overview of the study and results. Follow these guidelines:

- Summarize the original problem or research question.

- Summarize the major results.

- Identify the most significant research result or feature uncovered in the investigation.

- Explain the report's implications for future planning or research.

- Point out the need for any action.

In Model 13-3 the conclusion section is quite brief. The writer repeats the purpose of the national database on employment in U.S. families and suggests the usefulness it has for agencies concerned with families.

WRITING A PROGRESS REPORT

A *progress report* (also called a *status report*) informs readers about a project that is not yet completed.

Purpose

The number of progress reports for any project usually is established at the outset, but more might be called for as the project continues. Progress reports are often required in construction or research projects so decision makers can assess costs and the potential for successful completion by established deadlines. Although a progress report may contain recommendations, its main focus is to provide information, and it records the project events for readers who are not involved in day-to-day operations. The progress reports for a particular project make up a series, so keep your organization consistent from report to report to aid readers who are following one particular aspect of the project.

Organization

Model 13-4 is a progress report written by the on-site engineer in charge of a dam repair project. The report is one in a series of progress reports for management at the engineering company headquarters. The writer addresses the report to one vice president, but copies go to four other managers with varying interests in the project. The writer knows, therefore, that she is actually preparing a report that will be used by at least five readers. Progress reports include the following sections:

Introduction

The introduction reminds readers about progress to date. Explain the scope and purpose of the project, and identify it by specific title if there is one. Follow these guidelines:

- State the precise dates covered by this particular report.
- Define important technical terms for nonexpert readers.
- Identify the major stages of the project if appropriate.

- Summarize the previous progress achieved (after the first report in the series) so regular readers can recall the situation and new readers can become acquainted with the project.

- Review any changes in the scope of the project since it began.

The writer in Model 13-4 numbers her report and names the project in her subject line to help readers identify where this report fits in the series. In her introduction, she establishes the dates covered in this report and summarizes the dam repairs she discussed in previous reports. She also indicates that the project is close to schedule but that the expected costs have risen because another subcontractor had to be hired.

Work Completed

This section describes the work completed since the preceding report and can be organized in two ways: You can organize your discussion by tasks and describe the progress of each chronologically, or you can organize the discussion entirely by chronology and describe events according to a succession of dates and times. Choose the organization that best fits what your readers will find useful, and use subheadings to guide them to specific topics. If your readers are interested primarily in certain segments of the project, task-oriented organization is appropriate. If your readers are interested only in the overall progress of the project, strict chronological organization is probably better. Follow these guidelines:

- Describe the tasks that have been completed in the time covered by the report.

- Give the dates relevant to each task.

- Describe any equipment changes.

- Explain special costs or personnel charges involved in the work completed.

- Explain any problems or delays.

- Explain why changes from the original plans were made.

- Indicate whether the schedule dates were met.

The engineer in Model 13-4 organizes her work-completed section according to the repair stages and then lists them in chronological order. She also includes the date on which each event occurred or each stage was completed.

Work Remaining

The section covering work remaining includes both the next immediate steps and those in the future. Place the most emphasis on the tasks that will be covered in your next progress report. Avoid overly optimistic promises. Follow these guidelines:

- Describe the major tasks that will be covered in the next report.
- State the expected dates of completion for each task.
- Mention briefly those tasks that are further in the future.

The work-remaining section in Model 13-4 describes the upcoming stage of the repair work and states the expected completion date.

Adjustments/Problems

This section covers issues that have changed the original plan or time frame of the project since the last progress report. If the project is proceeding on schedule with no changes, this section is not needed. If it is necessary, follow these guidelines:

- Describe major obstacles that have arisen since the last progress report. (Do not discuss minor daily irritations.)
- Explain needed changes in schedules.
- Explain needed changes in the scope of the project or in specific tasks.
- Explain problems in meeting original cost estimates.

The writer in Model 13-4 includes an adjustments section in her report to explain unexpected costs and a short delay because of the deteriorating condition of the dam area.

Conclusion

The conclusion of a progress report summarizes the status of the project and forecasts future progress. If your readers are not experts in the technical aspects of the project, they may rely heavily on the conclusion to provide them with an overall view of the project. Follow these guidelines:

- Report any progress on current stages.
- Report any lack of progress on current stages.
- Evaluate the overall progress so far.
- Recommend any needed changes in minor areas of scheduling or planning.

- State whether the project is worth continuing and is still expected to yield results.

The engineer's conclusion in Model 13-4 assures her readers that, although she is three days behind schedule, she expects no further delays. She also identifies the final stage of the dam repairs and the expected completion date.

STRATEGIES—Support Coworkers

We all appreciate having our work recognized. Career success usually means others have helped you at some time. Remember to treat coworkers in an equally supportive way.

- Thank people who have helped you on a project, and give them credit in your written reports or oral presentations.
- Avoid interrupting people who are working overtime on a project unless you have an emergency.
- If you are not busy, offer to help a coworker who is busy.
- Respect your coworkers' schedules. Ask them if they can handle a job before assigning it.

WRITING A TRIP REPORT

A *trip report* provides a record of a business trip or visit to the field.

Purpose

A trip report is a useful record both for the person who made the trip and for the decision makers who need information about the subjects discussed during the trip. The trip report records all significant information gathered either from meetings or from direct observation.

Organization

Model 13-5 is a trip report written in the opening-summary organization pattern discussed in Chapter 12. The writer has visited a trade show to collect some ideas for clients who are rebuilding in downtown areas. The writer selects two main points for her report—the models she saw and a trade representative she talked to. Trip reports should contain the following sections.

Introductory Section

Use the subject line of your trip report to identify the date and location of the trip. Begin your report with an opening-summary that explains the purpose of the trip and any major agreements or observations you made. Identify all the major events and the people with whom you talked. Follow these guidelines:

- Describe the purpose of the trip.
- Mention special circumstances connected to the trip's purpose.
- If the trip location is not in the subject line, state the dates and the locations you visited.
- State the overall results of the trip or any agreements you made.
- Name the important topics covered in the report.

The opening summary in Model 13-5 states the reason for the trip and includes the writer's conclusion that she gathered useful ideas for the clients.

Information Sections

The information sections of your trip report should highlight specific topics with informative headings. Consider which topics are important to your readers, and group the information accordingly. Follow these guidelines:

- State the names and titles of people you consulted for specific information.
- Indicate places and dates of specific meetings or site visits.
- Describe in detail any agreements made and with whom.
- Explain what you observed and your opinion of it.
- Give specific details about equipment, materials, or systems relevant to company interests.

The writer in Model 13-5 divides her information into two relevant sections—the models she saw and the representative she talked to.

Conclusions and Recommendations

In your final section, summarize the significant results of your trip, whether positive or negative, and state any recommendations you believe are appropriate. Follow these guidelines:

- Mention the most significant information resulting from the trip.
- State whether the trip was successful or worthwhile.

- Make recommendations based on information gathered during the trip.

- Mention any plans for another trip or further meetings.

In Model 13-5, the writer uses her conclusion section to repeat her key recommendation—the company must hire consultants who specialize in rebuilding downtown areas.

WRITING A PROPOSAL

Proposals suggest new ways to respond to specific company or organization situations, or they suggest specific solutions to identified problems. A proposal may be internal (written by an employee to readers within the company) or external (written from one company to another or from an individual to an organization).

Purpose

Proposals vary a great deal, and some, such as a bid for a highway construction job on a printed form designed by the state transportation department, do not look like reports at all. In addition, conventional format varies according to the type of proposal and the situation. Proposals usually are needed in these circumstances:

1. A writer, either inside or outside a company, suggests changes or new directions for company goals or practices in response to shifts in customer needs, company growth or decline, market developments, or needed organizational improvements.

2. A company solicits business through sales proposals that offer goods or services to potential customers. The sales proposal, sometimes called a *contract bid,* identifies specific goods or services the company will provide at set prices within set time frames.

3. Researchers request funds to pay for scientific studies. The research proposal may be internal (if an organization maintains its own research and development division) or external (if the researcher seeks funding from government agencies or private foundations).

Proposal content and organization usually vary depending on the purpose and reader-imposed requirements. External proposals, in particular, often must follow specific formats devised by the reader. This chapter discusses the conventional structure for a proposal that is written to suggest a solution to a company problem—the most common proposal type.

In general, this type of proposal persuades readers that the suggested plan is practical, efficient, and cost-effective and that it suits company or research goals. Readers are decision makers who will accept or reject the suggested plan. Therefore, a successful proposal must present adequate information for decision making and stress advantages in the plan as they relate to the established company needs. In a high-cost, complicated situation, a proposal usually has many readers, all of whom are involved with some aspect of the situation.

Proposals are either solicited or unsolicited. If you have not been asked to submit a proposal for a specific problem, you must consider whether the reader is likely to agree with you that a problem exists. You may need to persuade the reader that the company has a problem requiring a solution before presenting your suggested plan.

Organization

Model 13-6 is an unsolicited proposal written by a manager of a city water treatment plant to his superior. The writer identifies a serious safety problem resulting from the current method of installing portable generators during rain storms. His proposal suggests changes to eliminate the safety hazard.

The writer of this formal proposal includes an informative abstract on the title page. His transmittal letter stresses the seriousness of the problem and notes the attached supplementary reports. He also requests authority to proceed with the project and offers to discuss the matter further. The body of the proposal represents the traditional organization for such a document.

Problem

The introductory section of a proposal describes the central problem. Even if you know that the primary reader is aware of the situation, define and describe the extent of the problem for any secondary readers and for the record, and also explain why change is necessary. If the reader is not aware of the problem, you may need to convince him or her that the situation is serious enough to require changes. The first section of the proposal also briefly describes the recommended solution. Include these elements:

- A description of the problem in detail. Do not, for instance, simply mention "inadequate power." Explain in detail what is inadequate about the generator.

- An explanation of how the situation affects company operations or costs. Be specific.

- An explanation of why the problem requires a solution.

- The background of the situation. If a problem is an old one, point out when it began, and mention any previous attempts to solve it.

- Deadlines for solving the problem if time is a crucial factor.

- An indication that the purpose of the proposal is to offer recommendations to solve the problem.

- Your sources of data—surveys, tests, interviews, and so on.

- A brief summary of the major proposed solution, such as to build a new parking deck. Do not attempt to explain details, but alert the reader to the plan you will describe fully in the following sections.

- The types of information covered in the proposal—methods, costs, timetables, and so on.

The writer in Model 13-6 begins by stating that there is a safety problem in the current method of installing portable generators and then describes the method in detail, pointing out where the hazards occur. He also warns the reader that the department is in violation of the Occupational Safety and Health Administration regulations.

Proposed Solution

This section should explain your suggestions in detail. The reader must be able to understand and visualize your plan. If your proposal includes several distinct actions or changes, discuss each separately for clarity. Follow these guidelines:

- Describe new procedures or changes in systems sequentially, according to the way they would work if implemented.

- Explain any methods or special techniques that will be used in the suggested plan.

- Identify employees who will be involved in the proposal, either in implementing it or in working with new systems or new equipment.

- Describe changes in equipment by citing specific manufacturers, models, and options.

- Mention research that supports your suggestions.

- Identify other companies that have already used this plan or a similar one successfully.

- Explain the plan details under specific subheadings relating to schedules, new equipment or personnel, costs, and evaluation methods in an order appropriate to your proposal.

The writer in Model 13-6 proposes four new installations—all designed to simplify the current method of installing portable generators. He uses subheadings to direct the reader to specific information about the schedule, costs, and coordination of the project.

Needed Equipment/Personnel

Identify any necessary equipment purchases or new personnel required by your proposal. Follow these guidelines:

- Indicate specific pieces of equipment needed by model numbers and brand names.

- Identify new employee positions that will have to be filled. Describe the qualifications people will need in these positions.

- Describe how employee duties will change under the proposed plan and how these shifts will affect current and future employees.

Schedules

This section is especially important if your proposal depends on meeting certain deadlines. In addition, time may be an important element if, for instance, your company must solve this problem to proceed with another scheduled project. Follow these guidelines if they are appropriate to your topic:

- Explain how the proposed plan will be phased in over time or on what date your plan should begin.

- Mention company deadlines for dealing with the problem or outside deadlines, such as those set by the IRS.

- Indicate when all the stages of the project will be completed.

The proposed changes in Model 13-6 will require the department to call for bids by contractors, obtain city permits, and review the design. Because of these requirements, the writer describes the necessary schedule in detail. For the reader's convenience, the writer separates the process into four stages.

Budget

The budget section of a proposal is highly important to a decision maker. If your plan is costly relative to the expected company benefits, the reader will need to see compelling reasons to accept your proposal despite its high cost.

Provide as realistic and complete a budget projection as possible. Follow these guidelines:

- Break down costs by category, such as personnel, equipment, and travel.
- Provide a total cost.
- Mention indirect costs, such as training or overhead.
- Describe project costs for a typical cycle or time period, if appropriate.

In Model 13-6, the writer includes a table to supplement the cost section. Because government units operate under budgets established by legislative bodies, the writer also explains how to cover the costs of this proposal under current budget allocations.

Evaluation System

You may want to suggest ways to measure progress toward the objectives stated in your proposal and for checking the results of individual parts of the plan. Evaluation methods can include progress reports, outside consultants, testing, statistical analyses, or feedback from employees. Follow these guidelines:

- Describe suggested evaluation systems, such as spot checks, surveys, or tests.
- Provide a timetable for evaluating the plan, including both periodic and final evaluations.
- Suggest who should analyze the progress of the plan.
- Assign responsibility for writing progress reports.

The writer in Model 13-6 offers to accept the responsibility of directing this project. As part of this responsibility, he will train employees in the new system, write monthly progress reports, obtain necessary permits, and monitor compliance with building regulations.

Expected Benefits

Sometimes writers highlight the expected benefits of a proposal in a separate section for emphasis. If you have such a section, mention both immediate and long-range benefits. Follow these guidelines:

- Describe one advantage at a time for emphasis.
- Show how each aspect of the problem will be solved by your proposal.
- Illustrate how the recommended solution will produce advantages for the company.
- Cite specific savings in costs or time.

Because this proposal involves a safety hazard, the writer in Model 13-6 uses the benefits section to emphasize how the proposed changes will solve the problem and bring the plant into compliance with existing safety regulations.

Summary/Conclusions

The final section of your proposal is one of the most important because readers tend to rely on the final section in most reports and, in a proposal, this section gives you a chance to emphasize the suitability of your plan as a solution to the company problem. Follow these guidelines:

- Summarize the seriousness of the problem.
- Restate your recommendations without the procedural details.
- Remind the reader of the important expected benefits.
- Mention any necessary deadlines.

The writer in Model 13-6 uses his conclusion to emphasize the safety hazard and stress the need for changes.

CHAPTER SUMMARY

This chapter discusses the conventional content and structure for the six types of reports most commonly written on the job. Remember:

- A feasibility study analyzes the alternatives available in a given situation.
- An incident report records information about accidents or other disruptive events.
- An investigative report explains how a particular question or problem was studied and presents the results of that study.
- A progress report provides information about an ongoing project.
- A trip report provides a record of events that occurred during a visit to another location.
- A proposal suggests solutions to a particular problem.

SUPPLEMENTAL READINGS IN PART 2

Caher, J. M. "Technical Documentation and Legal Liability," *Journal of Technical Writing and Communication*, p. 486.

Garhan, A. "ePublishing," *Writer's Digest*, p. 495.

Hart, G. J. S. "PowerPoint Presentations: A Speaker's Guide," *Intercom*, p. 501.

MODEL 13-1   Commentary

This feasibility study provides information to top executives who must select a site for the company headquarters in a new city. The report evaluates two available sites for their suitability for company operations and recommends one of them.

Discussion

1. The president of the company is the primary reader of this report, but other executives get copies. Officers of the company have already discussed moving to a new city, and the president is expecting this report. Review the "Introduction" section. What kinds of information would the writer have had to include if readers had never considered the possibility of moving the company to a new city?

2. Discuss the headings used in this report. How useful are they to a reader seeking specific information? What changes or additions might help the reader locate specific facts?

3. Discuss the overall organization in this report. How effective is it to discuss the Randolph Center ahead of the Lincoln Industrial Park in every section under the heading "Alternative Locations"?

4. Discuss how effectively the writer presents a case for choosing Randolph Center as the new location. Are there any points you would emphasize more than the writer does? What other kinds of information might support the writer's recommendation of Randolph Center?

To: Samuel P. Irving
          President

From: Isobel S. Archer
          Vice President—Administration

Date: April 16, 2005

Subject: Feasibility Study of Sites for Kansas City Offices

Introduction

This report assesses two available locations in greater Kansas City as possible sites for the new corporate offices of Sandrunne Enterprises. After reviewing the background, company needs, and features of the two available sites, I recommend we buy a building and lot in the Randolph Center.

Background

Sandrunne Enterprises has been considering moving the corporate headquarters to Kansas City from Springfield for about two years. Due to streamlining operations and expanding services, the current building in Springfield no longer fits the Company needs. Further, location in Kansas City will enhance our transportation connections and improve our ability to meet schedules. As you know, the Board of Directors has set strict parameters regarding the potential location for Sandrunne. The building/land must be large enough to accommodate our current requirements with enough room for reasonable expansion over the next five years, but overall acreage should not exceed two acres. The building/land must be in excellent condition, located in the greater Kansas City area in Missouri, and readily accessible to transportation carriers. Project cost should not exceed $900,000, and relocation should be completed by February 2006, when our current lease expires.

Possible Alternatives

I have surveyed the available building and land combinations with the assistance of Martin Realty Company, and the two locations that seem most appropriate are Randolph Center, a complex in North Kansas City, and Lincoln Industrial Park, also in North Kansas City. The two alternatives were chosen based on space suitability, condition of the

MODEL 13-1   Feasibility Study

S. P. Irving   -2-   April 16, 2005

building/land, transportation accessibility, cost, and availability within our time frame. Information for this study was obtained through meetings with William Martin, president of Martin Realty; Tracey Manchester, Director of Kansas City Regional Development Board; and John Blackthorn, Real Estate Administrator for the Department of Planning and Urban Development of Kansas City. I made two trips to each site, accompanied by Rafael Mendez from our transportation planning division.

Alternative Locations

Space Suitability

The Randolph Center building is on 1.2 acres of land and fits the basic corporate office/warehouse specifications required by the Board of Directors. Although the acreage is less than two acres, the office building, warehouse facility, and truck docks fit our exact requirements. Only minor modifications to the office building and the warehouse would be needed at Randolph Center. The warehouse space is actually 2,000 square feet over our current requirement and, therefore, allows for future expansion.

The Lincoln Industrial Park site is 2.4 acres. Most industrial parks are already plotted with specified acreage per plot, and the 2.4 plot available was one of the smaller locations close to our two-acre ideal.

Condition of Building/Land

The Randolph Center location contains a brick building built in 1992 in excellent condition with well-landscaped grounds. We would need inspections before purchase, but the current condition of the building seems in good repair. The offices have oak paneling, new carpeting, skylights in the reception area, windows and crown molding in every office, six-panel doors, and tiled bathrooms. Overall, the offices have a professional appearance in keeping with Sandrunne's image. The warehouse is equally well built with cement blocks throughout, and the layout, including the two dock areas, would fit Sandrunne's activities nicely.

The Lincoln Industrial Park location is newer (2002), and we would have to landscape the area as well as build offices. The land is flat and

S. P. Irving   -3-   April 16, 2005

dry, making construction fairly straightforward. Paved roads to the site
are complete, and other existing buildings are of high quality. Sandrunne
would have to break ground for construction, but, of course, we could
design the building exactly the way we want it.

Location and Transportation Accessibility

Since all of Sandrunne product is transported by truck, we must have
access to a highway system and a layout and docking area in which
the truckers can maneuver relatively easily. Both locations are within
the general area we had chosen as appropriate. Both have adequate
trucking access, and employees could reach either one via a four-lane
highway system. Rafael Mendez assured me that the truckers would be
pleased with the docking areas in Randolph Center. The basic docking
areas at Lincoln Industrial Park are excellent too, and our new building
could be constructed with the warehouse and docking facilities suitable
for our operation.

Cost

Only minor modifications are needed at the Randolph Center location. At
the Lincoln Industrial Park location, construction and landscaping will be
required. Total estimated cost for the two locations is as follows:

Randolph Center

Office Modifications	$ 40,000
Warehouse Modifications	70,000
Total Improvements	$110,000
Purchase Price	$500,000
Total Randolph Center Cost	$610,000

Lincoln Industrial Park

Office Construction	$350,000
Warehouse Construction	500,000
Land Purchase	100,000
Total Lincoln Costs	$950,000

Note: Cost estimates are from Gramm, Weismann, and McCall,
Associates, construction consultants.

S. P. Irving   -4-   April 16, 2005

The Lincoln Industrial Park location would be $50,000 over the established limit. Both location costs are estimated, and construction overruns could add considerably to the total cost for the Lincoln location. Costs could also be higher at the Randolph Center, but the risk is less.

Feasible Relocation Date

According to the realtors, the Randolph Center building is currently used for storage by the owners and could be vacated quickly upon closing. The closing time would be about 60 days, and an additional 60 days would be needed for modifications and cleaning. Sandrunne could probably move into the Randolph Center offices by September 1, 2005, well ahead of our February 2006 deadline.

The Lincoln Industrial Park location has more variables. The timetable must include design plans, site preparation, general contractor bids, license permits, and landscaping; therefore, completion is harder to predict. John Blackthorn estimates that Sandrunne could move in by January 1, 2006, but the risk of not making that deadline is somewhat high.

Conclusion and Recommendations

Sandrunne Enterprises has reached the point of moving to Kansas City to solidify its corporate position and expand its operation further. Both the alternative sites I investigated in North Kansas City have long-range advantages and meet the Company criteria. However, Lincoln Industrial Park involves a greater risk that both cost and deadline limits will be exceeded. Considering our need to move without hindrance, I recommend Randolph Center as our best choice. The location is excellent; the project cost is under budget; and the deadline can be met easily.

I have reports from our consultants that I would like to discuss further with you. My secretary will make an appointment for a discussion time.

IA:sd
c: M.L. Wilding
   M.P. Todd
   E.T. Fishman
   R.D. Micelli
   J.J. Warner

MODEL 13-2   Commentary

This incident report describes a chemical fire on company property. No one was seriously hurt, but nearly 60 people had to be treated for smoke inhalation, skin irritations, and minor burns. The writer describes the incident beginning with the discovery of the fire. The "Causes" section is based primarily on information from the fire investigator, and the "Recommendations" section focuses on improved company training for handling flammable materials.

Discussion

1. In groups, consider the "Recommendations" section of this report. Draft a memo to the crews in charge of loading and storage, and explain the need for a training session. Compare your draft with those of other groups.

2. In groups, draft a letter to the fire chief, and explain what steps the company is taking to prevent a future fire. Discuss payment for the volunteer fire equipment.

3. Check your student newspaper and the daily newspaper, and find a report of an accident. Outline the information for an incident report based on the newspaper story. Discuss what kinds of relevant information you are lacking when you rely on the newspaper account.

To: Thomas Cheng, Vice President

From: Carole Sazbatton $\mathcal{CS}$

Date: August 7, 2005

Re: Accident Report—Chemical Fire

Description of the Incident

On August 4, 2005, a fire started in our chemical mixing and storage facility, Building #10, at about 8:00 a.m. On site were 40–45 types of chemicals stored in 55-gal drums. Felix Consuelos reported that he saw the fire in one corner of the building, and he immediately sounded the alarm that automatically notified the Santa Maria Volunteer Fire Department. The firefighters arrived at 8:06 a.m. By that time the fire had spread to surrounding drums. Fire Chief Travis Burnett told me that they were afraid the 15 million-gal jet fuel tank might become involved, so they used all their equipment in a full effort. Several drums exploded, sending sulfur dioxide into the air. Heat and smoke were very strong. The fires were controlled and extinguished at 4:35 p.m. About 40 workers were treated at the scene for smoke inhalation and skin irritations. Nineteen people, including 14 firefighters and 5 children who had been playing in the park across the street, were sent to Mercy Hospital for treatment of chemical exposure, smoke inhalation, or minor burns. Everyone was released from the hospital after treatment. Chief Burnett reported that chemicals ate through hoses used in fighting the fire; burning chemicals seriously damaged one fire truck; and uniforms, masks, and oxygen tanks were ruined. The estimated loss to the fire department is $200,000. The damage estimate for Building #10 is not yet available.

Causes

Upon consulting with the fire investigator, Dean Stockton, I learned that the drums in the back of the building had been placed too close together to allow adequate air circulation. Further, the drums had not been sorted by chemical content to avoid storing incendiary chemicals near each other, thus causing the explosions as the fire spread.

Recommendations

I recommend mandatory training workshops on storing hazardous chemicals. We recently hired 11 new loaders, and they have not yet had a training session. Supervisors should also be directed to attend the training sessions and hold follow-up sessions at regular intervals.

MODEL 13-2   Incident Report

MODEL 13-3   Commentary

This investigative report is based on data available on the Web site of the Bureau of Labor Statistics. The report covers a survey of a sample of U.S. families to determine how many had at least one employed family member.

In research reports, the abstract may appear on the first page between the report title and the introduction, as shown here. Tables showing some of the employment statistics appear in an appendix at the end of the report.

Discussion

1. Discuss the abstract. Does it cover the central points in the report? Are there any changes you would make?

2. Discuss the "Introduction." How important is it to establish these definitions of family types and "children"? Would a reader instinctively understand the terms if the definitions were not included?

3. Discuss the "Method" section. Does the "Method" section explain enough about the data so the reader can understand the scope of the report?

4. Notice that the "Results" section repeats some of the data that appear in the tables in the appendix. Why would the writer repeat information from the tables in the "Results" section? How helpful are the divisions in the "Results" section?

5. The "Discussion" section speculates on reasons for the high employment rate. How helpful are these comments?

6. Notice that the "Conclusion" is brief. Discuss whether the writer needs to say more in the "Conclusion."

7. The investigative report pattern is designed for readers who are interested in how the data were collected as well as what the data signify. Discuss the usefulness of this format. Why would a general reader probably prefer a different format?

EMPLOYMENT CHARACTERISTICS OF U.S. FAMILIES

Abstract

This report covers the employment characteristics of U.S. families in 2003 from the annual survey conducted by the Bureau of Labor Statistics. Data indicate that over 82% of the 75.3 million families in the United States had at least one working member. Women had rates of more than 50% employment whether in a married-couple family or in a family without a spouse.

Introduction

The U.S. Department of Labor, Bureau of Labor Statistics reports that in 2003, 82% of the 75.3 million U.S. families had at least one employed member, down 0.4% from 2002. Employment information gathered from the end of the decade and end of the twentieth century indicates that American families continued to be a central part of the nation's work force.

The principal definitions used in this annual survey are as follows:

- Family—a group of two or more persons residing together who are related by birth, marriage, or adoption. The families are classified as married-couple families or as families maintained by women or men without spouses.
- Householder—the person in whose name the housing unit is owned or rented. The relationship of other persons in the house is defined in terms of the relationship to the householder.
- Married, spouse present—husbands and wives living together in the same household.
- Other marital status—persons who are never married, married with spouse absent, widowed, or divorced.
- Children—natural children, adopted children, and stepchildren of the husband, wife, or person maintaining the household.
- Employed—worked during survey week, worked 15 or more hours unpaid in family business, employed but absent that week because of illness, vacation, or other problems.

MODEL 13-3   Investigative Report

Method

The estimates in this report are based on the annual average data from the Current Population Survey (CPS), a national sample survey of about 60,000 households. The U.S. Census Bureau conducts this survey monthly for the Bureau of Labor Statistics. The information relates to people in the labor force age 16 years and older.

Sampling errors may occur if the sample differs from the popular values they represent. Bureau of Labor Statistics studies generally reach a 90% level of confidence that the sample will differ by no more than 1.6 standard errors. The CPS data also may be affected by nonsampling error resulting from an inability to obtain information for everyone in the sample or the inability or unwillingness of the participants to provide the correct information. The data for earlier years are not strictly comparable because of the steadily increasing U.S. population and revised population controls in the household survey in January 2003.

Results

Although the employment figures for families overall are high, differences related to race and ethnicity, marital status, and ages of children in the household were recorded.

Race and Ethnicity

In an average week in 2003, 6.1 million families had at least one unemployed member. The proportion of black families with an unemployed member was 13.7%; for Asian families, 9.4%; for Hispanic families, 11.1%; for white families, 7.1%. Of the 6.1 million families with at least one unemployed member, 70.5% also had an employed family member. Asian families had the highest percent (82.7%) in that category, followed by white families (73.6%), Hispanic families (70.1%), and black families (57.3%).

-3-

Beginning in 2003, families were classified by the race groups selected by the householder. In addition, people identified as Hispanic can be of any race. Therefore, details may not match totals due to rounding and "other races" not included.

Marital Status

Overall, 81.5% of married-couple families included an employed person in 2003, down by 0.6% from 2002. The proportion of married-couple families in which only the husband was employed was 19.2%, about the same as in 1999. The proportion of married-couple families in which both husband and wife worked declined slightly to 50.9% in 2003. The proportions of families with an employed person for households maintained by men (83.3%) and households maintained by women (71.9%), showed a decrease from 2002.

Parents in married-couple families with children under the age of 18 maintained a high rate of employment, with 97% of the families having a parent employed. This figure has been steady since 1997. (See Appendix, Table 1, for a breakdown of mother/father employment.)

Ages of Children

In 2003, families with children under age 18 had both parents working in 60.7% of married-couple families, a slight decrease. The father, but not the mother, was employed in 30.5% of these families, a slight increase. The mother was the sole working parent in 5.5% of the families.

When children were under the age of six, traditional married-couple families (the father, but not the mother, is employed) represented 29% of the total married-couple families. The mother was not employed 35.4% of the time in families maintained by mothers with children under the age of six. (See Appendix, Table 2

-4-

for breakdown of employed mothers with children under the age of three.) The father was not employed 15.9% of the time in families maintained by men with children under the age of six.

Discussion

These declines in employment figures probably reflect a general economy slump while new immigrants continued to arrive in the United States. In the 1990s, the United States had a high rate of employment with a strong economy in the last half of the decade. Jobs were plentiful and the unemployment rate was at a low level. Beginning in 2000, the economy slowed, and a number of companies closed operations or significantly cut back, increasing unemployment. As the economy began to improve in 2003, new jobs were created, but not in high enough numbers to reverse the trend of the preceding years.

In spite of increased unemployment, all families had a relatively high rate for employed family members. For mothers, employment rates were lowest when their children were under the age of three, but even then, the employment rate for mothers was 58.9%.

Conclusion

This annual survey by the Bureau of Labor Statistics provides valuable information about the working rates of major family categories, particularly about the working rates of parents. Although the data may not be entirely comparable from year to year, they do show trends over time and help government agencies and others assess the needs of families.

-5-

APPENDIX

Table 1

Employment Status of Married-Couple Families
with Children
(numbers in thousands)

	2002	2003
Married-couple families—total	25,191	25,308
Parents employed	24,372	24,590
Both parents employed	15,439	15,400
Mother employed	16,773	16,850
Mother employed/not father	1,334	1,490
Father employed/not mother	7,599	7,700
Neither parent employed	819	832

Table 2

Mothers—Employment Status
(numbers in thousands)

	Employed 2002	%	Employed 2003	%
Total w/children under 3 years	9,350	60.2	9,450	58.9
Total w/children 2 years	2,949	64.3	2,987	63.5
Total w/children 1 year	3,310	60.5	3,353	59.6
Total w/children under 1 year	3,091	56.1	3,110	53.7

MODEL 13-4   Commentary

This progress report, written by an on-site engineer, is one of a series of reports on a specific company project—dam repairs. The report will be added to the others in the series and kept in a three-ring binder at the company headquarters for future reference when the company has another contract for a similar job.

The writer addresses the report to a company vice president, but four other readers are listed as receiving copies.

Discussion

1. The writer knows that her readers are familiar with the technical terms she uses. If this report were needed by a nontechnical reader at the company, which terms would the writer have to define?

2. Discuss the information in the introduction and how effectively it supports the writer's assertion that work on the dam is progressing satisfactorily.

3. Discuss how the writer's organization and use of format devices support her report of satisfactory progress. Why did she list items under "Work Completed" but not under "Work Remaining"?

TO: Mark Zerelli   October 16, 2005
               Vice President
               Balmer Company

FROM: Tracey Atkins   *TA*
               Project Manager

SUBJECT: Progress Report #3–Rockmont Canyon Dam

Introduction

This report covers the progress on the Rockmont Canyon Dam repairs
from September 15 to October 15 as reported previously. Repairs to the
damaged right and left spillways have been close to the original
schedule. Balmer engineers prepared hydraulic analyses and design
studies to size and locate the aeration slots. These slots allowed Balmer
to relax tolerances normally required for concrete surfaces subjected to
high-velocity flows. Phillips, Inc., the general contractor, demolished and
removed the damaged structures. To expedite repairs, construction
crews worked on both spillways simultaneously. Construction time was
further reduced by hiring another demolition company, Rigby, Inc. The
project costs rose during the first month when Phillips, Inc., had to build
batching facilities for the concrete because the dam site had no facilities.

Work Completed

Since the last progress report, three stages of work have been
completed:

1. On September 18, aggregate for the concrete mix was hauled
230 miles from Wadsworth, Oregon. The formwork for the tunnel linings
arrived from San Antonio, Texas, on September 20.

2. Phillips developed hoist-controlled work platforms and man-cars
to lower workers, equipment, and materials down the spillways.
Platforms and man-cars were completed on September 22.

3. Phillips drove two 20-ft-diameter modified-horseshoe-shaped
tunnels through the sandstone canyon walls to repair horizontal
portions of the tunnel spillways. A roadheader continuous-mining
machine with a rotary diamond-studded bit excavated the tunnels in
three weeks, half the time standard drill and blast techniques would
have taken. The tunnels were completed on October 15.

MODEL 13-4   Progress Report

Mark Zerelli   -2-   October 16, 2005

Work Remaining

The next stage of the project is to control flowing water from gate leakage. Phillips will caulk the radial gates first. If that is not successful in controlling the flow, Phillips will try French drains and ditches. After tunnels are complete, both spillways will be checked for vibration tolerance, and the aeration slot design will be compared with Balmer's hydraulic model. Full completion of repairs is expected by November 15.

Adjustments

Some adjustments have been made since the last progress report. During construction work on both spillways, over 50 people and 200 pieces of equipment were on the site. Heavily traveled surfaces had to be covered with plywood sheets topped with a blanket of gravel. This procedure added $3500 to the construction costs and delayed work for half a day.

Conclusion

The current work is progressing as expected. The overall project has fallen three days behind estimated timetables, but no further delays should occur. The final stage of the project will require measurements at several areas within both spillways to be sure they can handle future flood increases with a peak inflow of at least 125,000 cfs. Balmer expects to make the final checks by November 25.

TA:ss
c: Robert Barr
   Mitchell Lawrence
   Mark Bailey
   Joseph Novak

MODEL 13-5   Commentary

This trip report describes a visit to a trade show. The writer went to the show looking for ideas about rebuilding city properties. She reports that she found several ideas that would be useful to clients, and she met with a major supplier. She recommends using consultants to develop plans for their clients.

Discussion

1. The writer begins with a summary that explains why she attended the trade show. How helpful is this opening summary to the readers?

2. The writer includes two information sections—one describing the model buildings she visited and the other discussing her meeting with a supplier representative. How helpful are the headings for these sections?

3. Consider the purpose of this report. Discuss whether the writer should have included more information about other people she met and other exhibits she saw.

4. In groups, draft a trip report using the following information: You work for the local public library system. Because the public library system is trying to cut back on expenses for periodicals subscriptions, you have been told to visit your school library and find out which technical periodicals are regularly used for class assignments. The plan is for the public library to cut back on those subscriptions if the school library already gets them and students use them there. You talk to Jon Milich, the head librarian, who shows you around the library and explains the periodical subscription list. Come to any conclusion you wish, and add specific information as needed.

To: Ken Germain, Vice President

From: Kirsten Swensen, Design Analyst   *Ks*

Date: July 23, 2005

Re: Trip to the Home Appliance Manufacturers Showcase 2005,
         Atlanta, Georgia, July 17–21, 2005

Summary

I attended the Home Appliance Manufacturers Showcase 2005 in
Atlanta, Georgia, for the full five-day convention because we have
several clients who have problems in redesigning their metropolitan
center property. The convention did provide a variety of ideas and plans
all designed to reattract builders, businesses, and residents back to the
central city. We will be able to use some of these ideas in proposals for
clients with interests in developing those districts.

Model Buildings

The model area just east of the convention hall showed three basic
designs for professional offices and retail shops with attractive
residences above or behind the businesses. The all-brick, semi-attached
buildings are all constructed on a low-rise, neo-Victorian scale and
designed with a quiet, early 20th century urban look. The architectural
plan created an uncrowded look overall. I brought back sample floor
plans to use in our client demonstrations.

Meeting with Copper Development Association Representative

On July 19, I met with Tricia Lawton of the CDA. She explained the new
low-pressure systems with small-diameter annealed copper tubing that
can deliver natural or bottled gas for appliances in these residences and
office complexes. These distribution systems will provide an effective,
high-quality feature for any plan we develop. I made arrangements to
consult with her again before we submit any final plans to clients.

Conclusion and Recommendations

My observations at Showcase 2005 convinced me that we need to work
with consultants who are specializing in commercial and residential
developments in central city areas. I have copies of relevant information,
but we need to arrange an ongoing relationship with a builder who has a
record of success in such developments.

c: M. Benamou

MODEL 13-5   Trip Report

MODEL 13-6   Commentary

This unsolicited proposal identifies a hazardous situation at a city water treatment plant. The writer prepares a formal proposal in this instance because he expects the reader to show it to others and he knows that requests for significant extra expenditures traditionally are presented in a formal report style in the city government offices.

The writer begins by describing the problem in detail, so the reader can understand the inherent dangers in the current situation. He then suggests a specific set of changes to eliminate the hazard to employees. In this case, the suggested changes are relatively uncomplicated. The process of soliciting bids and getting permits has specific stages, however, and the writer outlines these in a timetable.

Discussion

1. The reader of this proposal has not asked for it and may not be aware of the problem. Discuss how effectively the writer makes his point that a serious situation exists and must be resolved.

2. Discuss how effectively the writer presents the benefits section of his proposal.

3. The suggested timetable and projected costs are presented within the report. Discuss whether, as a reader, you would prefer these details in an appendix.

4. In groups, identify a campus problem that affects a large number of students. Draft a description of the problem that could be an opening section of a proposal for changes.

**PROPOSAL FOR STANDBY POWER IMPROVEMENTS
AT WATER TREATMENT PLANT**

Prepared for
Richard T. Sutton
Director of Operations

Submitted by
Michael J. Novachek
Water Department Manager

November 16, 2005

Abstract

Since 1996, the City Water Treatment Plant (WTP) has used two
generators to operate high-service pumps during electrical power failures.
One of these generators creates a dangerous safety condition because
electricians have to stand outside during rain storms to make connections.
The proposal is to install exterior outlets under shelter so standby power
can be safely connected.

MODEL 13-6   Proposal

CHESTERTON WATER DEPARTMENT
65 Portage Road
Chesterton, TN 38322

(423) 555-1000   Fax (423) 555-1100

November 16, 2005

Mr. Richard T. Sutton
Director of Operations
City of Chesterton
111 Civic Plaza
Chesterton, TN 38303

Dear Mr. Sutton:

I am attaching a proposal for installation of new electrical outlets and
outside shelter at the Water Treatment Plant to correct a current safety
problem. Our method for connecting standby power equipment during
rain storms is in violation of OSHA regulations and should be addressed
at once.

The recommended improvements are relatively minor. I am attaching a
report from Raymond and Burnside Engineering Associates, outlining
specifications and design features of the suggested improvements. Our
procedures for soliciting and reviewing construction bids will take about
seven months, so I would like to begin this project at once. May I have your
authorization to proceed?

Please call me at Ext. 5522 if you would like to discuss this matter in
greater detail.

Sincerely,

M. J. Novachek

Michael J. Novachek
Water Department Manager

MJN:sd
Attachment

PROPOSAL FOR STANDBY POWER IMPROVEMENTS
AT WATER TREATMENT PLANT

Safety Problem with Standby Power

The WTP uses four high-service pumps, which provide treated water to our customers. Three of the pumps have a rated capacity of 2 million gallons per day (MGD). The fourth pump has a rated capacity of 3 MGD. The average and peak daily demands at the WTP are 5 MGD and 7 MGD, respectively. Therefore, we typically run a combination of the pumps to meet our demand.

When an electrical failure occurs, we develop an unsafe situation for our electricians. The plant was originally designed and constructed to provide standby power only to the largest pump. The standby power came from the diesel-powered generator located alongside, and directly tied to, the 3-MGD pump. Therefore, when a power failure occurs, the power supply to the 3-MGD pumps can be switched from the standard supply to the generator. Under normal conditions, the plant is required to operate a 2-MGD pump along with the 3-MGD pump to meet customer demand. During power failures, we temporarily connect a portable standby power generator to one of the 2-MGD pumps. This temporary generator installation creates an unsafe condition.

The portable standby generator is stored on a trailer at Pump Station No. 11. During power failures, two electricians transport the portable generator to the WTP. The portable generator is parked outside, next to the doors of the pump room. The electricians run power cables from the generator to the Motor Control Center (MCC) of the pump, which is located in the pump room. Since power failures usually result from a severe storm, the electricians often make the generator connections while standing in the rain. The doors to the pump room remain open to provide an opening for the power cables to run from the generator, along the pump room floor, to the MCC. This setup exposes plant personnel to a live power cable and wet floor conditions and is in violation of Occupational Safety and Health Administration (OSHA) regulations. These conditions could create an enormous liability to the City if they result in an accident to an electrician or plant operator.

-2-

Proposed Solution

To eliminate hazardous conditions connected with installing the portable generators, I recommend the following:

- Install a male adapter plug to the portable standby generator.
- Install a female wall outlet at the exterior south wall of the pump room.
- Install an awning at the exterior south wall of the pump room.
- Install permanent wire and conduit from the outlet to the MCC of the pumps.

During a power failure, plant personnel can transport the portable standby power generator to a covered area located at the south exterior wall of the pump room. Personnel can connect standby power to the pump by inserting the generator plug into the permanent outlet. The awning covers the generator and wall outlet and keeps this area relatively dry. The pump room doors could be kept closed. No live electrical connections are required. Finally, no power cables are strewn along the floor.

Schedule

In order to incorporate the recommended improvements, the project will require four phases.

The design phase will require the services of a consultant engineer. I have discussed the project with engineers at Raymond and Burnside Associates, the company that designed the plant 15 years ago. The engineers assured me that the project is feasible and can be completed within the budgetary and time constraints set herein. The consultant will be responsible for the preparation of plans, specifications, and bid documents, as well as permits required by the Environmental Protection Agency. I recommend that Raymond and Burnside Associates handle the design services for this project since we have an existing agreement and they are familiar with the plant.

The bid phase will require the services of a consultant, as well as the City law director, to review the bids and recommend a contractor.

-3-

The construction phase will require an electrical contractor to perform the work. I recommend we use Raymond and Burnside Associates to verify that the work is being performed in accordance with the design plans and specifications.

The implementation phase comes after the improvements are completed. Once the system is in place, the two electricians on standby duty will be replaced by one on-duty plant maintenance worker.

I suggest the following schedule for the installation of the improvements:

Item	Completion Date
Service director approval to proceed	December 1, 2005
Procure agreement with engineer	December 14, 2005
Complete plans, specifications, and bid documents	January 3, 2006
City review of plans, specifications, and bid documents	January 17, 2006
Submit permits and plans to EPA	February 1, 2006
EPA review	April 1, 2006
Advertisement for bid	April 14, 2006
Bid opening	May 1, 2006
Award bid	June 1, 2006
Start construction	July 1, 2006
Complete construction	August 1, 2006

Cost

The estimated project cost to install a new standby power system improvement is $27,615. This cost includes engineering, construction, and contingencies as shown in Table 1.

Table 1
Probable Project Cost

Engineering

Design	$ 8,000
Services During Construction	4,000
Total Engineering Cost	$12,000

-4-

Construction

Awning	$ 2,500
Electrical Adapter	500
Electrical Outlet	500
Wire and Conduit	7,000
MCC Switches	3,800
Total Construction Cost	$14,300
Total Project Cost	$26,300
With 5% Contingency	$27,615

The engineering fee includes design, preparation of plans, specifications and bid documents, review of bid, and services during construction. Construction costs include all labor and materials required to construct the improvements. A contingency fee of 5 percent of the total engineering and construction costs is included to cover costs not foreseen at this time.

The expected 2007 annual operation costs with these improvements is $400. Without the improvements, cost is $2,500. This higher cost comes from the additional employees required to provide the service necessary to achieve the standby power.

The project probably can be funded out of the water fund. We had $20,000 left from last year's budget. A line item of $8,000 could be allotted from this year's budget. The combination of funds will cover the total cost.

Coordination

The improvements will require the coordination and cooperation of City forces to construct and operate the system. As Manager of the Water Department, I will take full responsibility for the following:

1. Instruct the employees regarding the new operation procedures once the improvements are installed.

2. Provide monthly progress reports to the Mayor and the Directors of Service, Law, and Finance.

-5-

3. Ensure that proper EPA and building permits are obtained.

4. Ensure that all City regulations involving capital improvements are followed.

5. Obtain proper approvals from the City Planning Commission and City Council.

Benefits

The main benefit of the proposed improvement is safety. As City leaders, we are responsible for the safety of our employees. The existing condition is unsafe and violates OSHA regulations. Also, the current permanent standby power system does not meet today's design standards. After improvements, we will be in full compliance with all safety regulations. The expected initial cost is reasonable given the seriousness of the situation, and the proposed improvements will lower personnel costs in the future.

Conclusion

Currently, the City water treatment plant operates its standby power system in an unsafe manner, thereby risking the health and safety of plant electricians and operators. With the installations proposed, these unsafe conditions can be eliminated. In addition, these improvements will bring our operation up to current design standards.

Chapter 13 Exercises

1. You have been hired as a computer consultant for the Edwin Burke Rehabilitation Center in Maine. The Center is open from 9 A.M. to 7 P.M. Monday through Friday and from 9 A.M. to 3 P.M. on Saturday. It provides out-patient physical therapy for people with mobility problems as they recover from surgery or accidents. The Center employs eight physical therapists, two full-time medical secretaries, and one part-time receptionist. The medical secretaries make appointments, maintain patient records, schedule procedures at other medical facilities, order supplies, and set up vendor contracts. Payroll is handled by the parent medical corporation in Houston. The Center also employs a cleaning crew of two people and hires local companies to do outside maintenance. Equipment purchases, such as treadmills and exercise bikes, include maintenance contracts. The Center's current computer system and equipment is eight years old. General Manager Cathy Cooper asks you to recommend a computer system (hardware and software) that will allow them to handle the daily tasks. She also wants to design and print eye-catching flyers and therapy instruction sheets for patients.

2. You have been hired as a local management consultant to Quick Auto Checkup, a chain of auto repair garages. The garages offer a full line of general service, such as oil changes, tire rotation, and tune-ups, and the company always blankets the area with free coupons when opening a new location. Select two potential locations in your community for a new Quick Auto Checkup garage, and write a formal feasibility study comparing these locations. Recommend one location to Jamal Tatum, President of Quick Auto Checkup.

3. Study a problem at the company where you work. The problem should be a situation that requires either a change in procedures or new equipment. Investigate at least two alternative solutions to this problem. Then, write a feasibility study in which you evaluate the solutions, and recommend one of them. Address the report to the appropriate person at your company, and assume that this person asked you to write the report. If your instructor wants you to prepare a formal report, use the guidelines in Chapter 11.

4. You are a construction-site supervisor for T. K. Construction. This morning, two of the workers were anchoring a plywood form in preparation for pouring a concrete wall. Jesse Dawson was using a power tool to nail the plywood. When he fired the tool, the nail penetrated the stud and the plywood and then struck and killed Miguel Miranda, who was working about 15 ft away. Both employees had been on the job only two days. When you investigate, Dawson tells you he was never trained to use the power nail gun. You also discover Miranda was not wearing any personal protective clothing or gear. The Occupational Safety and Health Administration (OSHA) requires

easily penetrated materials to be backed by something that would keep a nail from completely passing through. If no backing is available, workers do not use the power tool to drive nails. Write an incident report to David Weller, president of T. K. Construction.

5. You are the manager of the Scuba Deluxe Club. The club serves experienced divers and offers lessons to beginners. This afternoon, Ron Feldman, an experienced dive instructor, took out a group of eight beginning divers, all of whom had already completed five dives. The group went to Canyon Lake, a small but deep nearby lake. Ron directed the group to descend to about five or six feet and follow a compass heading until they reached the opposite shore. The group was to stay together and surface on the other side. When the group surfaced, Ron immediately saw that two divers were not yet there. After about five minutes, he became alarmed and began searching the surface for bubbles that indicated where the divers were. Suddenly, one diver, Brian McGee, shot to the surface in a panic. Ron yelled instructions to put McGee on oxygen as a precaution and continued to search for bubbles showing the whereabouts of Mike Chester. Chester's body was finally recovered at a depth of 65 ft, his gear in place and his tank empty. Ron tells you that the depth of the water and the diver's lack of navigational skills led to the disaster. He also admits that he should have set up a stronger control plan to guide the divers across the lake. Because Chester's gear was in place, you assume that he never knew he had gone too deep and probably suffered from a deep-water blackout before he died. When you talk to McGee later, he admits that he had trouble handling the rental equipment and neither he nor his friend Chester was familiar with the navigation by compass that was required. He said they did not want to hold up the group by asking for help. Write an incident report to Paul Calley, Executive Director of the Scuba Deluxe Club.

6. This assignment requires a visual inspection of a rest room on your campus. The selection of which room is up to you, but you must inspect the room *after 10:00 A.M. on a regular school day*. Observation or inspection requires that you examine the area under *existing conditions*. Make no attempt to control or "fix" the situation. Record what you see, hear, smell, and so on. Inspect the following: sinks, doors, walls, ceilings, floors, toilets/urinals, mirrors, soap containers, paper towel containers/air dryers, light fixtures. Take notes on the need for repairs and maintenance. Record the need for more equipment or more cleaning. In addition, observe the number of visitors during your inspection. Interview at least one person, either someone using the room or a maintenance worker, and include that person's comments in your investigative report. You do not have to recommend specific changes in this report, but you may include an overall assessment of the need for changes. Address your investigative report to the director of physical facilities at your school.

7. Based on the investigative report you wrote for Exercise 6, write an informal proposal for two or three specific changes in the rest room you observed. Address your proposal to the director of physical facilities at your school.

8. Write a progress report for a project you are doing in a science or technology class. Report your progress as of the midpoint in your project. Address the report to the instructor of the class, and say that you will submit another progress report when the project is nearly complete.

9. Write a proposal for a change at the company where you work. The change can include purchasing new equipment, adjusting work systems, new parking or other employee facilities, or other changes. If you do not have a current job, propose a change for your campus, such as increased parking, food services, or recreation facilities. Address the appropriate company manager or the director of student services. Assume that the reader did not ask for this proposal, and so you need to persuade your reader that a need or problem exists. If your instructor wants you to write a formal proposal, consult Chapter 11.

COLLABORATIVE EXERCISES

10. In groups, research current job prospects in a particular field or in a specific area of the country. Each group member should research five sources, including interviewing someone in the field or someone working in the location. Combine your information, and write an investigative report in which you report on job prospects for a field or a location. Address the report to the director of placement at your school, and assume that the director has asked you to prepare this investigative report.

11. In groups, research information about Internet Voice, also called Protocal or VoIP, which is an alternative to telephone service. Assume that your company, Acme Merchandising has offices in five other countries. Melissa Garcia, Director of Communications, needs to analyze the benefits of this new technology. She has heard that it can save 60% on international phone calls and allows great flexibility. Company business requires frequent calls among the offices in the five countries and in the United States. Write an information report to Ms. Garcia, and indicate whether you think this technology could be useful to Acme Merchandising. Assume that Ms. Garcia will show the report to other people.

INTERNET EXERCISES

12. Locate the Web site of a company that makes products you use. Review the site, and write a proposal for one change to the site organization, individual Web pages, or navigation elements. Address your proposal to the president of the company.

13. Search the Internet for five schools that offer academic programs like the one in which you are currently enrolled. Analyze the curriculum in these programs compared to the program at your school. Identify a change you would like to make in your current program, and write a proposal for this change to the chair of your major department.

14. Select a product that you use. Find the Web sites of three companies that make this product. Write a feasibility study comparing the brands and recommending one of them based on information from the Web sites. Include a comparison of the kinds of information and the amount of information available on the company Web sites. Address your informal report to a local retailer or dealer who is considering whether to add this brand to the inventory.

CHAPTER 14

Letters, Memos, and Email

UNDERSTANDING LETTERS, MEMOS, AND EMAIL

Correspondence includes letters, memos, and email. Although these frequently are addressed to one person, they often have multiple readers because the original reader passes along the correspondence to others, or the writer sends copies to everyone involved in the topic.

Give the same careful attention to reader, purpose, and situation in correspondence that you do in all writing. Because of its person-to-person style, however, correspondence may create emotional responses in readers that other technical documents do not. A reader may react positively to a bulletin in a company manual, but a memo from a supervisor on the same subject may strike the reader as dictatorial. Consider tone and organizational strategy from the perspective of how your readers will respond emotionally, as well as logically, to the message.

Letters

Letters are written primarily to people outside an organization and cover a variety of situations, such as (1) requests, (2) claims, (3) adjustments, (4) orders, (5) sales, (6) credit, (7) collections, (8) goodwill messages, (9) announcements, (10) records of agreements, (11) follow-ups to telephone conversations, and (12) transmittals of other technical documents.

Memos

Memos are written primarily to people inside an organization. Memos cover the same topics as letters. In addition, many internal reports, such as trip reports, progress reports, and short proposals, may be in memo form.

Email

Email (electronic mail) allows transmission of letters, memos, and other documents from one computer to another through a series of computer networks. Millions of people now use email daily because of the speedy transmission and the convenience, especially for short messages between people who have an established subject or who are communicating about routine matters.

Advantages of Email

Email has many clear communication advantages, such as the following:

- Employees can make quick decisions using email and avoiding a time-consuming meeting.

- Email creates a written record.

- Managers can reach dozens of employees quickly.

- Team members working on projects can communicate easily about small points without meeting.

- People can avoid playing "telephone tag," which is frustrating and time-consuming.

- People can communicate over long distances without regard to time zone differences.

- Companies can save postage and paper costs.

Disadvantages of Email

Email has some disadvantages as well, such as the following:

- Email can get lost in a daily sea of genuine email and spam (unwanted ads). People on the job can log into email in the morning and find more than 60 messages, most of which are spam. A genuine email can be overlooked or accidentally deleted as the person tries to clear the inbox.

- People sometimes write overly long messages because email does not provide an automatic image of length as a sheet of paper does. For most people, reading long documents on the screen is less comfortable than reading them on paper.

- Email is so easy to send that people sometimes dash off hastily written, badly planned, and negative messages without thinking.

- Email seems less formal than a printed document. Sometimes important communication is devalued if it arrives via email and not as a formal report.

Dangers of Email

Email must be used with caution. Because email does not pass through human hands and receivers use a password to access their email, senders often assume it is private. Email is not private.

By pressing one button, a receiver can send your message to thousands of people. The consequences can be serious. Only hours after Air Force pilot Scott O'Grady, shot down in Bosnia, was rescued by other U.S. pilots, one of the pilots sent a long, detailed message about the rescue to a few friends. The message included pilot code names, weapons loads, radio frequencies, and other classified information. Within minutes of reading the message, the pilot's friends forwarded it to other friends, who also passed it on. Finally, the message was posted on bulletin boards available to millions of people. The classified information was now in the hands of anyone interested in U.S. military operations.[1]

There are other dangers in casual use of email. Companies and individuals can be liable for messages sent by email. Lawsuits for sexual harassment, age discrimination, product liability, and other matters have resulted from email messages.[2] Many people do not realize that employers have the right to monitor email that is sent on a company system. A recent study showed that employers monitor about one-third of U.S. employees' email and Web use. In 2000, Dow Chemical fired 50 employees and disciplined 200 more for emailing pornographic materials from company computers.[3]

Some computer systems save all email even if the receiver deletes it, and computer experts can find files that were "deleted" months before. During the antimonopoly trial of Microsoft, the U.S. Department of Justice used more than 3 million pages from the company's email system to support the case against the company.[4] Consider carefully the way you use email. Do not send messages that could embarrass you, get you fired, cost you money, or send you to jail.

Finally, email has increased the opportunities for criminals to defraud and steal from the public. Spam and criminal scams via email have multiplied dramatically in the last few years. About 64% of email monitored in May 2004 was spam.[5] "Phishing" schemes use email messages to convince people that banks or other organizations need their personal and financial information to "protect" their accounts. Emails offering to share millions of dollars if the recipients will "help" funnel the money through their bank accounts arrive weekly. Such schemes result in identity theft or bank fraud for the victims. Businesses are adopting stricter email policies, and some companies set their computer systems to reject attachments automatically. These serious problems may reduce the use of email in spite of its convenience.[6]

DEVELOPING EFFECTIVE TONE

Tone refers to the feelings created by the words in a message. Business correspondence should have a tone that sounds natural and conveys cooperation, mutual respect, sincerity, and courtesy. Tone of this kind establishes open communication and reflects favorably on both the writer and the company. Because words on paper cannot be softened with a smile or a gesture, take special care in word choice to avoid sounding harsh or accusing. Remarks that were intended to be objective may strike your reader as overly strong commands or tactless insinuations, especially if the subject is at all controversial. Remember the following principles for creating a pleasant and cooperative tone in your correspondence.

STRATEGIES—Office Etiquette

Follow these office etiquette rules if you want to climb the promotion ladder:

- **Do** be on time for all meetings and appointments.
- **Do** make fresh coffee when you take the last cupful.
- **Do** clean up your desk occasionally. Clutter implies disorganized thinking.
- **Do not** eat during meetings, even if they run over lunchtime.
- **Do not** begin a separate conversation with someone next to you while a presenter or meeting leader is talking.
- **Do not** chat about your personal problems with groups of coworkers.
- **Do not** swear. Vulgarities imply a limited vocabulary and limited intelligence.

Natural Language

Use clear, natural language, and avoid falling into the habit of using old-fashioned phrases that sound artificial. A reader who has to struggle through a letter filled with out-of-date expressions probably will become annoyed with both you and your message, resulting in poor communication. This sentence is full of out-of-date language:

Per yours of the tenth, please find enclosed the warranty.

No one really talks this way, and no one should write this way. Here is a revision in more natural language:

I am enclosing the warranty you requested in your letter of March 10.

Keep your language simple and to the point. This list shows some stale business expressions that should be replaced by simpler, more natural phrases:

Old-Fashioned	Natural
Attached hereto . . .	Attached is . . .
We beg to advise . . .	We can say that . . .
Hoping for the favor of a reply . . .	I hope to hear from you . . .
As per your request . . .	As you requested . . .

It has come to my attention . . .	I understand that . . .
Prior to receipt of . . .	Before we received . . .
Pursuant to . . .	In regard to . . .
The undersigned will . . .	I will . . .

If you use such out-of-date expressions, your readers may believe you are as out of date in your information as you are in your language.

Positive Language

Keep the emphasis on positive rather than negative images. Avoid writing when you are angry, and never let anger creep into your writing. In addition, avoid using words that emphasize the negative aspects of a situation; emphasize the positive whenever you can, or at least choose neutral language. Shown here are sentences that contain words that emphasize a negative rather than a positive or neutral viewpoint. The revisions show how to eliminate the negative words.

Negative: When I received your complaint, I checked our records.

Positive: When I received your letter, I checked our records.

Negative: To avoid further misunderstanding and confusion, our sales representative will visit your office and try to straighten out your order.

Positive: To ensure that your order is handled properly, our sales representative will visit your office.

Negative: I am sending a replacement for the faulty coil.

Positive: I am sending a new coil that is guaranteed for one year.

Negative: The delay in your shipment because we lost your order should not be longer than four days.

Positive: Your complete order should reach you by September 20.

In these examples, a simple substitution of positive-sounding words for negative-sounding words improves the overall tone of each sentence, whereas the information in each sentence and its purpose remain the same.

No matter what your opinion of your reader, do not use language that implies that the reader is dishonest or stupid. Notice how the following revisions eliminate the accusing and insulting tone of the original sentences.

Insulting: Because you failed to connect the cable, the picture was blurred.

Neutral: The picture was blurred because the V-2 cable was not connected to the terminal.

Insulting: You claimed that the engine stalled.

Neutral: Your letter said that the engine stalled.

Insulting: Don't let carelessness cause accidents in the testing laboratory.

Neutral: Please be careful when handling explosive compounds.

Negative language, either about a situation or about your reader, will interfere with the cooperation you need from your reader. The emphasis in correspondence should be on solutions rather than on negative events.

You-Attitude

The *you-attitude* refers to the point of view a writer takes when looking at a situation as the reader would. In all correspondence, try to convey an appreciation of your reader's position. To do this, present information from the standpoint of how it will affect or interest your reader. In these sample sentences, the emphasis is shifted from the writer's point of view to the reader's by focusing on the benefits to the reader in the situation:

Writer emphasis: We are shipping your order on Friday.

Reader emphasis: You will receive your order by Monday, October 10.

Writer emphasis: To reduce our costs, we are changing the billing system.

Reader emphasis: To provide you with clear records, we are changing our billing system.

Writer emphasis: I was pleased to hear that the project was completed.

Reader emphasis: Congratulations on successfully completing the project!

By stressing your reader's point of view and the benefits to your reader in a situation, you can create a friendly, helpful tone in correspondence. Of course, readers will see through excessive praise or insincere compliments, but they will respond favorably to genuine concern about their opinions and needs. Here are guidelines for establishing the you-attitude in your correspondence:

1. Put yourself in your reader's place, and look at the situation from his or her point of view.
2. Emphasize your reader's actions or benefits in a situation.
3. Present information as pleasantly as possible.
4. Offer a helpful suggestion or appreciative comment when possible.
5. Choose words that do not insult or accuse your reader.
6. Choose words that are clear and natural, and avoid old-fashioned or legal-sounding phrases.

Tone is often a problem in email messages. Because so many emails are short, senders may not realize how abrupt or harsh they sound. Remember these tips for email tone:

- Avoid writing in all capital letters—called "screaming" in email.
- Avoid humor unless you know the receiver well.

- Proofread carefully. Writers tend to be sloppier in email than on paper. This casualness about correctness creates an impression that the sender does not respect the receiver.

- Never send an email when you are angry. The ease of clicking the "send" button makes email especially dangerous for creating a negative tone.

- Double-check word choice to be sure you have a pleasant and professional tone.

ORGANIZING LETTERS, MEMOS, AND EMAIL

Most correspondence is best organized in either a *direct pattern,* or an *indirect pattern,* depending on how the reader is likely to react emotionally to the message. If the news is good, or if the reader does not have an emotional stake in the subject, use the direct pattern with the main idea in the opening. If, however, the news is bad, the indirect pattern with the main idea after the explanation is often most effective because a reader may not read the explanation of the situation if the bad news appears in the opening. Most business correspondence uses the direct pattern. The indirect pattern, however, is an important strategy whenever a writer has to announce bad news in a sensitive situation and wants to retain as much goodwill from the reader as possible. A third pattern is the *persuasive pattern,* which also places the main idea in the middle portion or even the closing of a message. Writers use this pattern for sales messages and when the reader needs to be convinced about the importance of a situation before taking action.

Memos present a special writing challenge in that very often they have mixed purposes. A memo announcing the installation of new telephone equipment is a message to inform, but if the memo also contains instructions for using the equipment, the purpose is both to announce and to direct. Moreover, a memo can be either good or bad news, and if the memo is addressed to a group of employees, it can contain *both* good and bad news, depending on each individual's reaction to the topic. A memo announcing a plant relocation can be good news because of the expanded production area and more modern facility; however, it also can be bad news because employees will have to uproot their lives and move to the new location. You will have to judge each situation carefully to determine the most effective approach and pattern to use.

Avoid presenting bad news through email if possible. Email messages tend to be too short for well-developed negative messages. Bad news is best presented in letters or memos, which allow you to offer a detailed explanation of the situation.

Direct Organization

Model 14-1 shows a letter that conveys good news and is written in the direct pattern. In this situation, a customer has complained about excessive wear on the tile installed in a hospital lobby. Because the tile was guaranteed by the company that sold and installed it, the customer wants the tile replaced. The company will stand behind its guarantee, so the direct pattern is appropriate in this letter because the message to the customer is good news. The direct pattern generally has three sections:

1. The opening establishes the reason for writing the letter and presents the main idea.

2. The middle paragraphs explain all relevant details about the situation.

3. The closing reminds the reader of deadlines, calls for an action, or looks to future interaction between the reader and the writer.

In Model 14-1, the opening contains the main point, that is, the good news that the company will grant the customer's request for new tile. The middle paragraph reaffirms the guarantee, describes the tile, and explains why the replacement tile should be satisfactory. In the closing, the manager tells the customer to expect a call from the sales representative and also thanks the customer for her information about the previous tile order. The writer uses a pleasant, cooperative tone and avoids repeating any negative words that may have appeared in the customer's original request; therefore, the letter establishes goodwill and helps maintain friendly relations with the reader.

Indirect Organization

The basic situation in the letter in Model 14-2 is the same as in Model 14-1, but in this case, the letter is a refusal because the guarantee does not apply. Therefore, indirect organization is used.

The indirect pattern generally contains these elements:

1. The opening is a "buffer" paragraph that establishes a friendly, positive tone and introduces the general topic in a way that will later support the refusal or negative information and help the reader understand it. Use these strategies when appropriate:
   - Agree with the reader in some way:

     You are right when you say that . . .
   - State appreciation for past efforts or business:

     Thank you for all your help in the recent . . .

PRESCOTT TILE COMPANY
444 N. Main Street
Ottumwa, Iowa 52555–2773

April 13, 2005

Ms. Sonia Smithfield
General Manager
City Hospital
62 Prairie Road
Fort Madison, IA 52666–1356

Dear Ms. Smithfield:

We will be more than happy to replace the Durafinish tile in front of the elevators and in the lobby area of City Hospital as you requested in your letter of April 6.

When we installed the tile—model 672—in August 2004, we guaranteed the nonfade finish. The tile you selected is imported from Paloma Ceramic Products in Italy and is one of our best-selling tiles. Recently, the manufacturer added a special sealing compound to the tile, making it more durable. This extra hard finish should withstand even the busy traffic in a hospital lobby.

Our sales representative, Mary Atwood, will call on you in the next few days to inspect the tile and make arrangements for replacement at no cost to you. I appreciate your calling this situation to my attention because I always want to know how our products are performing. We guarantee our customers' satisfaction.

Sincerely yours,

Michael Allen

Michael Allen
Product Installation Manager

MA:tk

c: Mary Atwood

MODEL 14-1   A Sample Letter Using Direct Organization (in Semiblock Style)

PRESCOTT TILE COMPANY
444 N. Main Street
Ottumwa, Iowa 52555–2773

April 13, 2005

Ms. Sonia Smithfield
General Manager
City Hospital
62 Prairie Road
Fort Madison, IA 52666–1356

Dear Ms. Smithfield:

You are certainly correct that we guarantee our tile for 20 years after installation. We always stand behind our products when they are used according to the manufacturer's recommendations and the recommendations of our design consultant.

When I received your letter, I immediately got out your sales contract and checked the reports of the design consultant. Our records indicate that the consultant did explain on March 6, 2004, that the Paloma tile—model 672—was not recommended for heavy traffic. Although another tile was suggested, you preferred to order the Paloma tile, and you signed a waiver of guarantee. For your information, I'm enclosing a copy of that page of the contract. Because our recommendation was to use another tile, our usual 20-year guarantee is not in force in this situation.

For your needs, we do recommend the Watermark tile, which is specially sealed to withstand heavy traffic. The Watermark tile is available in a design that would complement the Paloma tile already in place. Our design consultant, Trisha Lyndon, would be happy to visit City Hospital and recommend a floor pattern that could incorporate new Watermark tile without sacrificing the Paloma tile that does not show wear. Enclosed is a brochure showing the Watermark designs. Ms. Lyndon will call you for an appointment this week, and because you are a past customer, we will be happy to schedule rush service for you.

Sincerely,

Michael Allen
Product Installation Manager

MA/dc
Encs.: Watermark brochure
           contract page
c: Ms. Trisha Lyndon

MODEL 14-2   A Sample Letter Using Indirect Organization (in Full-Block Style)

- State good news if there is any:

  The photographs you asked for were shipped this morning under separate cover.

- Assure the reader the situation has been considered carefully:

  When I received your letter, I immediately checked . . .

- Express a sincere compliment:

  Your work at the Curative Workshop for the Handicapped has been outstanding.

- Indicate understanding of the reader's position:

  We understand your concern about the Barnet paint shipment.

- Anticipate a pleasant future:

  The prospects for your new business venture in Center City look excellent.

Do not use negative words in the buffer or remind your reader about the unpleasantness of the situation. Buffers should establish a pleasant tone and a spirit of cooperation, but do not give the reader the impression that the request will be granted or that the main point of the message is good news.

2. The middle section carefully explains the background of the situation, reminding the reader of all the details that are important to the main point.

3. The bad news follows immediately after the explanation and in the same paragraph. Do not emphasize bad news by placing it in a separate paragraph.

4. The closing maintains a pleasant tone and, if appropriate, may suggest alternatives for the reader, resell the value of the product or service, or indicate that the situation can be reconsidered in the future.

The emphasis in an indirect pattern for bad news should be to assure the reader that the negative answer results from careful consideration of the issue and from facts that cannot be altered by the writer. As the writer of a bad news letter, you do not want to sound arbitrary and unreasonable.

In Model 14-2, the opening "buffer" paragraph does confirm the guarantee but mentions restrictions. In the next section, the manager carefully explains the original order and reminds the customer that she did not follow the company's recommendation—thus, the guarantee is not in effect. The manager encloses a copy of the waiver that the customer signed. The refusal sentence comes as a natural result of the events that the manager has already described. The final section represents a movement away from the bad news and suggests a possible solution—ordering different tile. The manager also suggests how some of the original tile can be saved. The final sentences look to the future by promising a call from the design consultant and noting the enclosure of a brochure that illustrates the suggested new tile. In this letter, the manager does not use such phrases as "I deeply regret" or "We are sorry for the inconvenience" because these expressions may imply some fault on the company's part where there is none. Instead, the manager emphasizes the facts and maintains a pleasant tone through his suggestion for a replacement tile and his offer to find a way to save some of the old tile.

Because you cannot give readers good news if there is none, use the indirect pattern of organization to help your readers understand the reasons behind bad news, and emphasize goodwill by suggesting possible alternatives.

Persuasive Organization

Model 14-3 shows a letter written in the persuasive pattern of sales messages. This letter was sent to company trainers who conduct seminars for employees on a variety of topics. The trainers are urged to give up traditional chalkboards and purchase porcelain marker boards for their training sessions.

The persuasive pattern generally includes these elements:

1. The opening in a persuasive letter catches the reader's attention through one of these strategies:
   - A startling or interesting fact:

     Every night, over 2,000 children in our city go to bed hungry.
   - A solution to a problem:

     At last, a health insurance plan that fits your needs!
   - A story:

     Our company was founded 100 years ago when
     Mrs. Clementine Smith began baking . . .
   - An intriguing question:

     Would you like to enjoy a ski weekend at a fabulous resort for only a few dollars?
   - A special product feature:

     Our whirlpool bath has a unique power jet system.
   - A sample:

     The sandpaper you are holding is our latest . . .

2. The middle paragraphs of a persuasive letter build the reader's interest by describing the product, service, or situation. Use these strategies when appropriate:
   - Describe the physical details of the product or service to impress the reader with its usefulness and quality.

SPEAKERS' SUPPLIES, INC.
1642 Ludlow Road
Portland, Oregon 97207-6123

Dear Trainer:

Would you like to eliminate irritating chalk dust and messy, hard-to-read chalkboards from your training sessions?

Let us introduce you to Magna Dry-Erase Boards. These heavy-duty, high-quality marker boards use special dry-erase color markers that glide smoothly over the porcelain-on-steel boards. One wipe of the special eraser, and you have a spotless surface on which to write again. If you prefer to use washable crayons or water-soluble markers, you can erase easily with a damp cloth. In addition, these versatile boards accommodate magnets, so you can easily display visuals, such as maps or charts, without extra equipment cluttering up your presentation area.

The Magna Dry-Erase Boards are suitable for a variety of training situations. Notice these special features:

- sizes from 2 ft × 3 ft to 4 ft × 12 ft
- beige, white, or silver-gray surface
- frames of natural oak or satin-anodized aluminum
- full-length tray to hold markers and eraser
- markers (red, blue, green, black) and eraser included
- hanging hardware included
- special board cleaner included

The enclosed brochure shows the beauty and usefulness of these durable boards—guaranteed for 20 years! Please call 1-800-555-3131 today and talk to one of our sales consultants, who will answer your questions and arrange for a local dealer to contact you. Or call one of the dealers in your area listed on the enclosed directory and ask about Magna Dry-Erase Boards. Get rid of that chalk dust forever!

Sincerely,

Nicole Fontaine

Nicole Fontaine
Sales Manager

NF:bp

P.S. Place an order for two 4 ft × 12 ft boards and get one of any other size at half price.

MODEL 14-3   A Sample Letter Using Persuasive Organization (in Full-Block Style)

- Explain why the reader needs this product or service both from a practical standpoint and from an enjoyment standpoint.
- Explain why the situation is important to the reader.
- Describe the benefits to the reader that will result from this product or service or from handling the situation or problem.

3. After arousing the reader's interest or concern, request action, such as purchasing the product, using the service, or responding favorably to the persuasive request.

4. The closing reminds the reader of the special benefits to be gained from responding as requested and urges action immediately or by a relevant deadline.

In Model 14-3, the first paragraph asks a question that probably will elicit a yes answer from the reader. The next paragraph describes the marker boards, pointing out how convenient and versatile they are. The list of special features highlights items that will appeal to the reader. The closing refers to the enclosed brochure and urges the reader to call a toll-free number and discuss a purchase. The final sentence returns to the topic in the first sentence and promises a solution to the chalk dust problems. Notice that a persuasive sales letter may differ from other letters in these ways:

- It may not be dated. Companies often use the same sales letters for several months, so the date of an individual sales letter is not significant.

- It may be a form letter without a personal salutation. The sales letter for marker boards is addressed to "Dear Trainer," identifying the job of the reader.

- It may include a P.S. In most letters, using a P.S. implies that the writer neglected to organize the information before writing. In persuasive sales letters, however, the P.S. often is used to urge the reader to immediate action or to offer a new incentive for action. The P.S. in the sales letter in Model 14-3 offers a special price on a third marker board if the reader buys two of the largest size.

- It may not include prices if other enclosed materials explain the costs. Sometimes the product cost is one of its exciting features and appears in the opening or in the product description. Usually, however, price is not a particularly strong selling feature. For example, the sales letter in Model 14-3 does not mention prices because they are listed in the enclosed brochure and the writer wants the reader to see the color photographs before learning the cost. Most sales letters, such as those for industrial products, magazine subscriptions, or mail-order items, include supplemental materials that list the prices.

Persuasive memos rarely follow this persuasive sales pattern completely, but they often do begin with an opening designed to arouse reader interest, such as

With a little extra effort from all of us, South Atlantic Realty will have the biggest sales quarter in the history of the company.

WRITING MEMOS AS A MANAGER

A mutually cooperative communication atmosphere is just as important between managers and employees as it is between employees and people outside a company. The tone and managerial attitude in a manager's memos often have a major impact on employee morale. Messages with a harsh, demanding tone that do little more than give orders and disregard the reader's emotional response will produce an atmosphere of distrust and hostility within a company. For this reason, as a manager, you need to remember these principles when writing memos:

1. *Provide adequate information.* Do not assume that everyone in your company has the same knowledge about a subject. Explain procedures fully, and be very specific about details.

2. *Explain the causes of problems or reasons for changes.* Readers want more than a bare-bones announcement. They want to know *why* something is happening, so be sure to include enough explanation to make the situation clear.

3. *Be clear.* Use natural language, and avoid loading a memo with jargon. Employees in different divisions of the same company often do not understand the same jargon. If your memo is going beyond your unit, be sure to fully identify people, equipment, products, and locations.

4. *Be pleasant.* Avoid blunt commands or implications of employee incompetence. A pleasant tone will go a long way toward creating a cooperative environment on the job.

5. *Motivate rather than order.* When writing to subordinates, remember to explain how a change in procedure will benefit them in their work, or discuss how an event will affect department goals. Employees are more likely to cooperate if they understand the expected benefits and implications of a situation.

6. *Ask for feedback.* Be sure to ask the reader for suggestions or responses to your memo. No one person has all the answers; other employees can often make valuable suggestions.

Here is a short memo sent by a supervisor in a testing laboratory. The memo violates nearly every principle of effective communication.

Every Friday afternoon there will be a department cleanup beginning this Friday. All employees must participate. This means cleaning your own area and then cleaning the complete department. Thank you for your cooperation.

The tone of this memo indicates that the supervisor distrusts the employees, and the underlying implication is that the employees may try to get out of doing this new task. The final sentence seems to be an attempt to create a good working relationship, but the overall tone is already so negative that most employees will not respond positively. The final sentence is also a cliché closing because so many writers put it at the end of letters and memos whether it is appropriate or not. Although the memo's purpose is to inform readers about a new policy, the supervisor does not explain the reason for the policy and does not provide adequate instructions. Employees will not know exactly what to do on Friday afternoon. Finally, the memo does not ask for feedback, implying that the employees' opinions are not important. Here is a revision:

As many of you know, we have had some minor accidents recently because the laboratory equipment was left out on the benches overnight and chemicals were not stored in sealed containers. Preventing such accidents is important to all of us, and therefore, a few minutes every Friday afternoon will be set aside for cleanup and storage.

Beginning on Friday, October 14, and every Friday thereafter, we'll take time at 4:00 p.m. to clean equipment, store chemicals, and straighten up the work areas. If we all pitch in and help each other, the department should be in good order within a half-hour. Please let me know if you have any ideas about what needs special attention or how to handle the cleanup.

This memo, in the direct organizational pattern, explains the reason for the new policy, outlines specifically what the cleanup will include, mentions safety as a reader benefit, and concludes by asking for suggestions. The overall tone emphasizes mutual cooperation. This memo is more likely to get a positive response from employees than the first version is. Since a manager's success is closely linked to employee morale and cooperation, it is important to take the time to write memos that will promote a cooperative, tension-free environment.

SELECTING LETTER FORMAT

The two most common letter formats are illustrated in Model 14-1, the semiblock style, and Model 14-2, the full-block style. In the *full-block style,* every line—date, address, salutation, text, close, signature block, and notations—begins at the left margin. In the *semiblock style,* however, the date, close, and signature block start just to the right of the center of the page. Business letters have several conventions in format that most companies follow.

Date Line

Since most company stationery includes an address, or letterhead, the date line consists only of the date of the letter. Place the date two lines below the company letterhead. If you do not use company letterhead, put your address directly above the date:

> 1612 W. Fairway Street
> Dayton, OH 45444-2443
> May 12, 2006

Spell out words such as *street, avenue,* and *boulevard* in addresses.

Inside Address

Place the reader's full name, title, company, and address two to eight lines below the date and flush with the left margin. Spell out the city name, but use the Postal Service two-letter abbreviations for states. Put one space between the state and the ZIP code. The number of lines between the date and the inside address varies so that the letter can be attractively centered on the page.

Salutation

The salutation, or greeting, appears two lines below the inside address and flush with the left margin. In business letters, the salutation is always followed by a colon. Address men as Mr. and women as Ms., unless a woman specifically indicates that she prefers Miss or Mrs. Professional titles, such as Dr., Judge, or Colonel, may be used as well. Here are some strategies to use if you are unsure exactly who your reader is.

Use Titles

When writing to a group or to a particular company position, use descriptive titles in the salutations:

> Dear Members of Com-Action:
>
> Dear Project Director:

Dear Customer:

Dear Contributor:

Use Attention Lines

When writing to a company department, use an attention line with no salutation. Begin the letter two lines below the attention line:

Standard Electric Corporation
Plaza Tower
Oshkosh, WI 54911–2855

Attention: Marketing Department

According to our records for 2006 . . .

Also use an attention line if the reader has not been identified as a man or a woman:

Standard Electric Corporation
Plaza Tower
Oshkosh, WI 54911–2855

Attention: J. Hunter

According to our records for 2006 . . .

Omit Salutations

When writing to a company without directing the letter to a particular person or position, omit the salutation and begin the letter three lines below the inside address:

Standard Electric Corporation
Plaza Tower
Oshkosh, WI 54911–2855

According to our records for 2006 . . .

General salutations, such as "Dear Sir" or "Gentlemen," are not used anymore because they might imply an old-fashioned, sexist attitude.

Use Subject Lines

Some writers prefer to use subject lines in letters to identify the main topic immediately. A subject line may also include specific identification, such as an invoice number, date of previous correspondence, or a shipping code. The subject line may appear in several different places, depending on the preference of the writer or on company style. First, the subject line may appear in the upper right-hand corner, spaced between the date line and the inside address:

September 24, 2006

Subject: Policy #66432–A6

Ms. Victoria Hudson
Marketing Analyst
Mutual Insurance Company
12 Main Street
Watertown, IL 60018–1658

Second, the subject line may appear flush with the left margin, two lines below the salutation and two lines above the first line of the letter:

Dear Ms. Valdez:

Subject: International Expo 2006

As you know, when the first contracts were . . .

Third, some writers center the subject line between the salutation and the first line of the letter:

Dear Ms. Valdez:

Subject: International Expo 2006

As you know, when the first contracts were . . .

The subject line may be underlined or typed in all capitals to help the reader see it quickly and to distinguish it from the first paragraph of the letter. Wherever you place a subject line, keep it brief, and use key terms or specific codes to help the reader easily identify the topic of the letter.

Model 14-4 shows a letter with a subject line. The company is sending a form letter, so there is no inside address. The subject line alerts the reader to the topic.

Body

The body of a letter is typed single-spaced and is double-spaced between paragraphs. Although computers can justify the right margins (end the lines evenly), most people are used to seeing correspondence with an uneven right margin. Justified margins imply a mass printing and mailing, while unjustified margins imply an individual message to a specific person.

WASHINGTON CITY WATERWORKS
Portage Plaza
Washington City, MO 64066
816–555–1678
For Account Inquiries: 816–555–3000

August 25, 2005

Dear Resident:

Subject: New Water Meters

Washington City Waterworks is pleased to announce installation of a new computerized water meter system effective November 1, 2005.

The system requires that new water meters be installed in the basement or outside the home. The free installation will take less than 45 minutes. Your new meter will send your actual reading to a nearby transmitter that will, in turn, send your reading to a central transmitter. That information will go via satellite to a processing unit in Kansas City and then back to our office.

Customers will notice several advantages of the new system:

- actual readings every month instead of estimates
- an "alert" if water usage dramatically increases, thereby catching water leaks promptly
- meter readings as frequently as the customer wishes
- a tamper-proof system to ensure correct readings

We believe this new state-of-the-art system will eliminate human error and eliminate confusion that could result from estimates. In the next two weeks, you will receive a notice of the date of installation. If you have any questions, please call our Customer Service representatives at 816–555–3000.

Sincerely,

Caleb Michaels

Caleb Michaels
Vice President of Operations

CM:sd

MODEL 14-4   A Sample Letter Using a Subject Line (in Semiblock Style)

Close

The close appears two lines below the last sentence of the body and consists of a standard expression of goodwill. In semiblock style, the close appears just to the right of the center of the page (see Model 14-1). In full-block style, the close is at the left margin (see Model 14-2). The most common closing expressions are "Very truly yours," "Sincerely," and "Sincerely yours." As shown in Model 14-1, the first word of the close is capitalized, but the second word is not. The close is always followed by a comma.

Signature Block

The signature block begins four lines below the close and consists of the writer's name with any title directly underneath. The signature appears in the four-line space between the close and the signature block.

Notations

Notations begin two lines below the signature block and flush with the left margin. In Model 14-1, the capital initials "MA" represent the writer, and the lowercase initials "tk" represent the typist. A colon or slash always appears between the two sets of initials.

If materials are enclosed with the letter, indicate this with either the abbreviation *Enc.* or the word *Enclosure.* Some writers show only the number of enclosures as "Encs. (3)"; other writers list the items separately, as shown in Model 14-2.

Second Page

If your letter has a second page, place the name of the addressee, page number, and date across the top of the page:

Ms. Sonia Smithfield   2   April 13, 2006

This heading may also be placed in the upper left-hand corner:

Ms. Sonia Smithfield
page 2
April 13, 2006

Envelope

Post Office scanning equipment can process envelopes most rapidly when the address is typed (1) with a straight left margin, (2) in all capitals, (3) without punctuation, (4) with an extra space between all words or number groups:

```
DR  RONALD  BROWNING
1234  N  SPRINGDALE  LANE
AKRON  OH  44313-1906
```

This format is not appropriate for the inside address of the letter.

SELECTING MEMO FORMAT

Many companies have printed forms so that all internal messages have a consistent format. If there is no printed form, the memo format shown in Model 14-9 often is used.

Subject Line. The subject line should be brief but clearly indicate a specific topic. Use key words so readers can recognize the subject and the memo can be filed easily. Capitalize the main words.

Close. Memos do not have a closing signature block as letters do. Write your initials next to your name in the opening, or write your name at the bottom of the page. The writer's and typist's initials appear two lines below the last line of the memo, followed by enclosure or copy notations.

Second Page. Format the second page of a memo like the second page of a letter with receiver's name, page number, and date at the top of the page.

SELECTING EMAIL FORMAT

Your software will format the mechanical elements of your email messages. You should add elements that create a business message appropriate for your reader and situation.

Subject Line. Put key words in your subject line so the reader can identify the content quickly. When you respond to an email, be sure to check that the subject line is still accurate. Using an old subject line with a new message on a different topic is a common error. If the company or department uses a general email address, put the name of the person you are trying to reach in the subject line.

Salutation.   Begin the message portion of your email with a formal salutation, such as "Dear Professor Jones:." If you normally use the reader's first name, use that in the salutation, such as "Dear Heather:." Some people prefer to drop the "Dear" and begin with the receiver's name only, as in "Professor Jones:" or "Heather:." Notice that all the salutations end in a colon. If you are uncertain, the formal salutation is always a safe choice. The email message in Model 14-5 uses the formal salutation because the writer is contacting a stranger in another workplace.

Subject: Guest Lecture

Date: Thu 17 Sep 2005 11:38:20 AM

From: Dana Kerr <dana9@uig.com>

To: skortze@hot1ph.com

Dear Mr. Kortze:

I am the program director for the annual July sales meeting held by Universal Insurance General. We are currently considering hotels for our 2008 meeting and are interested in the Phoenix Conference Center. We generally expect about 1000 sales representatives and 500 guests at the 4-day conference. I would appreciate receiving information on your hotel and conference facilities as well as menu selections for a banquet. Any materials you have that would help us make a decision about a conference site would be appreciated.

Sincerely,
Dana Kerr
Public Relations
Universal Insurance General
100 Eagle Plaza
Oklahoma City, OK 73112–0387
405.555.2611
Fax 405.555.2688
<dana9@uig.com>

MODEL 14-5   Sample Email Message

Close.   Because email addresses often do not reveal the company name or other details, use a full signature block for your closing, including (1) your full name, (2) your title or department, (3) the company name, (4) street address, (5) telephone number, and (6) fax number or email address. Model 14-5 shows a closing that gives the receiver useful information.

Email is so easy to produce that writers sometimes forget that it represents the company and the professionalism of the writer. Before sending email, check the following:

- Always review your spelling and grammar.

- Avoid using emoticons, such as :), or abbreviations, such as IMHO, because readers may not understand them and they detract from your professionalism.

- Do not erase the incoming message and send back a yes or no answer. The receiver may not remember the details of the original topic. Paraphrase the main points, or keep the original message as part of your answer.

- Avoid changing font sizes or adding features, such as boldface or italics. Not all email systems can read these, and your message may be garbled at the recipient's end.

USING THE FAX

Most companies now have fax (facsimile) machines to receive and transmit copies of correspondence and graphics to readers inside and outside the organization. The fax uses telephone lines and, therefore, arrives much faster than correspondence sent through the mail or even through interoffice mail in large institutions, such as a university.

Standard cover sheets usually are available to use when you send a fax. Be sure to include your company name, address, telephone and fax numbers, as well as your name, department, and telephone extension. Clearly write the name and title of the person receiving the correspondence along with the number of pages you are sending. Because the quality of fax machines varies and lines may experience interference, many people who fax correspondence in the interest of speed also send a printed copy through the mails so the reader receives a high-quality document.

Be aware that using the fax machine prevents confidentiality. You rarely can be certain who is standing at the receiving machine when you send a letter or memo. Faxed correspondence arrives unsealed as loose sheets. Marking the cover sheet confidential will not necessarily create any privacy. Fax machines usually are in high-traffic areas, and many people have access to them. If your correspondence is confidential or potentially damaging to someone's interests, do not send a fax, or do call the receiver and ask permission to send a confidential fax.

CHAPTER SUMMARY

This chapter discusses how to write effective letters, memos, and email. Remember:

- Letters are written primarily to people outside an organization, and memos are written primarily to people inside an organization.

- Effective tone in letters, memos, and email requires natural language, positive language, and the you-attitude.

- Correspondence can be organized in the direct, indirect, or persuasive pattern.

- Managers need to create a cooperative atmosphere in their memos to employees.

- The fax machine transmits messages rapidly, but is not confidential.

SUPPLEMENTAL READINGS IN PART 2

Frazee, V. "Establishing Relations in Germany," *Global Workforce,* p. 492.

Nielsen, J. "Ten Steps for Cleaning Up Information Pollution," *Alertbox,* p. 520.

ENDNOTES

1. Brigid Schulte, "Pilot's Computer Note Reveals Military Secrets," *Akron Beacon Journal* (July 12, 1995): A2.

2. Stephanie Stahl, "Dangerous E-Mail," *Informationweek* (September 12, 1994): 12.

3. Anick Jesdanum, "Employers Watching Workers' Web Use," *Akron Beacon Journal* (July 10, 2001): C16.

4. Ken Auletta, "Hard Core," *The New Yorker* (August 16, 1999): 42–69.

5. "Is the Future of E-Mail under Cyperattack?" (June 14, 2004). *USA Today Web site.* Retrieved June 15, 2004, from http://www.usatoday.com/tech/news/2004-06-14-email_x.htm.

6. Ibid.

MODEL 14-6   Commentary

This model shows a claim letter for defective screw tubes in a recent shipment. The writer addresses the reader by his first name because of the long business relationship they have had. Notice that the writer clearly states the order date and number in his opening sentence.

Discussion

1. Identify the organizational pattern used in this letter, and discuss why the writer would use this organization.

2. Why does the writer describe the problem in detail? Why not just state that the screw tubes are defective?

3. Discuss the tone in the letter. Does the writer seem to be angry? How effective do you think the closing paragraph will be in getting action?

4. Assume you are William Ruggles, and write a response to this claim. You will ship a new batch of screw tubes as soon as you finish checking production quality, and you are glad he sent the two samples. Your 15-year business relationship with this company is important, and you want to do everything possible to keep your customers happy. You estimate that it will take two days to match the new screw tubes against the specifications Brett Howard sent.

MARLEY ASSEMBLIES, INC.
1600 Spike Road
Mechanicsburg, PA 17066-4260
(412) 555-1313   FAX (412) 555-1300

March 15, 2005

Mr. William Ruggles
Production Manager
Acme Tubing, Inc.
750 Mull Road
Toledo, OH 43607–1389

Dear Bill:

I appreciate your promptness in getting our March 1, 2005, order
#A9968 to us within five days. The screw tubes #0867 were your usual
high quality, but screw tubes #0183 do not fit our assemblies properly.

I am enclosing two sample screw tubes #0183 from the order and a copy
of our specifications and line drawings. The problem with the screw
tubes is that the 20° bend is really 25°-27°, and the slot is not full length
as indicated on the specs. You will also notice that the two samples do
not measure the same width on the spring. Neither width is an exact fit
for our pans, which require the 3/16-in. width marked on the drawing.

I will keep the defective screws until you tell me what you want done
with them. In the meantime, we want a new shipment of screw tubes
#0183 after you have inspected them for conformity to our
specifications. I understand that these variations happen, and I know
you will want to take care of replacements immediately.

Sincerely,

Brett Howard

Brett Howard
Product Supervisor

BH:sd

Enc.

MODEL 14-6   Claim Letter—Marley Assemblies

MODEL 14-7   Commentary

This model shows a form letter sent to a specific group of customers at a large bank. Although envelopes will have individual addresses, the letter does not have an inside address. The writer is announcing a change in services that some customers will be disappointed to learn.

Discussion

1. Identify the specific organization pattern used in this letter, and discuss why the writer chose that pattern.

2. Discuss the tone of the letter and how the writer emphasizes the bank's desire to continue the customer relationship.

3. How does the writer imply that the change is not very significant?

4. In groups, draft a letter to all students at your school, and announce that there will be a $150 increase in general fees next semester to cover technology costs in the campus computer centers.

5. In groups, draft a memo to the 60 employees at Wilson Electronics Services, and explain that the company no longer can pay the full cost of the parking spaces in the Acme Parking Deck reserved for Wilson employees. Beginning next month, employees who wish to continue to park in the deck will have to pay half the regular rate ($145/month) charged by the parking deck management.

SOUTHEAST FIRST BANK AND TRUST COMPANY
One Atlantic Center
Coconut Beach, Florida 33578

Corporate Center 1–941–555–8989
Account Information 1–941–555–1133

November 12, 2005

Dear Southeast Account Customer:

Southeast First Bank and Trust appreciates your confidence in our
security and services over the years. We remain committed to providing
superior banking services that match your needs in today's ever-
changing financial conditions.

Many of our customers have shown an increasing interest in our
investment banking services, which include a full range of stocks and
mutual funds. As a member of our Select Banking Group, you may
already be using these services. Our Olympic Funds offer a selection of
investments that are especially designed for a retirement portfolio.
Beginning January 1, 2006, we will offer an expanded selection of bond
funds to allow you to further diversify your retirement financial
planning. Because of this expansion, we will no longer offer the Select
Banking Financial Management service after December 31, 2005. All
funds in that account will be transferred automatically to your checking
account unless you direct us to transfer them elsewhere.

Please call your personal Select Banker Representative at
1–941–555–1133 to discuss your options for transferring your Select
Banking Financial Management account. We appreciate the opportunity
to serve your financial needs now and in the future.

Sincerely,

LaTasha Williams

LaTasha Williams
Vice President—Banking Services

LW:sd

MODEL 14-7   Bad News Letter—Southeast First Bank and Trust Company

MODEL 14-8 Commentary

This email was sent to prospective customers of a telescope manufacturer. The mailing list was composed of people who had signed up to receive company announcements of new products. This email is a persuasive message designed to sell a new model of binoculars. Even if the email recipient does not want to buy the binoculars right away, the company hopes that the recipient will revisit the company's Web site to find more information.

Discussion

1. Analyze the persuasive appeals used in this email. How effective are these appeals for a reader who has visited the Web site only once? How effective are the appeals for a regular customer?

2. Discuss the subject line. How effective would the subject line be in attracting a casual reader?

3. Discuss the length of this persuasive email, and decide whether it is too short, too long, or just about right.

4. In groups, draft a persuasive email message to high school seniors telling them to consider coming to your school after graduation. Compare your draft with those of other groups.

5. Visit the Web site for one of the professional organizations listed in Appendix C, and gather membership information. Then, draft a persuasive email for appropriate potential members, and persuade them to join the organization. Compare your draft with those of your classmates.

Subject: Big Bargain Binoculars!

Date: Wednesday, October 15, 2005 10:42 AM

From: Star Monitoring

To: presley@uttw.edu

Beginning astronomers will be thrilled with these new high-magnification binoculars from Star Monitoring—the Deep Sky Angle Binoculars 678. Experienced sky observers will also want to test these large powerful binoculars.

The Deep Sky Angle Binoculars 678 will be a welcome addition to your stargazing equipment. The optics are fully multicoated on all air-to-glass surfaces to improve light transmission, and the binoculars feature interchangeable overhead eyepieces offering 26× and 32×. The standard 20× and optional 26× eyepieces have 52° apparent fields, and the 32× eyepieces have 66° apparent fields.

Star Monitoring is offering this high-test instrument for an astonishingly low price of $749.99. Click on www.starmonitor.com/678 to find more information about this remarkable bargain in binoculars. Or, email me directly to find out more and place an order.

Logan Canfield
Sales Director
logan @starmonitor.com

MODEL 14-8   Persuasive Email—Star Monitoring

MODEL 14-9   Commentary

This memo is from a supervisor to a division head, asking for maintenance support.

Discussion

1. Identify the organizational strategies the writer uses, and discuss why the writer probably chose these strategies.

2. Consider the tone of this memo. How does the writer place blame for the situation? How does the writer convey the seriousness of the problems?

3. In what ways does the writer show the reader that there are benefits in solving these problems? Is there any advantage to addressing the reader by name in the memo?

4. Write a brief memo to D. P. Paget from T. L. Coles announcing that R. Fleming, the maintenance supervisor, has been told to report on the problems within two days. Assure Paget that the problems are being taken seriously.

October 3, 2005

To: T. L. Coles, Division Chief

From: D. P. Paget, Supervisor   *DPP*

Subject: Problems in Cost Center 22, Paint Line

As I've discussed with Bill Martling, our department has had problems with proper maintenance for six weeks. Although the technicians in the Maintenance Department respond promptly to our calls, they do not seem to be able to solve the problems. As a result, our maintenance calls focus on the same three problems week after week. I've listed below the areas we need to deal with.

Temperature.   For the past two weeks, our heat control has not consistently reached the correct temperature for the different mixes. The Maintenance technician made six visits in an attempt to provide consistent temperature ranges. R. Fleming investigated and decided that the gas mixture was not operating properly. He did not indicate when we might get complete repairs. In the last ten working days, we've experienced nearly five hours of total downtime.

Soap.   The soap material we currently use in the waterfall booth does not break up the paint mixtures and, as a result, the cleanup crew cannot remove all the excess paint. Without a complete cleanup every twenty-four hours, the booth malfunctions. Again, we have downtime. We need to investigate a different soap composition for the waterfall booth.

Parking Lever.   The parking lever control has not been in full operation since January 14. Technicians manage to fix the lever for only short periods before it breaks down again. I asked for a replacement, but was told it wasn't malfunctioning enough to warrant being replaced. This lever is crucial to the operation because, when it isn't functioning, the specified range slips and paint literally goes down the drain. As a result, we use more paint than necessary and still some parts are not coated correctly.

I'd appreciate your help in getting these problems handled, Tom, because our cost overruns are threatening to become serious. Since all maintenance operations come under your jurisdiction, perhaps you could consider options and let me know what we can do. I'd be glad to discuss the problems with you in more detail.

MODEL 14-9   Memo

Chapter 14   Exercises

1. You are assistant to the communications director of your state's tourist bureau. Your office prepares all the printed brochures about various state attractions, such as state parks, historical sites, and recreational activities (e.g., fishing, boating, skiing). These brochures are directed at tourists and other visitors. Suzette Calvet writes to your office from Paris. She is planning a month-long visit to the United States and wants information about visiting your state. Select one location or activity in your state, and write a letter to Ms. Calvet that both provides information and persuades her to visit your area during her trip. Tell her you are sending brochures under separate cover. *Or,* if your instructor prefers, develop a one-page flyer or a brochure on the location or activity and write a letter to accompany it.

2. Yesterday, a female employee was attacked in the lobby of your company. Fortunately, two other employees chased the attacker out the door and down the street. Police officers picked him up in a few minutes. The female employee suffered only a few minor cuts, but the incident illustrates the dangers of keeping the company's four outside doors both unlocked and unattended. As Vice President of Operations, you decide that only the front door will be unlocked during the day. The two side doors and the back door will be locked all day to prevent outsiders from casually entering the offices. Employees will be able to leave by those doors, but they will have to go to the front door to reenter. A security guard will be stationed at the front door during office hours to monitor visitors and accept deliveries. Write a memo to all employees telling them of the new plan, which will start next Monday. Impress readers with the seriousness of the danger, but do not start a panic.

3. Select a "real-world" letter, memo, or email of at least one page. Make enough copies for your classmates. Present a critique of the document to the class, and point out effective and ineffective elements in the message. In addition, present a revision of the document. The revisions should reflect your ideas for improvement. Do not change the original situation in the document, but you may add details that increase the effectiveness of the message.

4. You are the president of your professional organization, Technical Editors International. You have decided to invite Dr. Stacy McDonald to speak at the group's next meeting—the annual banquet. Dr. McDonald is a well-known international scholar in ancient manuscripts, and she has a new book titled *Digging for Parchment: Old Scrolls and New Information*. This book contains information about ancient technical documents, and you think the group will be interested in her stories. You could arrange to have her book on sale the day of her speech if she would be willing to autograph the copies. You could also try to arrange interviews on local TV stations and radio shows.

Write Dr. McDonald, and invite her to speak at your annual banquet in two months.

5. Assume you are Dr. Stacy McDonald and have just received the invitation in Exercise 4. You would be glad to speak, but you have a few concerns. First, you are dreadfully allergic to nuts, and to ensure that you do not have a reaction, the dinner served cannot contain nuts in any dishes. Second, you live 40 miles from the banquet location, and you want a car and driver to take you to and from the banquet. Third, you do not mind posing for publicity photographs, but you do not permit taping your speeches. Also, you will not give interviews after your speech. Write a letter accepting the invitation and including your requirements. Ask if the group would prefer a scholarly speech on ancient manuscripts or a speech about the lighter side of hunting for manuscripts.

COLLABORATIVE EXERCISES

6. In groups, assume that you work for your state's Environmental Quality Office. Your task is to draft a letter to all landowners in the rural areas of the state. The letter should remind them that dirt and debris dams across streams are prohibited except in extreme circumstances. In rare instances where such dams might be needed, a permit must be obtained from the Environmental Quality Office. Building a dirt or debris dam violates Part 69 of the Inland Rivers and Streams Protection Act of 1997, Sections 6.294–6.298, annotated. The dams usually break down during rain storms and cause flooding at lower levels. The landowners need to keep streams on their property clear of dirt and debris dams. Draft a letter, and compare your results with those of other groups.

7. In groups, write a letter to the students at your school informing them of some changes. Several new buildings are under construction. A new recreation center and a new student services building will be under construction when the next term begins. These buildings will be important additions to campus. While construction is underway (about nine months), access to another important building will be severely restricted. You choose the building to which access will be restricted, and add relevant details as you write the letter.

INTERNET EXERCISES

8. Your supervisor has decided to get flat-screen monitors for the 15 clerks in the records department. Search the Internet for product and price information for flat-screen monitors. Write a memo to your supervisor describing the top three choices available and recommending one of them.

9. Your company, Tip-Top Foods, has had to recall a batch of peanut butter because ground-up peanut shells were found in several jars. The decision has been made to recall all the peanut butter produced over a three-day period to

be sure that no shells are in any batch produced during that time. Relevant batch numbers are PX-42-69, PX-43-70, and PX-44-71. The batch number appears on the bottom left of the front label. Customers can return the peanut butter to their grocery stores for a refund. The shells were ground so finely that the company does not believe there is any health hazard. Write a letter to consumers to be posted on the company's Web site that explains the situation and retains customer goodwill.

10. As the vice president of National Trust Bank, you have a problem. An employee (now arrested) sold the bank's list of customer email addresses to spam senders. These customers probably already have had an increase in spam. Fortunately, no financial information or home addresses were involved; only the customer's email address has been sold. Write an email that will go to all the customers on the bank's email list to explain the situation. Assure the customers that their financial accounts are safe. The bank is instituting high-level security measures to make sure this situation will not happen again.

CHAPTER 15

Career Communication

LOOKING FOR A JOB

Beginning a job search means that you actually have taken on the job of finding a job. Maintain a professional approach to all aspects of job hunting. Before applying for a job, follow these guidelines:

- Research the companies that employ people in your field. Find out about the job market and the kinds of skills employers are looking for. Most company Web sites list current job openings with details about the job requirements and instructions for applying. Libraries have business directories and guides. Check the *Wall Street Journal* for information about business in your field and companies you are interested in.

- Assess your own training, education, and experience. Consider how well you match the job requirements being requested.

- Analyze your interests and the skills you enjoy using. These elements will be the most important as you plan your career.

- Study positions advertised online or in the newspapers to determine the training that employers in your field are looking for. If you need more training in a particular skill, plan ways to get it as soon as possible.

- Because so many companies are international, do some research on the global expansion of companies in your field. Consider whether you should learn another language or increase your fluency in a second language.

- Attend career expos and workshops. These provide information and networking opportunities.

- Check all possible sources for jobs—newspapers, company Web sites, university placement services, job boards on the Internet, professional association Web sites, friends, professors, and college job fairs.

- Establish a way for prospective employers to reach you. Get an answering machine if you do not have one. If possible, make sure you are the only person to collect the messages. Remove any music or jokes you have on an answering machine, and record a business-like message for callers.

WRITING RÉSUMÉS

A serious job seeker today needs at least three styles of résumé—a traditional résumé with attractive formatting, a scannable résumé that employers can add to their databases, and an email résumé that can be sent over the Internet. Although the styles of these vary, they include most of the same information, and they all have the same purpose: to attract the attention of an employer and secure an interview for the job applicant. No matter which style you use to apply for a job, always bring a traditional résumé to an interview.

Traditional Résumés

A traditional résumé is printed on good white bond paper in black ink. The writer uses formatting elements (e.g., boldface, bulleted lists, underlining) that help to create an attractive, easy-to-read page. The usual recommendation for a new college graduate is to keep the traditional résumé to one page in length. A recent report indicates that a two-page résumé is equally acceptable and even preferable for those who have outstanding qualifications.[1]

Model 15-2 shows a traditional résumé for a new college graduate. Remember that the reader wants to find specific information quickly, so headings should stand out clearly. Because Model 15-2 is for a new graduate, education has a prominent position. There are many résumé styles, and this one is the basic chronological format. Include the following kinds of information in a traditional résumé:

Heading.   Put your name, address, and telephone number at the top of your résumé. You may include a business telephone number and email address. Appearances are important. If your email address is clever or funny, such as rockon@yahoo.com or amazeme@earthlink.net, get a professional email address for your job search. Email addresses that do not look professional may be automatically deleted as spam.

Some people include the address of a personal Web site, but this may not be a good idea. A potential employer might not be impressed by personal photos, diary-style comments, songs, or humorous cartoons. Consider the purpose of your Web site. If the Web site emphasizes professional work, such as graphic design, then including the link might be advantageous in your job search. If the Web site is purely personal, however, omit the link from your résumé.

Objective.   List a specific position that matches your education and experience, because employers want to see a clear, practical objective. Avoid vague descriptions, such as "I am looking for a challenging position where I can use my skills." Availability to travel is important in some companies, so the phrase "willing to travel" can be used under the objective. If you are not willing to travel, say nothing.

Education.   List education in reverse chronological order, your most recent degree first. Once you have a college degree, you can omit high school. Be sure to list any special certificates or short-term training done in addition to college work. Include courses or skills that are especially important to the type of position for which you are applying. List your grade-point average if it is significantly high, and indicate the grade-point scale. Some people list their grade-point average in their majors only, since this is likely to be higher than the overall average.

Work Experience.   List your past jobs in reverse chronological order. Include the job title, the name of the company, the city, and the state. Describe your responsibilities for each job, particularly those that provided practical experience connected with your career goals. In describing responsibilities, use action words, such as *coordinated, directed, prepared, supervised,* and *developed.* Dates of employment need not include month and day. Terms such as *vacation* or *summer* with the relevant years are sufficient.

Honors and Awards.   List scholarships, prizes, and awards received in college. Include any community honors or professional prizes as well. If there is only one honor, list it under "Activities."

Activities.   In this section list recent activities, primarily those in college. Be sure to indicate any leadership positions, such as president or chairperson of a group. Hobbies, if included, should indicate both group and individual interests.

References.   Model 15-2 shows the most common way to mention references. If you do list references, include the person's business address and telephone number. Be sure to ask permission before listing someone as a reference.

It is best to omit personal information about age, height, weight, marital status, and religion. Employers are not allowed to consider such information in the employment process, and most prefer that it not appear on a résumé.

Model 15-5 shows a résumé for a job applicant who is more advanced in his chemistry career than the applicant in Model 15-2. This résumé lists experience first, in the most prominent position. Other significant differences from the résumé for the new college graduate include the following:

Major Qualifications.   Instead of an objective, this applicant wants to highlight his experience, so he uses the *capsule résumé* technique to call attention to his years of experience and his specialty. He can discuss his specific career goals in his application letter.

Professional Experience.   This applicant has had more than one position with the same employer because he has been promoted within the company. The applicant highlights his experience by stacking his responsibilities into impressive-looking lists.

Education.   This applicant includes his date of graduation but does not emphasize it, and he omits his grade-point average because his experience is now

more important than his college work. He lists his computer knowledge because such knowledge is useful in his work.

Activities and Memberships.   After several years of full-time work, most people
no longer list college activities. Instead, they stress community and professional
service. This applicant lists his professional membership in the American
Chemical Society to indicate that he is keeping current in his field.

Model 15-6 shows another résumé style. This job applicant is also advanced
in his career, and he wants to emphasize the skills he has developed on the job
as well as highlight his successes. His actual dates of employment or of college
graduation are less important than the types of projects he has worked on.
This résumé style is used most by experienced job applicants who want to
change the focus of their careers or who have been out of the job market for a
while and are now reentering it. New college graduates with activities and
part-time work experience that developed skills relative to their career goals
may also find this style useful. In addition, the job applicant using this résumé
style may revise the lists of accomplishments and skills to emphasize different
areas for different positions. This résumé does list employment and education
but it differs from Model 15-2 and Model 15-5 in two important areas:

- *Accomplishments*—Rather than beginning with his work record, the
  applicant lists significant achievements and honors to draw the reader's
  attention to a pattern of success. Many people are in sales. The applicant
  here emphasizes his out-of-the-ordinary performance.

- *Special skills*—The applicant identifies the types of skills he has developed and lists them under a specific heading. In this case, the applicant
  is interested in a management position, and he knows that communication and organization skills are crucial for effective managers. Other
  types of special skills might include researching, public speaking, writing, counseling, analyzing, managing, or designing. This applicant's
  record of employment supports his list of accomplishments and skills.
  Do not list skills so vague you cannot provide specific details to support your assertion. Saying that you are "good with people" without
  any specific accomplishments to demonstrate your claim will not
  strengthen your résumé.

Scannable Résumés

Many large companies use computers to scan and store résumé information.
When employers want to interview prospective employees for a position, they
screen every résumé in their electronic files for key qualifications. For employers, the advantages of electronic screening are that it (1) eliminates the difficulty of handling thousands of résumés in a fair manner, (2) quickly identifies

candidates who have the primary job qualifications, and (3) processes more applications at less cost.

Electronic screening requires a special résumé style. After writing your traditional résumé, rewrite it in appropriate electronic style so you have both available. If you are applying to a company that uses electronic screening, you can inquire about a preferred format. If you cannot easily find out what the company prefers, send both a traditional and a scannable résumé, and state in your application letter that you are sending both as a convenience to the employer. Following are guidelines for developing a scannable résumé.

Name and Address.   Your name should be the first line of your résumé. Put your address, telephone numbers, and email address in a stacked list.

Key Words.   Nouns are more important in scannable résumés than action verbs because computers are set to look for key skills.

Immediately after your opening information, list key words that identify such items as your previous job titles, job-related experience, skills, special knowledge, degrees, certifications, and colleges attended. Be sure to use the terms listed in a job advertisement because these are the terms the scanner is seeking. For example, if the advertisement asks for "public speaking," use that term rather than "oral presentations." Some companies scan only a portion of a résumé, so be sure that your key words appear near the top of the page.

Single Columns.   Scanners read across every line from left to right. Do not use double columns to list information because it will look like gibberish in the computer files.

Paper and Typeface.   Scanners prefer clean and simple résumés. Use only smooth white paper with black ink and 12- or 14-point type. Do not use italics, underlining, boxes, shading, or unusual typeface. Capitals and boldface are acceptable as long as the letters do not touch each other.

Abbreviations.   Use abbreviations sparingly. Scanners can miss information if they are programmed to search for a whole word.

Folding and Stapling.   Do not staple or fold a scannable résumé. The scanner may have trouble finding words that are in the creases or that cover staple holes.

White Space.   Organize with lots of white space, so the scanner does not overlook key words. Use wide margins (60–65 characters per line) to avoid having text edges cut off if the recipient's equipment does not read the full line.

Model 15-3 shows the electronic version of the résumé in Model 15-2. Notice that some information, such as activities and references, is omitted.

The purpose of the application letter and résumé is to present an interest-attracting package that will result in an interview. Because the letter and résumé often are the first contact a new college graduate has with a potential employer, the initial impression from these documents may have a decisive impact on career opportunities.

Email Résumés

Many employers are now requesting that job seekers send in résumés via email. This method presents certain technical concerns because of the variety of word processing programs and hardware. Do not send your email résumé as an attachment. To avoid viruses, many companies set their email systems to reject attachments, and most managers are reluctant to open attachments from strangers. Use the information from your traditional résumé to prepare an electronic résumé.

Consider these guidelines:

- Do not use bullets, boldface, boxes, underlining, or italics.

- Prepare the résumé in ASCII (American Standard Code for Information Interchange), a text that includes no formatting elements but can be easily read by all computers. Save your résumé in "text only" or "Rich Text Format" (RTF).

- Use a single-column, left-justified format.

- Use the space bar instead of the tab.

- Use asterisks instead of bullets.

- Use a hard carriage return instead of the word wrap feature in your processing program.

As a test, send your email résumé to yourself to check readability and correctness.

Model 15-4 shows an email version of the traditional résumé in Model 15-2. Notice that the awards connected to Kimberly's education now appear with the education information. Because a potential employer might read the résumé online and have to scroll down, Kimberly wants all her educational information in one place.

Web-Based Résumés

Companies today are using both their own Web sites and Internet job sites to create a pool of applicants for a position. Web sites, such as www.careerbuilders.com and www.monster.com, offer convenient places for job seekers to look for jobs and to post their own résumés for potential employers.

If you decide to post your résumé with some of the Internet sites, check the costs carefully. Some charge a fee for posting a résumé for a specific length of time. Follow the directions on the Web site. You may be able to use your email version and just paste it in, or you may have to paste in sections under a prescribed format on the site. Eventually, online résumé postings will probably become more uniform so that employers can easily compare applicants from different sites.

Update your résumé at least every other month and add/change the keywords, so recruiters take another look at your posting.

STRATEGIES—Job Interviews

Preparation is essential for a successful job interview. Plan how to answer typical interview questions, such as the following:

- Why did you choose your major?
- What was your favorite course in college? Why?
- What are your greatest strengths and weaknesses?
- Tell me about yourself.
- Where do you see your career in five years?

You also should plan to ask questions about the company and the job, such as the following:

- Does this position have promotion possibilities?
- Does the company have plans to expand?
- Does the company support professional growth, such as attendance at workshops and conferences?

WRITING JOB-APPLICATION LETTERS

One of the most important letters you will write is a letter applying for a job. The application letter functions not only as a request for an interview but also as the first demonstration to a potential employer of your communication skills. An application letter should do three things:

1. *Identify what you want.* In the opening paragraph, identify specifically the position you are applying for and how you heard about

it—through an advertisement or someone you know. In some cases, you may write an application letter without knowing if the company has an opening. Use the first paragraph to state what kind of position you are qualified for and why you are interested in that particular company.

2. *Explain why you are qualified for the position.*   Do not repeat your résumé line for line, but do summarize your work experience or education and point out the specific items especially relevant to the position you are applying for. In discussing your education, mention significant courses or special projects that have enhanced your preparation for the position you want. If you have extracurricular activities that show leadership qualities or are related to your education, mention these in this section. Explain how your work experience is related to the position for which you are applying.

3. *Ask for an interview.*   Offer to come for an interview at the employer's convenience; however, you also may suggest a time that may be suitable. Tell the reader how to reach you easily by giving a telephone number or specifying the time of the day you are available.

In Model 15-1, the writer begins her application letter by explaining how she heard about the position, and she identifies her connection with the person who told her to apply for the job. In discussing her education, she points out her training in computers and mentions her laboratory duties that gave her relevant experience in product chemistry. Her work experience has been in research laboratories, so she explains the kind of testing she has done. In closing, she suggests a convenient time for an interview and includes a phone number for certain daytime hours. Overall, the letter emphasizes the writer's qualifications in chemistry and points out that the company could use her experience (reader benefits). An application letter should be more than a brief cover letter accompanying a résumé; it should be a fully developed message that provides enough information to help the reader make a decision about offering an interview.

WRITING OTHER JOB-RELATED LETTERS

Articles in Part 2 give advice about preparing for the job interview. The first interview, however, may be only one step in the job application process. Many employers conduct several interviews with the top candidates before filling a position. The first interview may be brief, perhaps a half-hour, allowing the interviewer to eliminate some candidates. Longer interviews then follow with the remaining candidates until the company offers the position to someone. If, after your first interview, you are very interested in the position, write a brief follow-up letter to remind the interviewer who you are and what special qualifications you have for the position. Because most job applicants do not take the time to write follow-up letters, you can make a stronger impression on the interviewer by doing so. Email, however, is not appropriate for a follow-up letter. Taking the time to write a letter and mail it shows that you not only are interested in the position but also sophisticated enough to make the extra effort to show appreciation for the interview.

Write your follow-up letter within three to four days after your interview. The letter should be short but include the following:

1. Express appreciation for the opportunity to discuss the position.

2. Refer specifically to the date of the interview and the position you applied for.

3. Mention something that happened during the interview (e.g., a tour of the plant) or refer to a subject you discussed (e.g., a new product being developed).

4. Remind the interviewer about your qualifications or mention something you forgot to bring up in the interview.

5. Express your continuing interest in the company.

Here is a sample follow-up letter that covers these points:

> I appreciated the opportunity to talk to you on September 3 about the position in cost accounting. During my tour of your offices, Ms. McNamara gave me a thorough introduction to the accounting operation at HG Logan Associates, and I remain very interested in your company.
>
> After hearing the details about the position, I believe my course work in cost accounting computer systems would be an asset in your organization.
>
> Thank you for your consideration of my application, and I look forward to hearing from you soon.

If you have another interview, write another follow-up letter to express your continuing interest.

If you receive a job offer and want to accept it, do so promptly and in writing. Use the letter to establish your eagerness to begin work and to confirm the details mentioned in the company's offer. Here is a sample acceptance letter:

> I am pleased to accept your offer of a position as financial analyst at Brady, Horton and Smythe at a salary of $42,000. As you suggested, I will stop in the Human Resources Office this week to get a packet of benefit enrollment forms.

I am eager to join the people I met during my visit to the Investment Sales Department and look forward to beginning work on June 15.

If you receive a job offer but decide not to accept it, write promptly and decline. Express your gratitude for the offer. You may or may not give a specific reason for your decision. Here is a sample letter declining a job offer:

Thank you for your offer of a position in the engineering division at Hughes Manufacturing. I was very impressed with the dynamic people I met during my visit, and I carefully considered your offer. Because of my interest in relocating to the Southwest, however, I have decided to accept a position with a company in New Mexico.

I appreciate your time and consideration.

You may receive a job offer but be uncertain about whether you want to accept it because you have other interviews pending and you want to complete those before making a decision. If so, write a letter asking for more time to respond to the company's offer. Express your interest in the position, and explain that you need a short extension of the deadline to be sure of your decision. Here is a sample letter asking for a delay in responding to a job offer:

I appreciate your offer of a nursing position in your clinic. Your facilities and staff made a strong impression on me, and I know that working at Family Health Care would be an exciting challenge.

Because I do have two interviews scheduled at hospitals in the city, I would appreciate your giving me until March 1 to respond to your offer. The extra ten days will allow me the opportunity to evaluate my career options and make the decision that is right for me and for my employer.

If you need my answer immediately, I, of course, will give it. If you can allow me the extra time to make a decision, I would be grateful. Please let me know.

All these letters may be part of your job-search process. Writing them carefully shows the company that you take the position seriously and are professional in your approach to your career.

CHAPTER SUMMARY

This chapter discusses how to write résumés and career correspondence. Remember:

- Résumés list a writer's most significant achievements relative to a specific position.

- Job seekers need three types of résumés: traditional, scannable, and email.

- A job-application letter presents a writer's qualifications for a position and requests an interview.

- A follow-up letter after a job interview will remind the interviewer about your qualifications.

SUPPLEMENTAL READINGS IN PART 2

Hagevik, S. "Behavioral Interviewing: Write a Story, Tell a Story," *Journal of Environmental Health,* p. 498.

Humphries, A. C. "Business Manners," *Business & Economic Review,* p. 505.

Messmer, M. "Prepare, Listen, Follow Up," *Management Accounting,* p. 517.

"ResumeMaker's 25 Tips—Interviewing," *ResumeMaker Web site,* p. 526.

Smith, G. M. "Eleven Commandments for Business Meeting Etiquette," *Intercom,* p. 533.

ENDNOTES

1. Elizabeth Blackburn-Brochman and Kelly Belanger, "One Page or Two?: A National Study of CPA Recruiters' Preferences for Résumé Length," *The Journal of Business Communication* 38.1 (January 2001): 29–57.

MODEL 15-1 through 15-7   Commentary

The following models demonstrate the job application letter and résumé styles:

- Model 15-1 is a job-application letter from a recent college graduate. The writer reviews her qualifications and emphasizes her strengths.

- Model 15-2 is a traditional chronological résumé to accompany the application letter.

- Model 15-3 is a scannable résumé for the same job applicant.

- Model 15-4 is an email résumé for the same job applicant.

- Model 15-5 is the traditional résumé of a job applicant who has significant professional experience. Notice he has had several positions with one employer and the format stresses his experience rather than his education.

- Model 15-6 is a functional-style, traditional résumé that emphasizes the job applicant's special accomplishments and skills.

- Model 15-7 is a traditional résumé of a job applicant who interrupted her education to marry and have children. She uses the functional style to emphasize the experience she gained from her jobs and her education. She is now working full-time and taking classes part-time. Notice that she lists her degree as expected at a future date and uses the term "general studies" to describe her one year of college work in Texas.

Discussion

1. Discuss the persuasive appeal the writer uses in her application letter in Model 15-1.

2. In groups, write scannable résumés for Models 15-5, 15-6, and 15-7.

3. In groups, write email résumés for Models 15-5, 15-6, and 15-7.

1766 Wildwood Drive
Chicago, Illinois 60666
July 12, 2005

Mr. Eric Blackmore
Senior Vice President
Alden-Chandler Industries, Inc.
72 Plaza Drive
Milwaukee, WI 53211–2901

Dear Mr. Blackmore:

Professor Julia Hedwig suggested that I write to you about an opening
for a product chemist in your chemical division. Professor Hedwig was
my senior adviser this past year.

I have just completed my B.S. in chemistry at Midwest University with a
3.9 GPA in my major. In addition to chemistry courses, I took three
courses in computer applications and developed a computer program on
chemical compounds. As a laboratory assistant to Professor Hedwig, I
entered and ran the analyses of her research data. My senior project,
which I completed under Professor Hedwig's guidance, was an analysis
of retardant film products. The project was given the Senior Chemistry
Award, granted by a panel of chemistry faculty.

My work experience would be especially appropriate for Alden-Chandler
Industries. My internship in my senior year was at Pickett Laboratory,
which does extensive analyses for the Lake County Sheriff's office. My
work involved writing laboratory reports daily. At both Ryan Laboratories
and Century Concrete Corporation, I have worked extensively in
compound analysis, and I am familiar with standard test procedures.

I would appreciate the opportunity to discuss my qualifications for the
position of product chemist. I am available for an interview every
afternoon after three o'clock, but I could arrange to drive to Milwaukee
any time convenient to you. My telephone number during the day
between 10:00 a.m. and 3:00 p.m. is (312) 555–6644.

Sincerely,

Kimberly J. Oliver

Kimberly J. Oliver

Enc.

MODEL 15-1   Application Letter

Kimberly J. Oliver
1766 Wildwood Drive
Chicago, Illinois 60666
(312) 555–6644
kjo@midwestu.edu

OBJECTIVE: Chemist in product development. Willing to travel.

EDUCATION

Midwest University, Chicago, Illinois
BS in Chemistry; GPA: 3.6 (4.0 scale), June 2005

Computer Skills: Lotus 1-2-3, HTML, Excel, Java, PowerPoint

WORK EXPERIENCE

Century Concrete Corp., Chicago, Illinois, 2004–present
  Research Assistant: Set up chemical laboratories, including purchasing
  equipment and materials. Perform ingredients/compound analysis. Produce
  experimental samples according to quality standards. (part-time)

Pickett Laboratory, Chicago, Illinois, Spring 2004
  Research Intern: Set up testing compounds. Wrote laboratory test reports
  for compound analysis.

Ryan Laboratories, Chicago, Illinois, Summers 2002–2003
  Laboratory Assistant: Assisted in testing compounds. Prepared standard
  test solutions; recorded test data; operated standard laboratory equipment;
  wrote operating procedures.

HONORS AND AWARDS

Chemical Society of Illinois Four-Year Scholarship, 2001–2005
Senior Chemistry Award, Midwest University, 2004
Outstanding Chemistry Major, Midwest University, 2004

ACTIVITIES

Member, American Chemistry Association, 2003–present
President, American Chemistry Association Student Chapter, 2004–2005
Member, Midwest Toastmaster Association, 2002–2004
Council Member, Women in Science Student Association, 2002–2005
Hobbies: tennis, piano

References available on request.

MODEL 15-2   Traditional Résumé

Kimberly J. Oliver
1766 Wildwood Drive
Chicago, Illinois 60666
(312) 555-6644
kjo@midwestu.edu

Key Words: Product chemistry. Bachelor's degree. Laboratory internship. Compound analysis. Laboratory test reports. Midwest University. Chemical Society of Illinois Scholarship. Written and oral communication skills. Lotus 1-2-3, HTML, Excel, Java, PowerPoint

Objective: Chemist in product development and testing.

Education

Midwest University, Chicago, Illinois, June 2005
B.S. in Chemistry; GPA 3.6 on 4-point scale

Computer Skills: Lotus 1-2-3, HTML, Excel, Java, PowerPoint

Work Experience

Century Concrete Corp., Chicago, Illinois, 2004–present
Research Assistant: Set up chemical laboratories, including purchasing equipment and materials. Perform ingredient/compound analysis. Produce experimental samples according to quality standards. (part-time)

Pickett Laboratory, Chicago, Illinois, Spring 2004
Research Intern: Set up testing compounds. Wrote laboratory test reports for compound analysis.

Ryan Laboratories, Chicago, Illinois, Summers 2002–2003
Laboratory Assistant: Assisted in testing compounds. Prepared standard test solutions; recorded test data; operated standard laboratory equipment; wrote operating procedures.

Honors and Awards

Chemical Society of Illinois Four-Year Scholarship, 2001–2005
Senior Chemistry Award, Midwest University, 2004
Outstanding Chemistry Major, Midwest University, 2004

KIMBERLY J. OLIVER
1766 Wildwood Drive
Chicago, Illinois 60666
(312) 555-6644
kjo@midwestu.edu

OBJECTIVE: Product Development Chemist

EDUCATION
**B.S. in Chemistry, Midwest University, Chicago, Illinois, June 2005
**Computer Skills: Lotus 1-2-3, HTML, Excel, Java, PowerPoint
**Chemical Society of Illinois Four-Year Scholarship, 2001-2005
**Senior Chemistry Award, Midwest University, 2004
**Outstanding Chemistry Major, Midwest University, 2004

WORK EXPERIENCE
**Research Assistant, Century Concrete Corp., Chicago, Illinois, 2004-present
Set up chemical laboratories; perform ingredient/compound analysis; produce experimental samples (part-time)
**Research Intern, Pickett Laboratory, Chicago, Illinois, Spring 2004
Set up testing compounds; wrote laboratory test reports
**Laboratory Assistant, Ryan Laboratories, Chicago, Illinois, Summers 2002-2003
Prepared standard test solutions; wrote operating procedures; recorded test data

ACTIVITIES
**Member, American Chemistry Association, 2003-present
**President, American Chemistry Association Student Chapter, 2004-2005
**Member, Toastmaster Association, 2002-2004
**Council Member, Women in Science Student Association, 2002-2005

References available

MODEL 15-4   Email Résumé

Jack E. Montgomery
21 Camelot Court, Skokie, IL 60622
(312) 555-5620 <jmont6@cdnrt.com>

Major Qualifications: 15 years in product development chemistry.
Specialty: Adhesives

PROFESSIONAL EXPERIENCE

CDX TIRE AND RUBBER CORPORATION, Chicago, IL

Formulation Chemist, 2000–present
- Develop new hot-melt adhesives, pressure sensitives, and sealants
- Perform analysis of competitors' compounds
- Determine new product specification
- Calculate raw materials pricing
- Conduct laboratory programs for product improvement

Analytical Chemist, 1993–2000
- Performed ingredient/compound analysis
- Developed and tested compounds (cured/uncured)
- Wrote standard operating procedures
- Developed improved mixing design
- Supervised processing technicians

STERLING ANALYTICAL LABORATORIES, Pittsburgh, PA

Research Laboratory Assistant, 1991–1993 (part-time)
- Prepared solutions
- Drafted laboratory reports
- Supervised students during testing procedures

EDUCATION

B.S.—Chemistry, Minor—Biology. University of Pittsburgh, 1993

COMPUTER SKILLS

PASCAL, Lotus 1-2-3, Minitab, Sigmastat, PowerPoint

ACTIVITIES AND MEMBERSHIPS

District Chairman, United Way Campaign, 2001
Member, Illinois Consumer Protection Commission, 1997–2000
Member, American Chemical Society, 1993–present

References furnished on request.

MODEL 15-5   Advanced Résumé

Jeffrey L. Hirsch
1630 W. Lynd Drive
Mount Troy, MI 44882
(517) 555-2327
jlhirsch@hallm.com

Objective: Marketing and Sales Management

Qualifications: Ten Years in Retail Marketing; MBA in Management

Accomplishments:
- Exceeded district objectives in 23 accounts, $2.6 million sales.
- Increased territory volume by 17%.
- Managed sales team of 12 people.
- Negotiated improved leases for retailers.
- Earned district, regional, and national awards for high performance including Market Development Award 2000, Sales Development Professional of the Year 2003.
- Published marketing advice in *Sales Times* and district newsletters.

Organizational and Communication Skills:
- Coordinated 2003 National Sales Meeting—Chicago.
- Developed regional sales programs and seminars.
- Presented district seminars for retailers.
- Conducted training seminars for new salespeople.
- Presented career seminars for high school programs.

Employment History:
Hallman Marketing Corporation
    District Sales Representative, 1999–present.
    Retail Sales Coordinator, 1997–1999.

Education:
MBA, emphasis in Management, University of Illinois, 1999.
BBA in Marketing, Illinois State University, 1997.

Professional Memberships:
Michigan Sales Associates, Program Coordinator, 2003.
Metropolitan Business Association.
Retail Business Association High School Mentor Program.

References: Available on request.

MODEL 15-6   Functional Résumé

Felicia Cummings
4236 N. Ryan Road
Port William, NY 10011
(315) 555-8457 (home)
(315) 555-1212 (business)

Objective: Museum Display/Historical Artifacts Displays

Qualifications: Six Years' Experience in Museum and Collectible Displays

Display Experience:
- Developed Southwest Artifacts Display for Museum
- Designed collectibles display for advertising photography
- Organized historical weaponry display for public exhibition
- Coordinated university displays of student art history research projects

Communication Skills:
- Wrote lectures and led museum tours for community groups
- Wrote brochures for museum fund-raising committee
- Use PowerPoint, Pagemaker, Photoshop

Research Experience:
- Catalogued historical collectibles and replications for artifacts dealer
- Organized research notes and wrote reports for archeology expedition to the Yucatan.

Employment History:
Assistant Manager, PW Antiques, Port William, NY, 2002–present
Assistant to curator, Museum of Indian Art, El Paso, TX, Summers, 1993–1995

Education:
B.A. in Art History, Port William College (expected June 2008)
General Studies, University of Texas-El Paso (1993–1995)

Professional Membership:
American Art History Association

References and portfolio are available on request.

MODEL 15-7   Functional Résumé

Chapter 15   Exercises

INDIVIDUAL EXERCISES

1. Select a company in your area, and apply for a summer job (or a part-time job during the school year) in your career area. Graduating seniors should apply for a full-time job in their career area. Prepare a traditional résumé based on the information in this chapter. Write an application letter that realistically discusses your interest in and qualifications for a position based on your *present* situation.

2. Prepare a scannable résumé based on the traditional résumé you prepared for Exercise 1.

3. Prepare an email résumé based on the traditional résumé you prepared for Exercise 1. If your instructor agrees, email the résumé to your instructor.

4. Prepare answers to the following typical interview questions. Discuss the effectiveness of your answers with the class. *Or,* if your instructor prefers, form groups and compare your answers. Decide what elements should be included in your answers.

- Why did you choose your major?
- Tell me about yourself.
- What is your greatest strength?

COLLABORATIVE EXERCISES

5. For panel presentations, your instructor will divide the class into groups. Each group will prepare one of the following subjects for a panel presentation. Each group member will be responsible for a subtopic of the main subject. Plan a five-minute presentation for each group member, unless your instructor assigns a different time limit. Read the articles on oral presentations in Part 2, and practice your presentation to be sure it fits the assigned time. Tailor your information to your audience of classmates. After making your presentation, submit an executive summary of your talk and a transmittal memo to your instructor. Guidelines for the executive summary and transmittal memo are in Chapter 11.

- *Job interviews*—Report on typical questions to expect during job interviews, how to answer questions effectively, role-playing interviews, handling a group interview, questions to ask the interviewer, and items to bring with you.
- *Body language and appearance*—Report on appropriate dress for interviews, definitions of "casual dress," appropriate accessories and grooming, appropriate body language for interviews, and handling luncheon or reception interviews.

- *Job prospects*—Report on opportunities for new graduates in specific career fields, types of positions available, expected salaries, promotion possibilities, geographic areas for opportunities, and sources for information about job prospects.

- *Business etiquette*—Report on office politics, arranging meetings, telephone manners, making introductions, handling business lunches, and sources for information about business etiquette.

- *International opportunities*—Report on searching for jobs in other countries, geographic locations and expected salaries, promotion opportunities, employer expectations for employees, and business meeting customs in other countries.

INTERNET EXERCISES

6. Evaluate one of the career sources listed in Appendix C. Follow the site's links to other Web sites, and prepare to discuss their usefulness in class. *Or,* if your instructor prefers, write an informal report reviewing how useful the links are on a particular site. Short reports are covered in Chapter 12.

7. Read "Behavioral Interviewing: Write a Story, Tell a Story" by Hagevik in Part 2. Search the Internet for more information about behavioral interviewing. Write a memo to your instructor, and share the most important advice you found about behavioral interviewing. Memo format is covered in Chapter 14.

CHAPTER 16

Oral Presentations

UNDERSTANDING ORAL PRESENTATIONS

You must be prepared to handle oral presentations gracefully. All the technology we use does not replace effective face-to-face communication skills. A survey of 725 managers reported that the ability to communicate in front of an audience is the number one skill needed for career advancement.[1] In a survey of personnel interviewers, 98% reported that communication skills have a strong effect on hiring decisions, but only 59% of the interviewers thought that job applicants showed adequate communication skills.[2] Engineers reported in a survey that 32% of their time was spent in formal and informal oral communication. One engineer commented, "A bad presenter is career-limited."[3] Oral presentations can be internal, such as for management, staff, or technicians, or external, such as for clients, prospective customers, or colleagues at professional conferences.

Purpose

Like written documents, oral presentations often have mixed purposes. A scientist speaking at a conference about a research study is informing listeners as well as persuading them that the results of the study are significant. Oral presentations generally serve these purposes, either separately or in combination:

- *To inform*—The speaker presents facts and analyzes data to help listeners understand the information. Such presentations cover the status of a current project, results of research or investigations, company changes, or performance quality of new systems and equipment.

- *To persuade*—The speaker presents information and urges listeners to take a specific action or reach a specific conclusion. Persuasive presentations involve sales proposals to potential customers, internal company proposals, or external grant proposals.

- *To instruct*—The speaker describes how to do a specific task. Presentations that offer instruction include training sessions for groups of employees and demonstrations for procedures or proper handling of equipment.

Advantages

Oral presentations have these advantages over written documents:

- The speaker can explain a procedure and demonstrate it at the same time.

- The speaker controls what is emphasized in the presentation and can keep listeners focused on specific topics.

- The speaker's personality can create enthusiasm in listeners during sales presentations and inspire confidence during informative presentations.

- The speaker can get immediate feedback from listeners and answer questions on the spot.

- Listeners can raise new issues immediately.

- If most of the listeners become involved in the presentation—asking questions, nodding agreement—those who might have had negative attitudes may be swept up in the group energy and become less negative.

Disadvantages

Oral presentations are not adequate substitutes for written documents because they have these distinct disadvantages:

- Oral presentations are expensive because listeners may be away from their jobs for a longer time than it would take them to read a written document.

- Listeners, unlike readers, cannot select the topic they are most interested in or proceed through the material at their own pace.

- Listeners can be easily distracted from an oral presentation by outside noises, coughing, uncomfortable seating, and their own wandering thoughts.

- A poor speaker will annoy listeners, who then may reject the information.

- Spoken words vanish as soon as they are said, and listeners remember only a few major points.

- Time limitations require speakers to condense and simplify material, possibly omitting important details.

- Listeners have difficulty following complicated statistical data in oral presentations, even with graphic aids.

- Audience size has to be limited for effective oral presentations unless expensive video/TV equipment is involved.

Types of Oral Presentations

An oral presentation on the job usually is in one of the following four styles:

Memorized Speech. Memorization is useful primarily for short remarks, such as introducing a main speaker at a conference. Unless it is a dramatic reading, a memorized speech has little audience appeal. A speaker reciting a memorized speech is likely to develop a monotone that will soon have the listeners glassy-eyed with boredom. In addition, in such presentations, if the speaker is interrupted by a question, the entire presentation may disappear from memory.

Written Manuscript. Speakers at large scientific and technical conferences often read from a written manuscript, particularly when reporting a research study or presenting complex technical data. The written manuscript enables the speaker to cover every detail and stay within a set time limit. Unfortunately, written manuscripts, like the memorized speeches, also may encourage speakers to develop a monotone. Usually, too, the speaker who reads from a manuscript has little eye contact or interaction with the audience. Listeners at scientific conferences frequently want copies of the full research report for reference later.

Impromptu Remarks. Impromptu remarks occur when, without warning, a person is asked to explain something at a meeting. Although others in the meeting realize that the person has had no chance to prepare, they still expect to hear specific information presented in an organized manner. Always come to meetings prepared to explain projects or answer questions about activities under your supervision. Bring notes, current cost figures, and other information that others may ask you for.

Extemporaneous Talk. Most oral presentations on the job are extemporaneous; that is, the speaker plans and rehearses the presentation and follows a written topic or sentence outline when speaking. Because this type of on-the-job oral presentation is the most common, the rest of this chapter focuses on preparing and delivering extemporaneous talks.

ORGANIZING ORAL PRESENTATIONS

As you do with written documents, analyze your audience and their need for information, as explained in Chapter 3. Consider who your listeners are and what they expect to get from your talk. The key to a successful presentation is knowing what facts will be most interesting to your audience. Emphasize these points in your presentation. Oral presentations also must be as carefully organized as written documents are. Even though you are deeply involved in a project, do not rely on your memory alone to support your review of the information or your proposal. Instead, prepare an outline, as discussed in Chapter 4, that covers all the points you want to make. An outline also will be easier to work with than a marked copy of a written report you plan to distribute because you will not have to fumble with sheets of paper.

Use either a topic or a sentence outline—whichever you feel most comfortable with and whichever provides enough information to stir your memory. Remember that topic outlines contain only key words, whereas sentence outlines include more facts:

Topic: Venus—Earth's twin

Sentence: Venus is Earth's twin in size, mass, density, and gravity.

Some people prefer to put an outline on note cards, with only one or two topics or sentences per card. Other speakers prefer to use an outline typed on sheets of paper because it is easier to glance ahead to see what is coming next. Oral presentations generally have three sections:

Introduction.   Your introduction should establish (1) the purpose of the oral presentation, (2) why it is relevant to listeners, and (3) the major topics that will be covered. Since listeners, unlike readers, cannot look ahead to check on what is coming, they need a preview of main points in the opening so that they can anticipate and listen for the information relevant to their own work. Your introduction also should define your terms. Even if your audience consists of experts in the topic, do not assume they understand everything you do. Explain the terms you use, and check your audience's comprehension before continuing.

Try to capture your listeners' attention with a dynamic opening statement,[4] such as the following:

- Make a negative statement to increase audience anticipation:

   "This drug will not cure cancer—at least not yet."

- Mention a startling fact to focus audience interest:

   "A killer asteroid heading toward Earth has a 75% chance of hitting Chicago this summer."

- Quote a well-known authority to establish credibility:

   "Dr. Richard Miller has confirmed that the universe is dominated by a constant force—a dark energy."

- Ask a question that your audience will want answered:

   "What makes microbursts so dangerous?"

- Suggest an unusual comparison or connection:

  "Diabetes and rotator cuff surgery can have the same effect on your ability to scuba dive."

Data Sections.   The main sections of your presentation are those that deal with specific facts, arranged in one of the organizational patterns discussed in Chapter 4. Remember these tips when presenting data orally:

- Include specific examples to reinforce your main points.

- Number items as you talk so that listeners know where problem number 2 ends and problem number 3 begins.

- Refer to visual illustrations, and explain their content to be sure people understand what they are looking at.

- Cite authorities or give sources for your information, particularly if people in the audience have contributed information for your presentation.

- Simplify statistics for the presentation, and provide the full tables or formulas in handouts.

Conclusion.   Your conclusion should summarize the main points and recommendations to fix them in listeners' minds. In addition, remind listeners about necessary future actions, upcoming deadlines, and other scheduled presentations on the subject. Also, ask for questions so that listeners can clarify points immediately.

Model 16-1 is a sentence outline of an oral presentation by a health officer to community groups. The officer is alerting groups to the need to act quickly in case the local water supply is infected.

As the outline shows, the introduction (I) states the potential problem of dangers to the water supply and identifies two limited options. Sections II, III, and IV explain the two major options for disinfecting contaminated water. Section V stresses the importance of community action in emergencies. In this situation, handouts that repeat crucial information would be useful to the audience.

The outline in Model 16-1 includes reminders to the speaker about when to show the slides and when to distribute the handout and ask for questions. Such reminders in your outline are a safeguard against your concentrating so much on what you are saying that you forget your visual aids. Some speakers use extra-wide margins in their outlines to allow room for extra notes.

EMERGENCY DISINFECTION OF DRINKING WATER

I. Disasters and accidents can create unsafe local water supplies.

[Slide 1—Options List]

A. Local health department information may differ from EPA recommendations.
B. Draining the hot water tank will produce limited amounts of water.
C. Well water is preferred source in emergencies.

II. Two methods of emergency disinfection.

[Slide 2—Methods List]

A. Boiling water for 1 minute will kill disease.
B. Chemical treatment uses chlorine and iodine.

III. Chlorine methods.

[Slide 3—Chlorine Methods]

A. Chlorine bleach bottles often have procedure written on the label.
B. Granular Calcium Hypochlorite in water disinfects.
C. Chlorine tablets have instructions on the package.

IV. Tincture of iodine methods.

[Slide 4—Iodine Methods]

A. Household iodine can be used to disinfect—5 drops to 1 quart.
B. Iodine tablets have instructions on the package.

V. Local health departments must take the lead in emergencies and offer immediate advice to residents.

[Distribute checklists and ask for questions.]

MODEL 16-1   Sentence Outline for Oral Presentation

PREPARING FOR ORAL PRESENTATIONS

Because oral presentations usually have time restrictions, concentrate on the factors that your audience is most interested in, and prepare written materials covering other points that you do not have time to present orally. In addition, for oral presentations, you must check on these physical conditions, if possible:

- *Size of the group*—Is the group so large that you need a microphone to be heard in the back, and will those in the back be able to see your graphic aids? Remember that the larger the group, the more remote the listeners feel from the speaker and the less likely they are to interact and provide feedback.

- *Shape of the room*—Can all the listeners see the speaker, or is the room so narrow that those in the back are blocked from seeing either the speaker or the graphic aids? Does the speaker have to stand in front of a large glass wall through which listeners can watch other office activities? An inability to see or an opportunity to watch outside activities will prevent listeners from concentrating on the presentation.

- *Visual equipment*—Does the room contain built-in visual equipment, such as a movie screen or chalkboard, or will these have to be brought in, thus altering the shape of the speaking area? Will additional equipment in the room crowd the speaker or listeners? Are there enough electrical outlets, or will long extension cords be necessary, thereby creating walking obstacles for the speaker or those entering the room?

- *Seating arrangements*—Are the chairs bolted in place? Is there room for space between chairs, or will people be seated elbow-to-elbow in tight rows? If listeners want to take notes, is there room to write? People hate to be crowded into closely packed rows. An uncomfortable audience is an inattentive one.

- *Lighting*—Is there enough light so that listeners can see to take notes? Can the light be controlled to prevent the room from being so bright that people cannot see slides and transparencies clearly?

- *Temperature*—Is the room too hot or too cold? Heat usually is worse for an audience because it makes people sleepy, especially in the afternoon. If the temperature is too cold, people may concentrate on huddling in jackets and sweaters rather than listening to the presentation. A cool room, however, will heat up once people are seated because of body heat. Good air circulation also helps keep a room from being hot, stuffy, and uncomfortable.

You may not be able to control all these factors, but you can make some adjustments to enhance your presentation. You can adjust the temperature control or change the position of the lectern away from a distracting window, for example.

STRATEGIES—Meetings

If you are in charge of organizing a meeting, you have certain responsibilities to make sure the meeting is worthwhile for participants.

- Distribute an agenda that includes the time, place, date, and directions for reviewing specific materials beforehand if necessary.
- Be sure that all the needed materials and equipment are available, such as name cards or tags, pens, paper, computers, clipboards, video players, flip charts, markers, water, carafes, and glasses.
- Establish break times if the meeting runs more than two hours.
- Attend the entire meeting even if you are not a major participant so you can handle unexpected requests and follow up on plans for another meeting.

DELIVERING ORAL PRESENTATIONS

After planning your oral presentation to provide your listeners with the information they need and organizing your talk, you must deliver it. Fairly or unfairly, listeners tend to judge the usefulness of an oral presentation at least in part by the physical delivery of the speaker. Prepare your delivery as carefully as you do your outline.

Rehearsal

Everyone who speaks in front of a group is somewhat nervous. Fortunately, nerves create energy, and you need energy to deliver your talk. Picture yourself performing well beforehand, and remember that no one is there to "catch" you in a mistake. In fact, the audience usually does not know if you reverse the order of transparencies or discuss points out of order. Rehearse your oral presentation aloud so that you know it thoroughly and it fits the allotted time. Rehearsing it will also help you find trouble spots where, for instance, your chronology is out of order. If possible, practice in the room you intend to use so that you can check the speaking area, seating arrangements, and lighting.

If you "go blank" during a presentation, pause and repeat your last point, using slightly different words. This repetition often puts a speaker back on track. Consider speaking in front of a group not as a dreadful ordeal that you must suffer through but as an essential part of your job, and develop your ability to make an effective presentation.

Notes and Outlines

Have your notes organized before arriving in the room, and number each card or page in case you drop them and have to put them in order quickly. If a lectern is available, place your notes on that. Avoid waving your notes around or making them obvious to your listeners. They want to concentrate on what you are telling them, not on a sheaf of white cards flapping in the air.

Voice

If you are nervous, your voice may reveal it more than any mistakes you make. Do not race through your talk on one breath, forcing your audience to listen to an unintelligible monotone, but do not pause after every few words either, creating a stop-and-start style. Avoid a monotone by being interested in your own presentation. Speak clearly, and check the pronunciation of difficult terms or foreign words before your talk. Be sure also to pronounce words separately. For example, say "would have" instead of "wouldof."

Professional Image

When you speak in front of a group, you are "on stage," and you want to project a professional image. Dress conservatively, but in the usual business style. Clothes that are too tight or that have unusual, eye-catching decorations will distract your audience from your message. Women should be aware that the higher the stage, the shorter skirts look. Long sleeves project authority and professionalism; short sleeves create a casual look.[5] Make sure the color you wear compliments you, and check, if possible, that your clothing does not blend into the background where you are presenting.[6] If you are traveling to a distant location to make a presentation, bring a spare outfit with you. If you are speaking after a meal, before presenting check to be sure your clothing does not have a drip of gravy on it. Many men keep a spare tie in their briefcases for such emergencies. Women often carry a scarf they can add to their outfits if necessary.

Do not hang on the lectern, sit casually on the edge of a table, or sway back and forth on your feet. Be relaxed, but stand straight.

Gestures

Nervous gestures can distract your audience, and artificial "on-purpose" gestures look awkward. If you feel clumsy, keep your hands still. Do not fiddle with your note cards, jewelry, tie, hair, or other objects. All your gestures should reflect confidence in your own expertise about the subject.

Some gestures can help focus the audience's attention on a specific point. If you suggest two solutions for a problem and hold up two fingers to emphasize the choice, the gesture will help your audience remember the number of alternatives you have explained.

Eye Contact

Create the impression that you are speaking to the individuals in the room by establishing eye contact. Although you may not be able to look at each person individually, glance around the room frequently. Do not stare at two or three people to the exclusion of all others, but do not sweep back and forth across the room like a surveillance camera. Eye contact indicates that you are interested in the listener's response, and a smile indicates that you are pleased to be giving this presentation. You should project both attitudes.

Questions

One of the advantages of oral presentations is that listeners can ask questions immediately. Answer as many questions as your time limitations permit. If you do not know the answer to a question, say so, and promise to find the answer later. You may ask if anyone else in the audience knows the answer, but do not put someone on the spot by calling his or her name without warning. If there is a reason you cannot answer a question, such as the information being classified, simply say so. If you are not sure what the questioner means, rephrase the question before trying to answer it by saying, "Are you asking whether . . . ?" In all cases, answer a question completely before going on to another.

USING VISUAL AIDS

Presentations usually benefit from visual aids. Handouts are the easiest to use because you prepare them ahead of time and pass them out when you want to. Using visual equipment and creating graphics present more challenges.

Equipment

If equipment breaks during your presentation, continue without it. Trying to manage repairs in the middle of a presentation will result in chaos. Follow these tips to avoid breakdowns:

- Check that all equipment is plugged in and turned on.
- Check that the projector is set up properly and that your laptop is selected as the source.
- Practice using equipment before you talk, so you know where the controls and features are and how to implement them.
- If you know you write slowly on chalkboards and flip charts, put your terms or lists on them before the presentation, so you can simply refer to them at the appropriate time.
- If you are using DVDs, VCRs, CDs, or audio cassettes, know which key word or image indicates where you want to start and stop.
- Check the microphone to be sure it is working before your audience arrives.
- To avoid computer freezeup, reboot just before you begin.[7]
- Turn off the laser pointer when you are not using it.

Presentation Software

PowerPoint and other presentation software have become increasingly popular for oral presentations; however, these programs create a new set of challenges for a speaker. Before tackling the mechanics of any presentation software, consider your audience and purpose, just as you would for any oral presentation. Plan your presentation based on what your audience wants to hear and on what you want to say to accomplish your purpose.

Advantages

Using a computer for visual presentations has several advantages:

- Audiences have become used to PowerPoint, and using it creates a sophisticated and expert image for you as a speaker.
- PowerPoint allows flexibility in presentations. You can make last-minute decisions about how to use your slides. You can skip ahead or back up to use a slide again. You can change the order of your slides at

any time. During the question-and-answer session, you can pull up relevant slides and review the information.

- Once you master the technology, you can eliminate fumbling with handouts, overheads, or flip charts or boards.

Disadvantages

Like all technology, PowerPoint has some disadvantages:

- Watching a parade of slides with nothing but key words on them can be very boring for an audience.

- Because the room is dim so the PowerPoint slides can be seen, you lose eye contact and interaction with your audience.

- Because so many presenters use it, PowerPoint makes all presentations seem to be generic, or at least very similar. This similarity often interferes with the audience's ability to remember your key points.

- Presenters sometimes become obsessed with creating visuals (e.g., moving graphs, shifting colors, animation) and forget that the key element in any presentation is clear and concise information.

Essentials

When you use PowerPoint or any presentation software, consider the following tips for creating effective visuals:

- Plan your presentation with storyboards. Storyboards are rough sketches of your slides in the order you want to use them. Some people prefer to plan their presentations using a two-column chart designed like a playscript, with the text on the left and the description of the slide on the right, opposite the appropriate text. This planning will save you time when you create your slides because you have already made the decisions about content and style.

- Use at least an 18-point font. A large font (40 point for your headings and 24–28 point for details) is best. If you want to use a smaller font, check the size of the room you will be using. If the text on the slide is not visible from the back of the room, choose a larger size.[8]

- Avoid italics, underlining, shadow lettering, embossing, or outlined letters because these features are hard to read quickly.[9]

- Use no more than five or six bullet points per slide.

- Practice your presentation to be sure you can handle the equipment properly.

- Have a backup plan. In addition to your PowerPoint, you might bring a set of transparencies or handouts covering key visuals in case the computer fails.

- Use "talking heads." Instead of "Sales," write "Falling Sales."

- Include the corporate logo in the corner of all slides, handouts, or overheads when presenting to clients.

- Distribute copies of complicated graphs or charts.

JOINING A TEAM PRESENTATION

In some instances, such as sales presentations to potential clients or training sessions, a team of people makes the oral presentation, with each person on the team responsible for a unit that reflects that person's specialty. Although each person must plan and deliver an individual talk, some coordination is needed to be sure the overall presentation goes smoothly. Remember these guidelines:

- One person should be in charge of introducing the overall presentation to provide continuity. This person also may present the conclusion, wrapping up the team effort.

- One person should be in charge of all visual equipment so that, for example, two people do not bring overhead projectors.

- If the question-and-answer period is left to the end of the team presentation, then one person should be in charge of moderating the questions that the team members answer to maintain an orderly atmosphere.

- All members should pay close attention to time limits, so they do not infringe on other team members' allotted time.

- All team members should listen to the other speakers, both as a courtesy and because they may need to refer to each other's talks in answering questions.

FACING INTERNATIONAL AUDIENCES

As companies expand their operations into worldwide markets and become partners with companies in other countries, you may have to make an oral presentation to an international audience. If so, you must consider language barriers. First, determine whether your audience understands English or you will be using a translator. If a translator will be repeating your words in another language after every sentence, your presentation will take twice as long, and you must adjust the length accordingly.

Rather than speak from an outline, write out your speech in detail. You want to be sure that nothing in your speech will offend or confuse your audience. Ask someone who speaks your audience's language to review your speech carefully, and cut out all confusing jargon, slang, acronyms, popular culture references, and humor. These elements do not translate well. When you give your speech, keep to the written script to avoid straying into comments that will present problems for the translator. As a courtesy, you might prepare visual aids in the language of your audience.

Even if your audience understands English, you will need to research the appropriate ways to open your presentation, how much time to allot for questions, and how to end without giving offense. For example, American presentation style is to begin by pointing out objectives. However, an Arab speaker may begin by expressing appreciation for hospitality and appealing for harmony. Germans typically do not interact with speakers, and the question-and-answer session may be short, but Mexicans enjoy a long question-and-answer session. The American closing typically calls for action, but other cultures may close presentations by emphasizing future meetings or group harmony.[10] Computer software presentations can be effective for international audiences because many cultures value structured and detailed presentations.[11]

It takes time and thought to prepare an appropriate oral presentation for an international audience. To be an effective speaker in such circumstances, you must develop techniques to ensure that your technical information is not changed in translation and your presentation style fulfills your audience's cultural expectations.

▬▬▬ *CHAPTER SUMMARY*

This chapter discusses making oral presentations on the job. Remember:

- Oral presentations may be (1) a memorized speech, (2) a written manuscript, (3) impromptu remarks, or (4) an extemporaneous talk.

- Speakers should (1) control nerves by preparing thoroughly, (2) rehearse aloud, (3) speak clearly, (4) dress professionally, (5) keep hand gestures to a minimum, (6) establish eye contact, (7) number notes and outline sheets, (8) practice with equipment, and (9) answer questions fully.

- PowerPoint can create interesting visuals, but it does interfere with interaction between the speaker and the audience.

- For a team presentation, one speaker should handle the general introduction and conclusion.

SUPPLEMENTAL READINGS IN PART 2

Frazee, V. "Establishing Relations in Germany," *Global Workforce,* p. 492.

Hart, G. J. S. "PowerPoint Presentations: A Speaker's Guide," *Intercom,* p. 501.

Robinson, J. "Six Tips for Talking Technical When Your Audience Isn't," *Presentations,* p. 530.

Smith, G. M. "Eleven Commandments for Business Meeting Etiquette," *Intercom,* p. 533.

Weiss, E. H. "Taking Your Presentation Abroad," *Intercom,* p. 535.

ENDNOTES

1. "Critical Link between Presentation Skills, Upward Mobility," *American Salesman* 36.8 (August 1991): 16–20.

2. Marshalita Sims Peterson, "Personnel Interviewers' Perceptions of the Importance and Adequacy of Applicants' Communication Skills," *Communication Education* 46 (October 1997): 287–291.

3. Pneena Sageev and Carol J. Romanowski, "A Message from Recent Engineering Graduates in the Workplace: Results of a Survey on Technical Presentation Skills," *Journal of Engineering Education* 90.4 (October 2001): 685–692.

4. Robert Goldbort, "Professional Scientific Presentations," *Journal of Environmental Health* 64.8 (April 1, 2002): 29–31.

5. Dawn E. Waldrop, "What You Wear Is Almost as Important as What You Say," *Presentations Web site.* Retrieved July 31, 2000, from http://www.presentations.com/speakingtips.

6. "What to Wear (or What Not to Wear) at Your Next Business Presentation." *PresentationHelper Web site.* Retrieved August 24, 2004 from http://www.presentationhelper.co.uk/what_to_wear.htm.

7. Jim Endicott, "If Disaster Strikes Onstage, Stay Focused and Be Creative," *Presentations Web site.* Retrieved June 12, 1999, from http://www.presentations.com/speakingtips.

8. Steve Kay, "It's Showtime! How to Give Effective Presentations," *Supervision* 60.3 (March 1999): 8–10.

9. Mary Munter and Lynn Russell. *Guide to Presentations* (Upper Saddle River, NJ: Prentice-Hall, 2002), p. 79.

10. Farid Elashmawl, "Multicultural Business Meetings and Presentations: Tips and Taboos," *Tokyo Business Today* 59.11 (November 1991): 66–68.

11. Michael Conner, "Working through PowerPoint: A Global Prism for Local Reflections," *Business Communication Quarterly* 67.2 (June 2004): 228–231.

Chapter 16 Exercises

INDIVIDUAL EXERCISES

1. Prepare a 2–4-minute oral presentation based on one of the written assignments for this class. Prepare a sentence outline of your talk to submit to your instructor.

2. Prepare a 2–4-minute presentation on a topic from another class. Prepare a sentence outline of your talk to submit to your instructor.

3. Watch a formal talk or an interview on PBS or one of the news channels. Take notes on how well the speaker is prepared and how effective the presentation is or the interview answers are. Prepare an outline of your evaluation for class discussion and submit the outline to your instructor.

COLLABORATIVE EXERCISES

4. In groups, select a speech from a government Web site, such as the U.S. Department of Commerce or U.S. Department of the Interior. Print out the speech, and design a set of slides to accompany it. Share your results with the other groups. *Or,* if your instructor prefers, develop a team presentation for the information in the speech and create appropriate graphics.

5. In groups, select a campus activity, and gather information about it. Develop a team presentation for the class in which you persuade your audience to participate in some way in this activity. Prepare appropriate graphic aids.

INTERNET EXERCISES

6. Find three Web sites of U.S. universities or colleges. Prepare a 3–5-minute PowerPoint presentation in which you compare the visual design and text of the sites.

7. Search the Web for information about a specific community problem, such as air pollution, and prepare a 5–10-minute PowerPoint presentation based on the information you find.

8. Search the internet for copies of historic speeches by one of the following: Martin Luther King, Jr., John F. Kennedy, Winston S. Churchill, Dwight D. Eisenhower, William Jennings Bryan, or Franklin D. Roosevelt. Print a copy of the speech, and prepare an outline of the speech as if you were the presenter. Next, prepare a PowerPoint presentation of six to eight slides to accompany the speech. Finally, prepare a two-column outline, placing the text outline in the left column and the notations of the slides in the right column in their proper place. Submit the speech outline, copies of the slides, and the two-column outline to your instructor with a transmittal memo. Transmittal memos are covered in Chapter 11.

PART 2

Technical Writing: Advice from the Workplace

Ethics in Technical Communication

Lori Allen and Dan Voss

This excerpt from the book Ethics in Technical Communication *discusses unethical use of statistics, graphics, and photographs. Ethical communication involves illustrations as well as words. Consider these guidelines when you are preparing graphics.*

LYING WITH STATISTICS

Misuse of numbers for purposes of deception is analogous to misuse of words for deception. Actually changing numbers is commission. Conveniently leaving out numbers that don't serve one's interests is omission. Manipulating numbers to send a distorted message is like using circumlocution or slanted language. All are unethical. Of the myriad opportunities for deception with statistics, three of the most common are invalid survey techniques, misuse of percentages, and misleading use of averages.

Surveys

Surveys are tricky animals. To be valid, they must represent a sufficiently large and representative segment of the population. Questions must be worded to clearly differentiate between opinions, not to deliberately lead respondents down a primrose path to desired responses.

Consider this situation: A group of high school students is asked to survey residents of a wealthy and predominantly elderly community known to oppose tax increases on whether they would support a bond issue for a new football stadium with Astroturf. The students knock on 100 doors. In 95 cases, as soon as the homeowners learn the subject of the survey, they chase the students down the driveway wielding frying pans and broomsticks. In the other five cases, the students find one genuine supporter, one who is adamantly opposed, and three who are opposed but whom the students badger for hours until they finally give in and check "Yes" as the only way to get the students off their property. The result is "Four out of five residents surveyed recommended grassless football stadiums for those high schools that have football stadiums." The other 95? Oh, that's right—they weren't surveyed.

Percentages

Percentages are meaningful only if they are presented along with the base to which they are applied and they can be very misleading. Consider the executive who tells her employees that she will be receiving only a 3 percent raise this year while they all receive 6 percent. That sounds great until you allow for the fact that the executive's raise is 3 percent of $100,000 and the workers' raise is 6 percent of $20,000. Or consider this one: If somebody offered to give you a 50 percent raise this month if you agreed to accept a 40 percent cut the following month, would you take it? Your first impulse might be to say, "Sure." Try it out. Let's say you're making $10 an hour. With a 50 percent raise, you'd be making $15 an hour. Now apply the 40 percent cut. You drop to $11 an hour, so you're still $1 ahead, right? Think again. The 40 percent cut applies to your *new* base salary, not your original base salary and 40 percent of $15 is $6. The net result is: You'd wind up working for $9 an hour, $1 less than you started out with.

Averages

Averages come in three flavors: mean, median, and mode. When most people use the word "average," they are referring to the mean, which is computed by adding all the individual values and dividing by the total number of items. *Mean* gives an honest expression of average when there is a fairly even distribution of values. When the values are clustered at the extremes of the range, the mean presents a distorted picture. *Median* refers to the midpoint among the values: the point at which there are an equal number of items of lesser and greater value. It is more valid than the mean for cases involving extremes; less so for even distributions. *Mode* refers to the individual value that occurs most often. It would be useful if you were trying to present the average height of a class which contained four 6'8" basketball players and one 5'2" cheerleader.

Of the three types of averages, the mean offers the most opportunity for deception. Consider the robber baron who claims that the employees in his dilapidated sweatshop of a factory live in comfort. His promotional brochure boasts: "Our employees work in a climate-controlled environment with an average temperature of 75 degrees." This deception is particularly masterful, because it demonstrates both verbal and statistical persiflage. The factory is, indeed, controlled by the climate. Since it has neither heating nor air conditioning, it is 90 degrees in the summer and 60 degrees in the winter. What is the average? You guessed it—75 degrees!

Or suppose you run into a state trooper on your way back from your bachelor's or bachelorette's party. The conversation might run something like this:

"Gee, officer, I was only averaging 50 miles an hour!"

True statement. Unfortunately, you achieved that average by going from 0 to 100.

Lying with Graphics

If one picture is worth a thousand words, then one lie with graphics is worth a thousand lies with words. Examples of deliberate deception with charts and figures abound. Pie charts can lie if the slices do not accurately reflect the percentages (which, of course, are in one-point type). Line graphs can lie by distorting

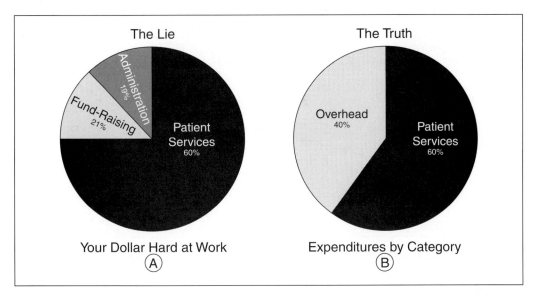

FIGURE 4-1   Pie charts lie when the numbers don't match the slices.

the axes. Organization charts can present a misleading picture of the chain of command.

Figure 4-1a might come from the annual report of an unprincipled charitable organization seeking to raise funds. It differs from an honest depiction of the same data (Figure 4-1b) in two important respects. First, it divides the overhead costs into two subcategories to conceal the amount of expenses unrelated to patient services. Second, it visually distorts the slices to well below the actual 40 percent overhead figure. The real numbers are there to avoid fraud charges, but they're about as legible as the Surgeon General's warning on cigarette packages.

Figure 4-2 "loads the dice" by manipulating the units of measure on the ordinate or Y-axis. At first glance, the reader might conclude that the workers' wages are climbing much more rapidly than those of the executives. Look again. The units of measure on the vertical axis of the executive pay chart are much larger than they are on the vertical axis of the similar looking employee pay chart, thereby making it appear as if the executives' pay increases are smaller than those of the employees when, in fact, they are much larger.

Figure 4-3 might be the version of the organization chart that Employee B sends out with his or her résumé. It is carefully designed to convey the impression that Employee B is second in command to A. It also tends to imply that Employees C, D, E, and F are below B but above Employees G, H, I, and J in the pecking order—which further reinforces B's "vice-emperor" message. However, if you follow the reporting lines closely, you will find that Employees B through J are actually all equal; all bear the same relationship with the boss.

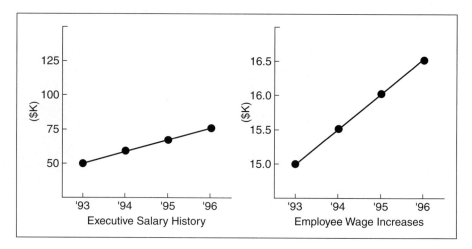

FIGURE 4-2   Who's been getting bigger raises? Look again.

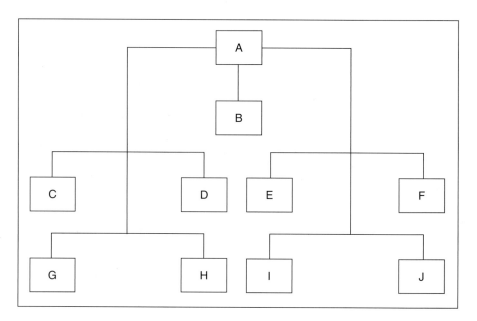

FIGURE 4-3   Who's more important, J or B?

LYING WITH PHOTOGRAPHS

Nearly everyone in our generation recalls the use of obviously doctored photographs in tabloid newspapers back in the 1950s and 1960s: "Kitten Born with Dog's Head" screamed the headline. To document this noteworthy biological event, under the headline appeared a lurid photograph with a puppy's head clumsily air-brushed onto a kitten's body. Sometimes if you looked closely, you could even discern the traces of the airbrush.

Well, that happens no more. No more clumsiness, that is. In today's world of dizzily onrushing technology, any entry-level illustrator with the appropriate software can, in a matter of minutes, give a kitten a dog's head. Or slap an electronic test set on a piece of equipment where it has never been used before. Or deftly whisk away any telltale plumes of pollution from the smokestacks of a factory when designing the company's environmental awareness brochure.

Given the present shift from optical to digital technology in photography, within a few years nearly all business and technical photography will be computer based. And since any digital file can be subtly—or not so subtly—altered without a trace, it will become almost impossible to know whether a photograph is genuine or retouched.

This technological advance has upped the ethical ante for technical communicators. Increasingly, if our audiences are to trust the integrity of the photographs in the documents and visual presentations we produce, they will have to depend on our honesty.

Consider this scenario. Suppose the manufacturer of a CAT scan imaging system asks a technical illustrator in its publications department to enhance the clarity of its images for a promotional brochure. Suppose, further, that the perception of higher resolution than the product can actually provide sways a hospital purchasing agent to order one of the systems. And suppose, six months later, that a neurosurgeon depends on that equipment to pinpoint a blood clot so she can operate on a child who has been gravely injured in a car–bike accident. The neurosurgeon goes in, the clot isn't where the machine says it was, and before she can get to it, the child dies on the table. What is the ethical responsibility of the technical illustrator to the parents of the child?

How to Avoid Costly Proofreading Errors

Carolyn Boccella Bagin and Jo Van Doren

*Proofreading errors cost businesses money. This article tells you how to pro-
duce error-free copy. Try finding all the mistakes in the sample copy in the
article, and check your results against the corrected version.*

Are proofreading errors costing you money? Some organizations have sad stories to tell about the price of overlooked mistakes.

- One insurance firm reported that an employee mailed a check for
  $2,200 as a settlement for a dental claim. Payment of only $22.00 had
  been authorized.

- One executive wasted $3 million by not catching a hyphen error when
  proofreading a business letter. In originally dictating the letter to his
  secretary, the executive said, "We want 1,000-foot-long radium bars.
  Send three in cases." The order was typed, "We want 1,000 foot-long
  radium bars."

- A magazine accidentally ran a cake recipe in which "¾ cup" was
  printed as "¼ cup." Irate readers sent complaint letters and cancelled
  their subscriptions.

Your company's image could be marred by unfortunate mistakes that find
their way into the documents you produce. Developing good proofreading
techniques and systems can save time, money, and embarrassment.

How Can You Produce Error-Free Copy?

- Never proofread your own copy by yourself. You'll tire of looking at
  the document in its different stages of development and you'll miss new
  errors. (If you must proof your own copy, make a line screen for yourself or roll the paper back in the typewriter so that you view only one
  line at a time. This will reduce your tendency to skim your material.)

- Read everything in the copy straight through from beginning to end:
  titles, subtitles, sentences, punctuation, capitalizations, indented items,
  and page numbers.

- Read your copy backward to catch spelling errors. Reading sentences
  out of sequence lets you concentrate on individual words.

- Consider having proofreaders initial the copy they check. You might find that your documents will have fewer errors.

- If you have a helper to proof numbers that are in columns, read the figures aloud to your partner, and have your partner mark the corrections and changes on the copy being proofread.

- If time allows, put your material aside for a short break. Proofreading can quickly turn into reading if your document is long. After a break, reread the last few lines to refresh your memory.

- Read the pages of a document out of order. Changing the sequence will help you to review each page as a unit.

- List the errors you spot over a month's time. You may find patterns that will catch your attention when you proofread your next document.

- If you can, alter your routine. Don't proofread at the same time every day. Varying your schedule will help you approach your task with a keener eye.

- Not everyone knows and uses traditional proofreading marks. But a simple marking system should be legible and understandable to you and to anyone else working on the copy.

WHERE DO ERRORS USUALLY HIDE?

- Mistakes tend to cluster. If you find one typo, look carefully for another one nearby.

- Inspect the beginning of pages, paragraphs, and sections. Some people tend to skim these crucial spots.

- Beware of changes in typeface—especially in headings or titles. If you change to all uppercase letters, italics, boldface, or underlined copy, read those sections again.

- Make sure your titles, subtitles, and page numbers match those in the table of contents.

- Read sequential material carefully. Look for duplications in page numbers or in lettered items in lists or outlines.

- Double-check references such as, "see the chart below." Several drafts later, the chart may be pages away from its original place.

- Examine numbers and totals. Recheck all calculations and look for misplaced commas and decimal points.

- Scrutinize features that come in sets, such as brackets, parentheses, quotation marks, and dashes.

TRY YOUR HAND (AND EYE) AT THIS TEST

Mark the mistakes and check your corrections with our marked copy below.

It is improtant to look for certain item when proofing a report , letter, or othr document. Aside from spelling errors, the prooffer should check for deviations in format consitent use of punctuation, consistent use of capitol letters, undefined acronyms and correctpage numbers listed in the the Table of contents.

After checking a typed draft againts the original manuscript one should also read the draft for aukward phrasing, syntactical errors, and subject/ verb agreement and grammatical mistakes. paralell structures should be used im listings headed by bullets or numbers: ie, if one item starts with the phrase "to understand the others should start with to plus a verb.

The final step in proofing involves review of the overall appearance on the document. Are the characrters all printed clearly Are all the pages there? Are the pages free of stray marks ? Is the graphics done? Are bullets filled in? All of the above items effect the appearance of the the document and determine whether the document has the desired effect on the reader.

HOW DID YOU DO?

Check your markup against ours. (We only show corrections in the copy and not the margin notes that most proofreaders typically use.)

It is improtant to look for certain item when proofing a report, letter, or othr document. Aside from spelling errors, the prooffer should check for deviations in format consitent use of punctuation, consistent use of capitol letters, undefined acronyms and correctpage numbers listed in the the Table of contents.

After checking a typed draft againts the original manuscript one should also read the draft for aukward phrasing, syntactical errors, and subject/ verb agreement and grammatical mistakes. paralell structures should be used im listings headed by bullets or numbers; ie, if one item starts with the phrase "to understand the others should start with to plus a verb.

The final step in proofing involves review of the overall appearance on the document. Are the characrters all printed clearly Are all the pages there? Are the pages free of stray marks ? Is the graphics done? Are bullets filled in? All of the above items effect the appearance of the the document and determine whether the document has the desired effect on the reader.

Ten Tips on Writing White Papers

Darren K. Barefoot

Product information, in printed form or available on the company's Web site, is central to a company's success in reaching potential customers. This article gives advice about writing product descriptions for company Web sites.

Technical writers are often asked to contribute to their company's white papers (documents that educate industry customers about products or services). Sometimes a product manager will write the paper and a technical writer will edit it; other times the writer will generate the text. These completed papers are usually posted on the company's Web site and distributed to potential customers at trade shows and meetings. Having written and reviewed my fair share of these documents. I offer the following ten suggestions for improving results.

1. ANALYZE YOUR AUDIENCE

Before you begin writing, define your audience. You might even want to include a section called "Intended Audience" in the body of the paper. An audience analysis will reveal the extent to which you must explain technical concepts. The results of this analysis will determine not only the level of technical content but also your writing style. For example, if your white paper is pitching a brand new technology to business development experts (as opposed to engineers), you should take care to explain the technology in language your audience can understand.

2. CREATE AN OUTLINE

Your outline will vary depending on your subject, but generally it's a good idea to structure your white paper so that it presents a solution to a problem. You could use an introductory section to define the problem, and then describe your solution in the body of the document. Ideally, you should have a subject matter expert, a marketing representative, and a member of your potential audience review the outline. This spares you the pain of removing sections because "that window didn't make it into the release" or "we're not pitching that feature right now."

3. USE DIAGRAMS CAREFULLY

While a picture may be worth a thousand words, a diagram often requires a thousand-word explanation. Clear, professional-looking diagrams are invaluable, but overly complex diagrams tend to look silly and can often confuse readers. There's nothing worse than the "hockey puck object models" and "Dadaist program architectures" that we find in industry documents.

You should also make sure that all diagrams in a document are of similar quality. The authority of a white paper is diminished when a colorful, professional "marketecture" diagram is followed by a technical diagram whipped up by one of the developers in Microsoft *Word*.

4. EXPLORE A LESS FORMAL WRITING STYLE

Unfortunately, most white papers are less compelling than most user manuals—and that's not a compliment to user manuals. Because white papers are marketing documents, they should be not merely readable but riveting. Don't be afraid to use metaphors, contractions, humor, and other devices. This doesn't necessarily mean writing in dialect or using other verbal fireworks to get attention: It simply means minimizing the technical lingo and inserting the occasional analogy. One of the best white papers I've read used a can of black beans to explain the basics of XML.

5. SOLVE A PROBLEM

Too many white papers are extended brochures that offer minute detail about features but don't explain why the reader might want the product. Your white paper should tell a simple but compelling story about how your company, product, or feature particularly and uniquely unravels the reader's Gordian knot. For example, instead of writing "Buy a DVD player because it has 128-bit oversampling and advanced virtual surround sound," you'd be better off with "Buy a DVD player because you need better sound and a crisper picture."

6. BEWARE OF MARKETING-SPEAK

As technical writers, we're the first readers who can sniff out the vague adjectives that sometimes accumulate in white papers. If potential customers have gone to the trouble of actually procuring your document, they want to know what your product does and why they should buy it. One of my professors used to say that "big truths are for posters, small truths are for short stories." Tell small truths in your white papers, and leave discussions of "leveraging synergy" and "re-imagining Weblications" to the press releases. Also, if you want to be taken seriously, do not fabricate new terms.

7. Supply Supporting Evidence

Quotations from your customers are valid, but everyone knows they are often solicited and tweaked to fit a company's message. It's more valuable to provide third-party information that supports your proposed solution. For example, if your company's DVD player has an embedded clock, you might state that a reputable survey reports that "80 percent of VCR owners cannot set their VCR clock."

8. Introduce and Summarize

An effective, compelling introduction is crucial, and should obviate need for an "executive summary"—the synopsis that precedes the table of contents in most white papers. These summaries have little value: Your audience may not consist solely of suit-wearing bigwigs with too little time to read the whole document, and the non-executives in your audience may resent the implication that these summaries are not for them.

At the end of the document, be sure to repeat the problem and summarize how your product addresses it. Too many white papers peter out after the last feature.

9. Spiff It Up (Appearance Is Important)

Many of the white papers I read look like freshman history essays: page after page of nothing but text. In terms of style and length, a white paper has a lot in common with a magazine article, so it should receive similar graphic treatment. Use all of the usual designer's tools—text boxes, columns, pull quotes, graphics, and so forth—to create an attractive, readable document.

10. Trim It Down (It's Probably Too Long)

Even if you've followed all this advice, your white paper isn't going to get more attention than a Stephen King novel, or even the reader's e-mail. The shorter your paper, the better. The length of the paper depends on the complexity of your product and method, but I certainly don't care to read anything longer than, say, fifteen pages. If you can't say what is necessary in that space, perhaps you need to divide your message into more than one document. For example, my company has a product suite of developer tools that could be presented in one long white paper. Instead, we plan to produce future documents along two themes related to our products—development and deployment.

Most people drastically underestimate the time and effort it takes to produce a decent, readable white paper. While this list is not exhaustive, it may help you avoid a few common pitfalls.

Technical Documentation and Legal Liability

John M. Caher

This article describes a New York state legal case in which the court analyzed information included with a drug prescription to determine company liability after a patient committed suicide. In making a decision, the court examined the writer's prose for completeness and clarity.

If a technical writer's prose on a prescription drug data sheet is unclear, and a patient blows his brains out in a medicinally-caused fit of depression, is the pharmaceutical company liable?

That is exactly the question recently addressed by New York's highest court, the Court of Appeals, in a case underscoring the responsibilities and liabilities of the technical writing staff. The court's unanimous decision in *Martin v. Hacker* made two things perfectly clear: one, companies are definitely on the hook for financial damages when their product documentation is imprecise or inaccurate; and two, in this case the data sheet was sufficiently clear.

What makes this decision particularly interesting is the court's painstaking analysis of the verbiage contained in the data sheet and its effort to evaluate not only the content, but the context of the documentation. Although the decision is binding only in New York, opinions by the Court of Appeals traditionally carry great influence around the country and its legal precedents, procedures, and methodologies are often followed by other tribunals. Therefore, the ruling, dated Nov. 23, 1993, has indirect implications for technical writers and their employers across the country.

THE FACTS OF THE CASE

Eugene Martin was a retired New York State trooper under a doctor's care for high blood pressure. In 1981, Martin's physician prescribed a prescription drug, hydrochlorothiazide (HCT). A year later, a second drug, reserpine, was prescribed by the same doctor. The physician advised Martin of various side effects of reserpine, including the fact that the drug can produce depression, although usually would not do so in the dosage prescribed. It is unclear from the record whether the doctor informed Martin that HCT can exacerbate the side effects of reserpine.

On Feb. 13, 1983, Martin—who had no history of mental illness or depression—shot and killed himself in a drug-induced despondency. His widow sued various parties, including the doctor, Chelsea Laboratories (the manufacturer), and Rugby Laboratories (the distributor), alleging in part that

486

the written warnings supplied with reserpine and HCT were insufficient as a matter of law.

THE LAW OF LIABILITY

Under the law, there are essentially three theories—contract, due care, and strict liability—under which a technical writer or manufacturer may be held responsible for an injurious result. Under the contract theory there is a binding agreement to perform or provide a service. The due-care theory holds that a manufacturer knows more about a product than the consumer and therefore has a weightier responsibility. Strict liability imposes the greatest burden of all. Under that theory, a manufacturer or employer can be held responsible for damages regardless of fault. The lawsuits against Chelsea and Rugby were brought as strict-liability claims, predicated on the assumption that a prescription drug is by nature inherently unsafe. The legal issue in *Martin v. Hacker*, as framed by Judge Stewart F. Hancock, Jr., centered on a drug manufacturer's obligation to fully reveal the potential hazards of its products.

Since prescription drugs must be prescribed by a medical doctor, the information and warnings contained on package inserts are directed toward physicians. They must be written in accordance with labeling specifications of the Food and Drug Administration, which requires the following information in this order:

- Description
- Actions
- Indications
- Contraindications
- Warnings
- Use in pregnancy
- Precautions
- Adverse reactions
- Dosage and administration
- Overdosage
- How supplied

Eugene Martin's widow contended the package inserts were inadequate in at least three respects. She alleged: that the Warnings section was ambiguous as it relates to the type or category of patient at risk for suicide; that the Adverse Reactions section diminished the suicide caveat found in the Warning section; and that there was insufficient warning of the increased suicide risk when reserpine is prescribed in conjunction with HCT.

THE COURT'S APPROACH

The court's analysis is instructive and, perhaps, revealing.

Unsatisfied to merely scrutinize the package inserts from arm's length, the panel dissected the work of the technical writer and considered at the micro and macro levels not only what was said but what may have been implied or suggested in whole and in part.

"Whether a given warning is legally adequate or presents a factual question for resolution by a jury requires careful analysis of the warning's language," Judge Hancock wrote for the court as he began the analysis. "The court must examine not only the meaning and informational content of the language but also its form and manner of expression." Judge Hancock's prescription called for a surgical, tripartite examination of the package insert, with a focus on the warning's accuracy, clarity, and consistency. The court went to considerable lengths to spell out precisely the standard of review, carefully defining its own terms, as follows:

Accuracy—"For a warning to be accurate, it must be correct, fully descriptive and complete and it must convey updated information as to all of the drug's known side effects."

Clarity—In the context of a drug warning, the language of the admonition must be "direct, unequivocal and sufficiently forceful to convey the risk." The court went on to state:

"A warning that is otherwise clear may be obscured by inconsistencies or contradictory statements made in different sections of the package insert regarding the same side effect or from language in a later section that dilutes the intensity of a caveat made in an earlier section. Such contradictions will not create a question of fact as to the warning's adequacy, if the language of a particular admonition against a side effect is precise, direct, unequivocal and has sufficient force. The clarity of the overall warning may in such instances offset inconsistencies elsewhere in an insert."

Consistency—Essentially, the court said the whole is greater than the sum of its parts: "While a meticulous examination and parsing of individual sentences in the insert may arguably reveal differing nuances in meaning or variations in emphasis as to the seriousness of the side effect, any resulting vagueness may be overcome if, when read as a whole, the warning conveys a meaning as to the consequences that is unmistakable."

THE DOCUMENTATION AND THE ANALYSIS

The court analyzed and scrutinized the package inserts, pursuant to the allegations of insufficiency raised by the plaintiff. Judge Hancock specifically addressed the various sections of the inserts that gave rise to this lawsuit.

The Warnings Section

The Warnings section contains three declaratory sentences:

1. Extreme caution should be exercised in treating patients with a history of mental depression.

2. Discontinue the drug at the first sign of despondency, early morning insomnia, loss of appetite, impotence, or self-deprecation.

3. Drug-induced depression may persist for several months after drug withdrawal and may be severe enough to result in suicide.

Cynthia J. Martin, the widow, asserted that the first sentence set the parameters for the following sentences in the Warnings section. Namely, she argued that the caveats of the second and third sentences apply, or appear to apply, only to the patients described in the first sentence, those "with a history of mental depression." Mrs. Martin's contention was that the sequence of information—the pattern in which it was presented—made it unclear.

The court rejected her argument, described the third sentence as "direct, unqualified, and unequivocal," and added that it would "defy common sense and subvert the clear intendment of the third sentence to read it as limited by the first sentence of this section."

The Adverse Reactions Section

Under the FDA labeling specifications, the Adverse Reactions section should follow the Warnings section of drug package inserts. Mrs. Martin contended that a portion of the Adverse Reactions section served to dilute the third caveat in the Warnings section. The relevant portions read as follows:

> Central nervous system reactions include drowsiness, depression, nervousness, paradoxical anxiety, nightmares. . . . These reactions are usually reversible and usually disappear after the drug is discontinued.

Mrs. Martin contrasted the final sentence in the above quote with the third admonition in the Warnings section, which advised that medication-induced depression may well persist after drug use is discontinued.

A contradiction?

The court said no.

It noted that the last section of the Adverse Reactions section pertains to the usual duration of the side effect, not to its seriousness. "Thus, the last sentence does not contradict the unequivocal and straightforward statement in the Warnings section that drug-induced depression may be severe enough to result in suicide."

Furthermore, the court looked to another section, the Actions portion, that precedes both the Warnings and Adverse Reactions sections. The Actions section states: "Both cardiovascular and central nervous system effects may persist for a period of time following withdrawal of this drug."

So, what we have here is the Actions section advising that the side effects may continue even though use of the product has ceased, the Warnings section admonishing that effects may persist after withdrawal and may be serious enough to result in suicide and the Adverse Reactions section offering that the reactions typically subside when drug use is curtailed.

Taken as a whole, the court said, those sections "are sufficient to convey to any reasonably prudent physician an unambiguous and consistent message" which "comports exactly with the risk of taking reserpine."

The Dosage and Administration Section

Mrs. Martin's final claim, that the reserpine insert does not sufficiently warn of the dangers inherent in ingesting both reserpine and HCT, was rejected by the court as meritless.

The Dosage and Administration section states that "concomitant use of reserpine with (other drugs) necessitates careful titration of dosage with each agent." Also, the HCT package insert includes a forewarning that the drug "may add to or potentiate the action of other antihypertensive drugs."

THE DECISION

The court dismissed the lawsuit against Chelsea and Rugby, holding that the package inserts "contained language which, on its face, adequately warned against the precise risk in question, i.e., depression-caused suicide." It upheld a lower court and found "no triable issues of fact regarding the warning provided to physicians by the package inserts accompanying these prescription drugs."

In other words, the tech writing staff did its part in adequately conveying information to the prescribing physician. The remaining question for a trial court jury is whether the physician did his part and adequately warned Eugene Martin of the danger of drug-induced despondency.

CONCLUSION

Gerald M. Parsons of the University of Nebraska, Lincoln, has written in this journal about the growing number of lawsuits focusing on the work of technical writers, and the increasing willingness of courts to carefully analyze, and perhaps second-guess, the writer's phraseology. *Martin v. Hacker* is progeny to this trend.

The potential exposure to the pharmaceutical companies involved in this case was enormous. But they escaped liability, precisely because the court was willing to take extraordinary care in analyzing the prose of the technical writer. It seems self-evident, for legal and moral reasons, that if the courts are willing to put that much effort into reading and evaluating a tech writer's work, then the author and employer must be at least as diligent in their construction of technical documentation. The stakes are substantial, and not just in a legal sense. When a technical writer's work is unclear, and an operator inadvertently reformats a hard drive, that's unfortunate. But if a technical writer's inaccuracy or imprecision claims a life, that's another matter altogether.

BIBLIOGRAPHY

Enterprise Responsibility for Personal Injury, Vol. II, *The American Law Institute*, Philadelphia, April 15, 1991.

Holzer, H. M., Product Liability Law: The Impact on New York Businesses, *Brooklyn Law School,* 1990.

Markel, M. H., *Technical Writing: Situations and Strategies,* St. Martin's Press, New York, 1992.

Martin, Cynthia J., Individually and as Executrix of Eugene J. Martin, Deceased, v. Arthur Hacker, et al., and Chelsea Laboratories, Inc., et al., *83 NY2nd 1,* Nov. 23, 1993.

Parsons, G. M., A Cautionary Legal Tale: The Bose v. Consumers Union Case, *The Journal of Technical Writing and Communication,* 22:4, pp. 377–386, 1992.

There Ought to Be a Law: Product Liability in New York State, *The Public Policy Institute, Special Report,* June 1991.

Establishing Relations in Germany

Valerie Frazee

Communicating across cultures can present difficulties and take extra time. German business culture tends to be formal and structured. This article discusses the German emphasis on data and the need to know someone's credentials before conducting business.

Consider the case of one American-German partnership that started off on the wrong foot. Terri Morrison, president of Getting Through Customs based in Newtown Square, Pennsylvania and coauthor of the book *Kiss, Bow or Shake Hands,* shares the story of an American manager with a U.S. company purchased by a German firm. This manager made the trip overseas to meet his new boss.

Morrison explains: "He gets to the office four minutes late. The door was shut, so he knocked on the door and walked in. The furniture was too far away from the boss' desk, so he picked up a chair and moved it closer. Then he leaned over the desk, stuck his hand out and said, 'Good morning, Hans, it's nice to meet you!'"

The American manager was baffled by the German boss' chilly reaction. As Morrison reveals, in the course of making a first impression he had broken four rules of German polite behavior: punctuality, privacy, personal space and proper greetings. This first meeting ended with both parties considering the other rude, a common result of cross-cultural misunderstandings.

A LOVE OF STRUCTURE

The most important thing to understand about Germans, according to both Morrison and Dean Foster, director of the cross-cultural training division of Princeton, New Jersey-based Berlitz International Inc., is that they have a high regard for authority and structure. "From our perspective, the Germans appear to us as people who are very compartmentalized, heavily emphasizing the structure, much more concerned about the process than what they're doing," Foster explains. "Germans perceive Americans as being far too fluid, far too mushy, far too unfocused."

Germans' love of structure can mean that communicating through their organizations will take a little longer, as employees participate in consensus-building conversations and check to make sure everything is in order before moving ahead to the next phase. This sense of structure extends to the physical world and influences even personal appearances. Morrison notes Germans tend to stand straight up, rarely putting their hands in their pockets and never slouching in a meeting. German greetings are formal, always employing the use of titles such as doctor or professor. And German companies are full of offices with closed doors.

The easygoing, familiar demeanor of an American businessperson clashes with these German values. Morrison warns: "You don't want to take the attitude of the laid-back American. . . . Being an entrepreneur is wonderful and is respected around the world, but when you go to Germany, [the Germans] respect authority."

Among other things, Germans respect big names and big numbers. If you work for a company with name recognition and you have an impressive title, play these things up on your business cards. Also emphasize the number of years a company has been in business or the number of workers your organization employs.

HOW TO PREPARE

So how can you put all this information to use? First off remember that Germans like to work with a lot of data. So proving to them that you have found a better way of doing something will take more than a demonstration of how well your way works. Germans are likely to ask: How did you reach that conclusion? What was your method? Foster recommends being prepared to present your evidence. And part of that is going into the discussion knowing what the German way of doing the same thing is.

If expatriates will be giving presentations in Germany, Morrison advises they have all sorts of documentation with them and that their presentation materials are thoroughly researched. And HR should advise employees not to start out with a joke or a funny story. Germans don't appreciate humor in a business setting.

Germans prefer not to mix business with pleasure. Creating a friendly work environment to encourage productivity seems to be an American concept. Advise expatriates not to be disheartened when they find this isn't a universal work style. "The warm and friendly atmosphere may develop over time, but at first you have to establish respect and you do that by acknowledging the level, the status, the achievements and the rank of your colleagues—and they, in turn, [will do so] with you," Foster explains.

This doesn't mean the Germans don't form close relationships, or that they are a less emotional people, as some stereotypes would suggest. In fact, Germans would say that Americans are too casual in their offhand manner of forming friendships. Foster says, "The complaint I've heard over and over from Germans is that you can't get close to Americans. They appear friendly when they shouldn't be—there's no place for that in business. But when you finally get to know [Americans], they never want to make that deep commitment."

THE GLASS CEILING

The glass ceiling is a little lower in Germany than in the United States, meaning women have to work harder to establish that highly regarded sense of respect from work colleagues. Morrison explains: "Women have pretty high positions in government—and that's all. Women generally don't have big-deal jobs in private industry."

She shares an anecdote from a senior-level American woman on a U.S. team that met with a German team in the course of the merger of their two companies. The woman was extremely frustrated. The Germans wouldn't address her in the course of the discussion.

Fortunately this is fixable. Remembering that credibility is a key issue, managers need to put in extra effort to establish the authority of women team members. If a woman is in charge of a team, the men on the team need to support her. Morrison says: "When a question is addressed to the U.S. group, [all the men in the group] need to look back at the [woman manager] and say: 'Well, what do you think?' If the team won't do that, the [women managers] can't win."

Foster says that women who are known authorities or experts in their field will be treated as respected work colleagues. So the trick is communicating and establishing that credibility. This is true of men too, just to a lesser degree. "I think [the Germans] need to know before the meeting who you are and why you're the one selected to be there," he says.

He adds that in many cases it depends on the individual woman—and on the particular German: "As an American woman, it's understood that you don't necessarily have to follow the same rules." He continues: "But if you're working with older and traditional German men, it still may be difficult for them to understand."

There's much that binds the German and American cultures together. The people dress similarly, they live in democracies and they have an equal interest in the bottom line. But the challenge is to uncover the differences. Being aware of these differences greatly improves your odds for a successful business relationship.

ePublishing

Amy Garhan

Readers on the Web expect to get the information they want quickly. Web readers usually decide within a few seconds whether to stay on a site. This article presents eight strategies for writing useful and readable Web documents.

On the Web, snap judgments are the norm. It doesn't matter how compelling or on-target your writing is—if your online readers can't tell inside of a few seconds what your page is about and whether it meets their needs, they'll click away. And often, even if they see that you're giving them what they want, they can't or won't take the time to plow through your work.

If you don't make your basic points instantly, chances are you won't be able to communicate anything at all. Smart online writers and editors accept this limitation, and succeed anyway. Here are eight basic techniques that can make your Web writing easy to read, or scan:

1. Keep It Short and Break It Up

Web users tend to be in a hurry, so keep your Web writing short. Aim for about half the word count you would use in print. Break your text into fairly self-sufficient chunks of about 300 to 500 words each.

If you must present long documents, consider placing the main sections on separate Web pages. Just be sure that each page stands well on its own, and provide navigation throughout so your online audience understands how the piece is organized. However, avoid taking a piece that's meant to be read from beginning to end and chopping it onto separate pages. Sure, your readers won't have to scroll as much but making them click more just for the heck of it isn't a better solution.

2. Use Intuitive Headings

Begin every section of a Web page with a visually prominent heading that sums up the main point of that section. Avoid cute headings or teasers, where further reading is needed to get the true meaning. Keep headings short and make them understandable to a general audience. Consider offering an index of these sections at the top of the page; this serves as an outline as well as aids navigation of long pages.

3. START WITH A SYNOPSIS

It's usually best to lead with a couple of sentences that sum up the entire page in a nutshell. Indicate what that page is about, why it matters and who should care. Story-style leads and suspense tactics generally don't work as well on the Web.

4. CRAFT LINK TEXT CAREFULLY

Hyperlinks stand out visually on a Web page, so Web readers view them as signposts as well as connections. Keep your links very short (one to three words if possible). Link text should indicate where online readers would end up and what they would find there. (This is why "click here" links are inferior.) Well-crafted links are so important that you should edit your sentences specifically to yield good links. Also, don't put too many links too close together; they'll lose their impact.

5. HIGHLIGHT KEYWORDS

Ideally, every section should contain one or two words or phrases that visually stand out from the main text and serve to expand on the section heading. Well-crafted links can accomplish this goal, but highlighted keywords are another approach. Try to keep them to three words or less and choose text that encapsulates the main point of the paragraph. Be willing to edit sentences specifically to yield good keywords.

6. USE BULLETED LISTS

Bulleted lists are much easier to scan than narrative text, so don't hesitate to use them much more than you would in print. If you're ending up with a long, complex page, look for opportunities to use bulleted lists as a way to cut the word count and simplify the structure. Include a blank line between list items. If list items are more than a few words long, begin each with a very short heading in bold type (like the list you're reading now). Only use numbered lists to indicate a sequence or hierarchy.

7. OFFER A PRINTER-FRIENDLY VERSION

If you're presenting a longer piece that is meant to be read from beginning to end, consider offering a "printer-friendly" version of the document. This is still a Web page (HTML file), but it offers the same text stripped of images and design elements. This way, online readers who scan your writing and find it worth reading have the option of reading your document on paper, away from the computer (which not only is more convenient, but also easier on the eyes).

8. INCLUDE USEFUL PAGE TITLES

You can—and should—create a page title for every Web page on which your writing appears. This is a line of text displayed above the browser's menu bar. The page title should succinctly encapsulate the purpose of that page. Although few people will read the page title while browsing your page, if they decide to bookmark that page, the page title is what will appear in their list. And if a bookmark listing doesn't make sense, it's unlikely that they'll return to that page.

If you're writing for your own site, you can implement all of these techniques yourself. If you're writing for someone else, be proactive about suggesting link text, headings, highlighted keywords and how to break up longer pieces. Many online publishers welcome this level of input from writers, as long as you're not trying to dictate design issues.

In fact, if you can file your text in the form of an HTML document rather than a word processing or text-only file, you can handle these matters yourself rather than relying on someone else to interpret your instructions.

Behavioral Interviewing:
Write a Story, Tell a Story

Sandra Hagevik

One of the techniques interviewers use to assess job candidates is to ask questions about knowledge and experience related specifically to the skills needed for the position. Successful candidates will use the opportunity to relate stories about their accomplishments. This article discusses the interviewer's typical questions and gives advice about how to develop answers.

Whether you're seeking a job or seeking new employees, your ability to write or tell good stories could make an enormous difference in the success of your undertaking. That's especially true with a structured form of interviewing developed by Dr. Paul Green of Behavioral Technology, Inc., which increases the potential for matching a person's skills to the requirements of a job. This type of screening is far less subjective than more informal methods, and it relies on good story-telling techniques.

PREPARATION FOR INTERVIEWERS

Interviewers using this method ask good questions. They rate candidates' job or technical skills and performance levels using measurements aligned with specific job responsibilities and a rubric that determines how well those tasks are performed. Here's how it works. Before screening applicants, the interviewers carefully define job responsibilities to identify representative skills and capabilities. Questions are then designed to probe for depth of knowledge and experience relating to each skill, as well as for insight into personal characteristics—just as story writers look for underlying motives and circumstances. Finally, a grading scale is developed to quantify answers, usually on a scale of three or five points that measures responses as everything from "exceeds expectations" to "does not meet expectations." Interviewers must agree upon the specifications for each category, and receive training on how to identify them in candidates' responses. Only then, after the groundwork is laid, can an interview be conducted—it's a bit like revealing the plot of a story.

Interviewers search for answers that reflect thinking skills, problem-solving strategies, working habits, ability to learn, flexibility, and other personal characteristics relevant to job success in their corporate culture. They may ask questions to clarify their understanding of answers that are vague or too general. If they discover a positive, they may seek a negative, or vice versa. They may probe for strengths and weaknesses, successes and failures, challenges and problems. The questions are often asked about hypothetical or real-life situations one may face on the job. Or the candidate may be asked to complete a job-related task on site in a measured time frame. Computer programmers may be asked to write several lines of code, writers to synthesize conflicting sources of data, managers to design a budget or project plan, instructors to teach a class or make a presentation. Behavioral interview questions or tasks are preplanned, structured, and consistent; they focus on job responsibilities, seek specific examples, and are open-ended. Like good story writers, good interviewers are as curious about character and motivation as they are about intelligence and the ability to learn.

Following the interview or behavioral sample, the candidate will be rated according to an agreed-upon scale. The score is tallied, and a decision is made to continue the process or end the story. The basis for that decision is how clearly, concisely, and precisely the interview questions were answered.

If You're the Job Seeker

Prepare a wide range of brief stories about your accomplishments to illustrate specific skills. Gain a thorough understanding of job specifications, tasks, or requirements. Frame your responses to questions as described below.

Demonstrate Results with Short Answers, Vivid Examples

Typically, an interviewer will start with general questions to review your work history (refer to the extra copy of your résumé you have brought along). Go light on history, then cite an example or two of recent accomplishments that parallel expectations for this job. Because your interviewers will be looking for behavioral examples, expect questions that lead with "Give me an example of . . . ," "Tell me about a time when . . . ," and "Describe a situation in which you . . . " *Always* focus your answer on your role, the steps you took, the strategies you invoked, and the results that ensued. If the results were disappointing, tell a story about what you learned from the experience. Other questions you might encounter include the following:

- Tell me about a time when you were proud of your decision-making skills.

- Give me an example of a problem you solved and what your role was.

- Tell me about a time when you failed.

- What activities in your previous job tapped into your creative capabilities?

- Describe a situation when you had a conflict with a supervisor.

Avoid answering such questions with responses that are vague, abstract, redundant, incomplete, or off target. Interviewers report annoyance with people whose answers miss the question, especially in this type of interview. Also, don't spend much time beating yourself up or reviewing your painful past. Concentrate on short, vivid stories that demonstrate learning.

To prepare for behavioral interviews, practice telling brief stories about your accomplishments. Make them concise, interesting, focused, and purposeful. Preparation is key, and in the process of constructing responses for behavioral interviews, you will have identified a model for other types of communication—for any situation in which being listened to is important. We all want our stories to be heard.

PowerPoint Presentations:
A Speaker's Guide

Geoffrey J. S. Hart

PowerPoint, like all visual aids, should support your oral presentation, not overwhelm it. This article provides guidelines for using PowerPoint in ways that enhance your presentation.

Vinton Cerf, one of the founders of the Internet, reportedly parodied the well-known quote about the cost of attaining power, observing that if power corrupts, "*PowerPoint* corrupts absolutely." Pointed though Cerf's statement is, it places far too much blame on the software. After all, speakers must take some responsibility for their presentations. As in any other form of communication, you must decide what you're going to say and how you plan to say it. But once that's done, you need to use all the skills at your disposal to make the chosen medium work for you.

Let's assume that you've chosen to make an oral, computer-based slide presentation, whether with *PowerPoint* or any of its competitors, and that you've practiced the material well enough to speak comfortably on it. Now your task is to create a presentation that will support your talk. Although a well-crafted, well-delivered speech can often succeed without visual accompaniment, presenting a message in two media (spoken words combined with onscreen text, for example) can improve retention of your message. Here are ten things to keep in mind as you develop your presentation.

IT'S ALL ABOUT YOU

People are present to hear you speak, not to read slides; if you feel they would be satisfied reading slides, stay home and send them a printout or a self-running presentation file. The true value of a presentation is what *you* have to say, plus your ability to expand upon your original message if the audience seems to be missing the point or raises questions. Use the slides to support your talk, not replace it. Visuals should help the audience understand and remember what you're saying, so design the onscreen information to support that goal.

Since reading slides is as deadly as reading a prepared speech, use the slides primarily to remind you of what you wanted to say and to focus the viewer's attention. It's natural for viewers to want to read the text you present, so take advantage of that desire and use the projected images to provide context for what you're about to say.

501

Reduce Your Text to the Minimum

The more time people spend reading text, the less attention they can devote to you, so keep text as short and focused as possible. Onscreen text should summarize the point you're about to make, not provide all the details. A well-designed screen typically contains a title that provides the overall context, an introductory line or heading that specifies what you'll be talking about in that context, and no more than four bullets of five words each. If you have more to say, use several slides rather than cramming the information into a single slide.

If you want the presentation to include the text of your speech rather than just the bullet-point summary, include that extra text using the "speaker's notes" that *PowerPoint* and its competitors provide. This information will be visible when you print out the presentation, but not when you display the individual slides. If you don't known how to use this feature, provide the information in the form of a printed handout. PDF files available for downloading are another option.

Whether to provide the handout before or after your talk is hotly debated. If the audience receives the handout in advance, they'll be less inclined to try to write down all of the information on your screens while you talk; on the other hand, they may spend more time reading your handout than listening to you. One suitable compromise involves providing the handout far enough in advance of the talk that your audience has time to skim through it before you begin speaking. That's particularly important if you'll be displaying large or complex graphics that will be difficult to see onscreen; you can describe things on the handout that would be impossible to show on the screen. If you choose to provide the handouts after your talk, explain to the audience that it won't be necessary to record each screen.

Don't Do Everything Simultaneously

If you present multiple points on each screen (most commonly in the form of bullets), bring these points onscreen one at a time, not simultaneously. Audiences will try to read everything on the screen, and while they're reading, they're not listening to you. So present the information in bite-sized chunks, pausing long enough between chunks for the audience to read the five or so words that I recommended—typically a few seconds.

Never trust that what you expect to appear onscreen has actually appeared. In an ideal setup, you can see what's onscreen by pressing the "next slide" button on your laptop—you don't have to turn your back to the audience, and the audience has time to read each line of text. (Be aware, however, that it's easy to accidentally tap the "next slide" button twice. Learn how to retrace your path when this happens.) Less ideal setups may force you to stand sideways so you can keep one eye on the screen and the other on your audience.

MAKE IT LEGIBLE

For a presentation to be legible, the signal must stand out from the noise; that is, the text and graphics must be legible against their background. Black text on a light background (for example, ivory or pale yellow) provides maximum contrast without relying on a glaring white background that strains the viewer's eyes, particularly in a darkened room. But white or off-white on a dark blue background and yellow on a dark green background also work well, and are easier on the viewer's eyes.

Try to limit yourself to two typefaces if you're not producing a presentation on typography, since unnecessary typographic changes draw attention to the type at the expense of the message. One typeface for titles and one for the main text should suffice for most purposes. Whether to use serif or sans serif type is less of an issue than it used to be, since serif typefaces designed for on-screen use can be every bit as legible as typical sans serif typefaces; in fact, the serif font is sometimes *more* legible, since many sans serif fonts make it difficult to distinguish between lowercase L, l, and the number one.

The type must always be large enough to be read at a distance. To test legibility, stand about ten feet from your monitor and ask yourself whether you can read the text easily: if you can't read it, neither can your audience. Please note that ten feet is only a crude estimate: You should stand far enough away that your computer monitor appears to be the same width as the projection screen will appear to viewers sitting at the back of the conference hall.

USE GRAPHICS JUDICIOUSLY

It seems obvious that you should use graphics to communicate graphical concepts ("a picture is worth a thousand words"), but you'd be surprised how often this isn't done—or is done poorly. (Of course, if you've sat through as many presentations as I have, maybe you're not surprised.) Unfortunately, most graphics appear in the form of weird lines or shapes superimposed on the slide background. Although these graphics improve the esthetics of your presentation—in those rare cases when the audience actually shares your taste in *objects d'art*—more often they simply clutter the screen, make it harder to read, and distract the audience.

Instead, use graphics that illustrate what you want to say, and use a pointer to direct attention to features of the graphic that you want to emphasize while you talk. Be wary of laser pointers, since they often dazzle the viewer's eyes; a yardstick or a telescoping metal stylus often work far better. Better still, emphasize the key parts of each graphic directly in your presentation software by adding graphical pointers (arrows that point to or circles that surround the point of interest) one at a time as you talk. If you don't know how to automate this process so that new pointers appear and old ones disappear each time you tap the keyboard, use a new screen for each image. Avoid overwhelming viewers with a myriad of highlighted information.

DON'T GET TOO COMPLEX

If you do present graphics, simplify them as much as possible. The principle of "abstraction" means that you should show only the visual features that most directly concern the viewer, not all possible features. For that reason, an illustration is often simpler and clearer than a photograph. In technical communication, software screenshots are a mainstay of our graphics both because they relate strongly to our typical topic and because, if you're like me, you can't draw a straight line with a ruler, let alone create a convincing illustration. But nobody can read an unenlarged *Excel* spreadsheet onscreen. Where possible, crop and enlarge the image to focus only on the details you're discussing.

CUT THE MULTIMEDIA

As a general rule, eliminate all multimedia effects, particularly the fancy whizbang sounds and transitions that most presentation software now offers. It *is* possible to use animations and sound effectively, just as it's possible to run a mile faster than four minutes—but few of us can do either. The main exception, of course, is when you can't communicate a concept in any way other than using sound or animation. It's not possible to communicate music well without actually playing the music, and the motion of objects is best handled with moving video. Just remember that, although skillful use of multimedia can improve communication, unskilled use can distract the audience into paying so much attention to the media that they forget the point you're trying to make. (Usability expert Jared Spool reported exactly this effect in one example of "talking heads" videos used in online help: Viewers really liked the help file, but retained less information than in a traditional version with only text and static graphics.)

STRIVE FOR CONSISTENCY

Learn to use your software's template features rather than building each screen from scratch. Consistency is important indeed in a presentation, since visual objects such as logos that jump from position to position between slides draw the eye; the viewer ends up playing "Where's Waldo?" instead of concentrating on your presentation. That's also true of colors, fonts (typeface and size), and layout. Don't change *anything* randomly—change it to make a point by drawing the eye intentionally to something you want viewers to focus on.

AND IN CONCLUSION . . .

Novelist Terry Pratchett once remarked that "wisdom is one of the few things that look bigger the further away it is." That's particularly true of presentations. If you want your presentation to hold up to close scrutiny and still make you appear as wise as you undoubtedly are, keep it simple and straightforward by following the tips presented here.

Business Manners

Ann Chadwell Humphries

Successful professionals must understand appropriate etiquette for typical business situations. Knowing proper business etiquette will instill confidence and help you respond effectively to others in the workplace. This article reviews common etiquette mistakes and gives advice on how to avoid making them.

COMMON ETIQUETTE BLUNDERS

1. Being rude on the telephone. This, the most common business etiquette blunder, includes not returning phone calls promptly, not identifying oneself, and screening calls arrogantly. Always return calls within 24 hours—and preferably the same day. Identify yourself and the nature of your business up front; anticipate resistance and be helpful. And help people get the information they need. Avoid overly protecting the boss with questions too directly asked, such as, "What is the nature of your call?"

2. Interrupting. Let people finish their sentences, and ask permission if you need to interrupt their work. Don't barge into conversations or offices without giving a signal.

3. Introducing people incorrectly. Say the most important person's name first, then the secondary person's. In the workplace today, the most important persons—regardless of sex or age—are outside guests, officially titled people, and superiors. "Outside visitor, this is our company President." "Mayor, this is our Vice-President of Marketing." "Boss, this is an employee of 35 years."

4. Wimpy or vise grip handshakes. Wimpy handshakes or overpowering grips are no-noes. Grab your handshake high, around the thumbs, and shake in kind. Don't swoop on delicate folks, or withdraw from an acrobatic handshaker. And women—get your hand out there! Men hesitate to initiate this action. They've been taught to wait for a lady to extend her hand first. Handshakes are expected in business life. So learn to shake hands well, and offer yours easily.

5. Incorrect eye contact. In the United States, looking people in the eye means you have nothing to hide, that you are listening, that you are interested. In certain other cultures, however, direct eye contact is considered confrontational and disrespectful. Vary eye contact from short glances to longer holds of 3–5 seconds, but don't stare to make yourself look good. You will be overbearing.

6. Poorly managed business meetings. Start and end them on time. Distribute agendas ahead of time so people can prepare. If you are the leader, control the action. Don't suppress confrontation, or it will emerge elsewhere. But you can defer it to a more appropriate time. Last, limit telephone calls to the leader, so the group will not be held up.

7. **Poor or inappropriate appearance.** To the business executive, appearance is important. It's not the only measure of a person, but it does give visual interest to doing business and indicates your knowledge of, and respect for, the rules of the game. You must dress with understated distinction to be considered seriously in business. Invest in quality clothing, and be impeccably groomed.

8. **Forgetting names.** Ask anybody about his or her most uncomfortable business etiquette dilemma and forgetting names will emerge in the top five. To help you remember names, one positive strategy is to repeat the person's name as soon as you hear it, and to use the name at least once if you have a long conversation. Don't over-use a person's name, however.

But when you don't remember someone's name, there are several simple strategies to follow. Use your judgment about which is most appropriate for, and least embarrassing to, you and the other people involved.

a. **Introduce yourself first.** This is for the occasion—either in a one-on-one conversation or if you're part of a group—when you only vaguely remember the person. Introducing yourself helps take people off the spot.

b. **Stall for time.** When you know the person, but it takes a few minutes for your mind to compute the name, act friendly and don't withdraw. Don't sport the "Charlie Brown" nervous smile. Focus on them, not yourself.

c. **Introduce the people you do know.** Pause to let those whose names you don't remember introduce themselves. Or, say lightly, "I'm so sorry. I know who you are. I've just gone blank." Or, "I've just forgotten your name," with an implied, "Silly me." Keep the conversation going with what you do remember.

d. **Help out someone who looks as if he or she is struggling to remember *your* name.** This is not pompous or arrogant; it's considerate. Many people slur their names so they're unintelligible, or present themselves in an ordinary manner, so they aren't memorable. Take care when you introduce yourself.

Code of Ethics for Professional Communicators

International Association of Business Communicators

Companies, organizations, and individuals need to follow ethical guidelines that match their activities. This article reprints the code of ethics available on the Web site of the International Association of Business Communicators (IABC) [http://www.iabc.com]. The association stresses ethical behavior for communicators as part of its professional development programs.

PREFACE

Because hundreds of thousands of business communicators worldwide engage in activities that affect the lives of millions of people, and because this power carries with it significant social responsibilities, the International Association of Business Communicators developed the Code of Ethics for Professional Communicators.

The Code is based on three different yet interrelated principles of professional communication that apply throughout the world.

These principles assume that just societies are governed by a profound respect for human rights and the rule of law; that ethics, the criteria for determining what is right and wrong, can be agreed upon by members of an organization; and, that understanding matters of taste requires sensitivity to cultural norms.

These principles are essential:

- Professional communication is legal.

- Professional communication is ethical.

- Professional communication is in good taste.

Recognizing these principles, members of IABC will:

- engage in communication that is not only legal but also ethical and sensitive to cultural values and beliefs;

- engage in truthful, accurate and fair communication that facilitates respect and mutual understanding; and,

- adhere to the following articles of the IABC Code of Ethics for Professional Communicators.

Because conditions in the world are constantly changing, members of IABC will work to improve their individual competence and to increase the body of knowledge in the field with research and education.

ARTICLES

1. Professional communicators uphold the credibility and dignity of their profession by practicing honest, candid and timely communication and by fostering the free flow of essential information in accord with the public interest.

2. Professional communicators disseminate accurate information and promptly correct any erroneous communication for which they may be responsible.

3. Professional communicators understand and support the principles of free speech, freedom of assembly, and access to an open marketplace of ideas; and, act accordingly.

4. Professional communicators are sensitive to cultural values and beliefs and engage in fair and balanced communication activities that foster and encourage mutual understanding.

5. Professional communicators refrain from taking part in any undertaking which the communicator considers to be unethical.

6. Professional communicators obey laws and public policies governing their professional activities and are sensitive to the spirit of all laws and regulations and, should any law or public policy be violated, for whatever reason, act promptly to correct the situation.

7. Professional communicators give credit for unique expressions borrowed from others and identify the sources and purposes of all information disseminated to the public.

8. Professional communicators protect confidential information and, at the same time, comply with all legal requirements for the disclosure of information affecting the welfare of others.

9. Professional communicators do not use confidential information gained as a result of professional activities for personal benefit and do not represent conflicting or competing interests without written consent of those involved.

10. Professional communicators do not accept undisclosed gifts or payments for professional services from anyone other than a client or employers.

11. Professional communicators do not guarantee results that are beyond the power of the practitioner to deliver.

12. Professional communicators are honest not only with others but also, and most importantly, with themselves as individuals; for a professional communicator seeks the truth and speaks that truth first to the self.

Ethos

Charles Kostelnick and David D. Roberts

Companies try to build a consistent image through their logos and signs. This article from the book Designing Visual Language *discusses how different designs for logos and signs can reflect a specific ethos (basic values or character) for a company.*

The success of signs and logos often depends on the ethos they build—or fail to build. To do their work effectively, these images have to be credible and relevant. Like other forms of visual language, an unprofessional-looking image can devastate the trust it's supposed to build. For example, the image in Figure 9-36 would be inappropriate for representing a bank or other financial institution—it's too simplistic and doesn't project qualities customers look for in a bank, such as strength and prestige. It simply lacks ethos for this purpose.

Sometimes ethos can be enhanced by conventional forms that meet reader expectations for the entity the symbol represents. Typical ways to represent a financial or insurance company would be to use the picture of a famous historical figure (a president or other patriot) or a picture of an impressive natural or artificial object like a tree, a mountain, or a home office building.

Our roller blade sign (Figure 9-37a) builds credibility through a more rigid and technical convention—a circle with a slash through it. Most readers will associate this symbol with other proscriptive signs telling them where not to park, walk, or ride their bikes. The convention of the circle and slash gives the sign instant credibility. We could create the same message with another version of the sign, like the one in Figure 9-37b, but it wouldn't be as credible. In fact, readers might disregard this "no roller blading" message because it flouts the accepted convention and thereby lacks authority.

We can also assess the credibility of a sign, then, by how well it embodies the character of the group it represents and which aspects of that character it emphasizes. Given that groups are seldom of one mind, creating a logo that reflects a group's collective character can be a real challenge. Suppose we're

FIGURE 9-36 Bank Logo Lacking Ethos.

FIGURE 9-37   Following Conventions Can Build Credibility.

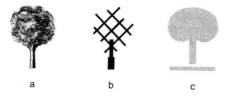

FIGURE 9-38   Ethos May Depend on the Readers.

choosing a logo for a regional group that promotes tree planting as a way to beautify communities and preserve the natural environment. What are the relative strengths and weaknesses of the logos in Figure 9-38 in representing that mission?

Logo *a* looks more natural than the others because it seems to be drawn freehand, whereas the others are geometric and mechanical. For those members who see themselves as protectors of natural things, who find intrinsic beauty in such things, and who deplore the encroachment of the artificial, the logo in Figure 9-38a will build trust. It's also more realistic, which might engender credibility with more traditional members who see their role as protectors and conservators rather than as advocates of change. On the other hand, its rustic, homemade appearance may reduce its credibility with readers who want the organization to project a more professional, businesslike ethos.

By contrast, the sharp, angular branches of the logo in Figure 9-38b give it a contemporary, dynamic appearance. It might find most credibility with readers who see the group as an advocate for aggressive action, a lightning rod for change. The image might also appeal to urban members because of its hard-edged design. On the other hand, more traditionally-minded members might find this highly abstract form odd and unsettling and perhaps even antithetical to the goals of the organization—to restore and maintain nature.

The simple, geometric forms of the logo in Figure 9-38c give it a universal look; they also soften and subdue the logo, rendering it more approachable as well as more passive. This logo expresses an ethos of acceptance and political neutrality. The leadership within the group might find this logo weak and insipid, while volunteer workers might find comfort in its calm, steadfast inclusiveness.

As you can see, creating an image to represent the identity of an organization takes a good deal of thought and revision, especially when the primary audience includes the group members themselves. However, creating a visual identity can be a two-way street: An image can gain credibility if it accurately reflects the ethos of an organization, but in doing so, it can also help *build* that ethos.

It's All in the Links:
Readying Publications for the Web

Mindy McAdams

This article provides guidelines for creating useful hypertext links in electronic documents. The author includes cautions about common mistakes writers make.

If there were a single door leading to the World Wide Web, someone might hang a giant sign on it reading **EDITORS WANTED**. Adapting existing documents for use on the Web (or on an Intranet, an in-house-only version of the Web) is a distinctly editorial process that today is often handled by graphic designers or, in some cases, by automation.

Editors who want to work with Web documents should start by learning HTML (hypertext markup language), the code on which the World Wide Web is built. HTML isn't a programming language; it's a system of codes for tagging structural elements in a document. Motivated people can master basic HTML from a book in three or four days—it's no more complex than the editing codes that most newspaper copyeditors are required to learn.

HTML codes tag elements, such as headings and paragraphs, according to their purpose in the document. They can also define links between related documents like a set of pamphlets on employee benefits or the five parts of a serial newspaper article.

READABILITY CAN BE BUILT IN

The biggest difference between reading on-line and reading printed text is that on-line the text moves. You're scrolling up and down. As you scroll, you tend to scan the document. The environment urges you to speed-read, so if you don't see anything but plain text, you're likely to lose interest.

Two simple practices greatly increase the readability of on-line texts:

- Use headings and boldface more than you would in print.

- Keep paragraphs short.

Boldface headings enhance the reader's ability to scan text rapidly. In print, too many headings make a page look clunky and disjointed, but on-line headings disappear off the top of the screen as the reader scrolls down. As a rule of thumb, one heading per screen of text within a document works very well.

Headings aren't just window dressing. Each should provide a clear predictor of the content of the paragraphs below it. Try to spare speed-hungry on-line users from reading anything they aren't absolutely interested in.

Boldface or italics used to highlight key phrases in the text function much as pull-out quotes do in print, but without the redundancy. Boldface can eliminate the need for headings where they would be intrusive or inappropriate, as in brief documents (400 words or less).

Using short paragraphs with a line space between them also facilitates scanning. Just as you would break up a long printed article in which the writer had neglected to use any paragraphs, so in an on-line article you must break long paragraphs into two or three shorter ones.

Hypertext Links Are Key

Hypertext, the most powerful feature of HTML, is fundamental and also unique to electronic publishing. A word or phrase in the text is underlined or highlighted. When you mouse-click on that word, a new document comes onto your screen. The highlighted text is similar to a footnote numeral—it refers you elsewhere—but it usually links to another large block of text, not a small note.

The text of a hypertext link can be a title (the link goes to a document with that title), a person's name (the link may go to a biographical sketch or résumé), a word (the link may go to a dictionary definition)—or any phrase or clause. Choosing or writing a good phrase or clause to use as a link requires editorial talent.

The cardinal sin of link-making is to use the words "click here." (You'll see those words all over the Web. Many people sin.) A link exists to be clicked, and to say so in the link text is a sorry waste of words.

Writing Strong Link Text

The text used for the link should, first and foremost, produce a reasonable expectation in the user. If the link is a button that reads **HOME**, the user will expect to go to the site's home page (or first page).

If a link said, "Find out more about our product," would you be disappointed if that link went to an e-mail form that provided nothing new but invited you to write for additional information? A better link would say, "Contact us to find out more."

Consider this text:

Find out <u>who we are</u> and let <u>us</u> know what you'd like to see.

Can you guess, without following the two links, how "who we are" and "us" might be different? They actually lead to two different documents: "About

This Site" and a page of Web staff bios. This would be a better way to write the links:

> Find out <u>about our Web site</u> and let <u>us</u> know . . .

Link text can't always include active verbs, but a keen editor can often figure out a way to enliven a passive clause or dull phrase: "Our annual report is now on-line" can become "Read our annual report on-line," and "Membership form" can become "Apply for membership today."

AVOIDING CARELESS LINKS

It's a shame to waste the power of links by creating them randomly. Consider their cumulative effect on your on-line readers and avoid the following kinds of pitfalls:

Excessive Use of Links

When links appear in almost every line of a document, users are less likely to follow *any* link at all (who has the time?). By eliminating links that are irrelevant, extraneous, or redundant, you do a great favor for busy, impatient users. Reference lists and search engines exist for anyone who wants to find every site related to a given topic.

Irrelevant or Extraneous Links

My favorite example is the word *Washington,* which appeared as a link in a news article about an action in Congress. Clicking it took me to a page about tourist attractions in the U.S. capital. Yes, it was a page about Washington, but it wasn't directly related to the article where the link appeared. Such links waste users' time and erode their faith in your ability to lead them to something good.

Gratuitous Use of External Links

You can fill a document with links that go to Web sites all around the world, but keep in mind that a person who leaves your site may never return. The more you send them away, the greater the impression that your own site has little to offer. Provide links to exemplary external sites as appropriate, especially when they add useful information that your site can't provide. But don't send people away without a good reason.

A Dearth of Link Options

If a Web page has only one link, it may as well be printed on paper. Never try to force a single path on the user. Making choices is what interactivity is all about.

BREAKING UP LARGE ARTICLES

In print a reader can flip a few pages and read the last paragraph (or read all the headings, or look at all the photos). A long article formatted as a Web page takes more time to load—to come on screen—than a shorter one. Because of the linear scrolling path from top to bottom, the user's ability to scan a very long article is reduced.

Careful use of hypertext removes this limitation and makes a longer article more pleasant to read on-line—and more work for an editor to prepare. The relationship is a direct one: The more effort an editor puts into breaking up a long article and linking the pieces intelligently, the more engaging that article will be.

BASIC TASKS OF HYPERTEXT EDITING

The four basic tasks a hypertext editor undertakes contribute much to the effectiveness and appeal of an on-line document.

1. **Finding appropriate places to break.** The writer may have already divided the article into sections, but the editor usually will have to split up the article even further. Base your decisions on content, on presenting a whole idea. Try to make each piece as independent as possible. Sometimes this means you'll delete or rewrite transition sentences.

2. **Deciding how long each piece should be.** The broken-out pieces don't have to be of equal length, but the goal is to keep them all relatively short (500 words is a good average, although it's too short for some kinds of material). In some cases, one piece will be much longer or shorter than all the others.

3. **Creating the link structure.** If there are very few pieces (three or four), they might be best handled as sidebars, with the titles of all related pieces listed (as links) at the end of each one. (That structure provides access to any part of the article from any other part, which is good.) A larger number of pieces can be more intricately interwoven. Provide contextual links so that users can skip around, choosing their own paths, but also give them the option to follow a traditional path through the article (using "next" and "previous" as links, if appropriate).

4. **Providing an overview.** When an article is broken into a dozen or more parts, users appreciate the option of glimpsing the whole cloth before following a thread. The overview may be structured as a table of contents (list), a map (graphics), or an abstract or introduction (text) that has links to all the pieces on one page. Each piece of the article should include a link to the overview.

THE ON-LINE READER IS A MOVING TARGET

Construct your set of documents in such a way that users can easily skip the things they don't care about but won't miss the things they like.

Remember that users are always moving—scrolling up and down and jumping from link to link. They don't like to be still. They don't like to sit with their hands folded in their laps. This doesn't mean they are impossible to grab. They'll stop and read when something catches their eye. The trick is to hang onto them long enough.

Prepare, Listen, Follow Up

Max Messmer

For an effective job search, you need to gather information about opportunities in your field. This article provides guidelines for arranging and conducting an information-seeking interview with someone in your field.

Begin your informational interview campaign by writing down the names of people, organizations, and industries that interest you. Define the topics you want to explore. Clarify how these subjects relate to your work experience and career objectives. As you build your contact list and start making phone calls, remember to be persistent and maintain a positive attitude.

DEFINING YOUR OBJECTIVES

Many job seekers make the strategic error of phoning someone and immediately asking if there are any available positions. It's not likely a job will be open precisely when you call. It's more likely that your abrupt inquiry will place the person you're speaking with in the uncomfortable position of having to say, "No, I'm sorry, I can't help you."

Instead, let the contact know you're on a fact-finding mission to learn more about the industry. Productive informational interviews require listening. In most cases, people are flattered and receptive when asked to talk freely about their profession and experience.

A KNOWLEDGE-DRIVEN INQUIRY

In preparing for your informational interview, you should go through the same rigorous process you would undertake prior to a job interview. Be as thorough as you can in defining the topics you intend to discuss. Study and take notes on business publications, books, and corporate literature. Utilize resources on the Internet, CD-ROMs at libraries, and other databases for relevant information about the industries and companies you're contacting. This knowledge-driven approach will enable you to demonstrate both intellect and enthusiasm—two assets that must come across in any interview setting.

Another effective technique is to bring with you information of direct value to the person with whom you're meeting. It could be a book, an article, or even a report that you developed in the past. This is a no-lose opportunity. Be prepared to discuss your current situation—your skills, accomplishments, and current job-seeking objectives. You also should have available several copies of your résumé and business cards.

517

GATEKEEPERS ARE YOUR ALLIES

As you make phone calls requesting informational interviews, you are likely to run into "gatekeepers" who will screen calls. Treat these professionals with the utmost courtesy. Introduce yourself and politely try to establish a relationship. In the course of these short conversations, the gatekeeper might even tell you about others in the company who might be willing to meet with you. If you try to bypass gatekeepers, you run the risk of creating an adversarial relationship, thus damaging your cause.

CLARIFY TIME AND PURPOSE

Once the meeting begins, verify exactly how much time the person has available. It's also valuable to have an established agenda. Remind the person you're talking to that your objective is informational: You want to ask questions and learn more so that your job search is as productive as possible.

LISTEN CAREFULLY

Provided it is acceptable to the other person, you should take notes during your meeting so that you can retain as much information as possible. However, your note taking should not interfere with the flow of the conversation. Listen attentively, write down major ideas, and underline key points. You'll have time after the interview to transcribe your notes.

Typical questions to ask are:

- What are some of the challenges of working in this field?
- What skills or experience do you feel are most important for success in this industry/profession?
- What type of background is most suitable?
- How long have you been in this industry/profession? What did you do before?
- Why did you choose this field?
- What are some new developments in your industry? (You should have some knowledge of trends prior to the meeting.)

FOLLOW THROUGH IS CRITICAL

Monitor the length of your meeting. As it nears the agreed-upon finish, it is your responsibility to respect the other person's time by politely inquiring if he or she can talk further.

At the close of the interview, ask for names of others you could contact for information. Then within a day or two of your meeting, send a short thank-you note. In the weeks that follow, keep in touch with your new contacts by sending them relevant news articles that relate to their industry or profession, or phoning them to give them progress updates on your job search.

The time you invest conducting informational interviews will prove invaluable in expanding your career network. You will gain insight into various corporations, industries, and hiring trends, and you'll have a chance to practice talking about your own talents and experience in a less formal setting than an official job interview.

Ten Steps for Cleaning Up Information Pollution

Jakob Nielsen

Employees waste hours every year handling irrelevant email correspondence on the job. This flood of email has created "information pollution" in the workplace. This article offers tips on how to control your work time and reduce the interruptions caused by email and instant messages.

Our knowledge environment is getting ever more contaminated by information pollution. Things we need to know are drowning in irrelevant information. Symptoms include:

- In most companies, employees squander **an hour or more each day** simply "doing email."

- Employees fritter away **48 hours each year** trying to unearth job-related information on bad intranets compared to the time they would need on an intranet with usability in the top 25%. The resulting productivity loss amounts to millions of dollars for mid-sized companies.

- Many websites alienate users by burying answers to basic questions in useless corporatese.

- Email messages that customers actually want, such as useful newsletters or customer-service confirmations, don't survive overflowing inboxes—often because senders ignore the principles of good email design.

WHAT INDIVIDUALS CAN DO

All time-management courses boil down to one basic piece of advice: **set priorities** and allocate the bulk of your time to tasks that are crucial to meeting your goals. **Minimize interruptions** and spend big chunks of your time in productive and creative activity.

Unfortunately, current information systems encourage the opposite approach, leading to an **interrupt-driven** workday and reduced productivity. Here are six steps to regaining control of your day:

1. **Don't check your email all the time.** Set aside special breaks between bigger projects to handle email. Don't let email interrupt your

projects, and don't let the computer dictate your priorities. Turn off your email program's "Biff" feature (the annoying bell or screen flash that notifies you every time an email message arrives). If you're using Microsoft Outlook, go to Tools > Options > Preferences > E-mail Options and uncheck *"Display a notification message when new mail arrives."*

2. **Don't use "reply to all"** when responding to email. Abide by the good old "need to know" principle that's so beloved by the military and send follow-up messages only to those people who will actually bene-fit from the reply.

3. **Write informative subject lines** for your email messages. Assume that the recipient is too busy to open messages with lame titles like "hi."

4. **Create a special email address** for personal messages and newsletters. Only check this account once per day. (If you're geekly enough to mas-ter filtering, use filters to sort and prioritize your email. Unfortunately, this is currently too difficult for average users.)

5. **Write short.** J. K. Rowling is not a good role model for email writers.

6. **Avoid IM** (instant messaging) unless real-time interaction will truly add value to the communication. A one-minute interruption of your colleagues will cost them ten minutes of productivity as they reestab-lish their mental context and get back into "flow." Only the most important messages are worth 1,000 percent in overhead costs.

What Companies Can Do

At the corporate level, we need to implement four more steps:

7. **Answer common customer questions on your website** using clear and concise language. This will save your customers a lot of time—thus making you popular—and will keep them from pestering you with time-consuming phone calls and emails.

8. **User test your intranet.** Clean it up so that employees can find stuff faster, and make the intranet homepage their entry point for keeping up on company news and events.

9. **Don't circulate internal email to all employees;** instead put the infor-mation on the intranet where people can find it when they need it. (This obviously assumes that you've fixed the intranet's usability.)

10. **Establish a company culture in which it's okay not to respond to email immediately.** This frees employees from the pressure of incessantly checking email and lets them get more work done.

As individuals and as organizations, we can all do our share both to cope with the existing information pollution and emit fewer new pollutants. Ignoring this problem will only make it worse every year. If we act now, we can get it under control.

Top Ten Guidelines for Homepage Usability

Jakob Nielsen

The home page is the first page an Internet user sees when visiting a Web site. This article gives ten tips for creating an effective, appealing, and usable homepage.

Homepages are the **most valuable real estate** in the world. Each year, companies and individuals funnel millions of dollars through a space that's not even a square foot in size. For good reason. A homepage's impact on a company's bottom line is far greater than simple measures of e-commerce revenues: The homepage is your company's **face to the world.** Increasingly, potential customers will look at your company's online presence before doing business with you—regardless of whether they plan to close the actual sale online.

The homepage is the **most important page on most websites,** and gets more page views than any other page. Of course, users don't always enter a website from the homepage. A website is like a house in which every window is also a door: People can follow links from search engines and other websites that reach deep inside your site. However, one of the first things these users do after arriving at a new site is go to the homepage. Deep linking is very useful, but it doesn't give users the site overview a homepage offers—if the homepage design follows strong usability guidelines, that is.

Following are ten things you can do to increase the usability of your homepage and thus enhance your website's business value.

MAKE THE SITE'S PURPOSE CLEAR: EXPLAIN WHO YOU ARE AND WHAT YOU DO

1. Include a One-Sentence Tagline

Start the page with a tagline that summarizes what the site or company does, especially if you're new or less than famous. Even well-known companies presumably hope to attract new customers and should tell first-time visitors about the site's purpose. It is especially important to have a good tagline if your company's general marketing slogan is bland and fails to tell users what they'll gain from visiting the site.

2. Write a Window Title with Good Visibility in Search Engines and Bookmark Lists

Begin the TITLE tag with the company name, followed by a brief description of the site. Don't start with words like "The" or "Welcome to" unless you want to be alphabetized under "T" or "W."

3. Group All Corporate Information in One Distinct Area

Finding out about the company is rarely a user's first task, but sometimes people do need details about who you are. Good corporate information is especially important if the site hopes to support recruiting, investor relations, or PR, but it can also serve to increase a new or lesser-known company's credibility. An "**About <company-name>**" section is the best way to link users to more in-depth information than can be presented on the homepage. (See also my report with 50 guidelines for the design of "about us" areas of corporate websites.)

Help Users Find What They Need

4. Emphasize the Site's Top High-Priority Tasks

Your homepage should offer users a clear starting point for the main one to four tasks they'll undertake when visiting your site.

5. Include a Search Input Box

Search is an important part of any big website. When users want to search, they typically scan the homepage looking for *"the little box where I can type,"* so your search should be a box. Make your search box at least 25 characters wide, so it can accommodate multiple words without obscuring parts of the user's query.

(Update added 2004: Based on findings in my Web Usability 2004 project, my recommendation is now to make the search box 27 characters wide.)

Reveal Site Content

6. Show Examples of Real Site Content

Don't just describe what lies beneath the homepage. Specifics beat abstractions, and you have good stuff. Show some of your best or most recent content.

7. Begin Link Names with the Most Important Keyword

Users scan down the page, trying to find the area that will serve their current goal. **Links are the action items** on a homepage, and when you start each link with a relevant word, you make it easier for scanning eyes to differentiate it from other links on the page. A common violation of this guideline is to start all links with the company name, which adds little value and impairs users' ability to quickly find what they need.

8. Offer Easy Access to Recent Homepage Features

Users will often remember articles, products, or promotions that were featured prominently on the homepage, but they won't know how to find them once you move the features inside the site. To help users locate key items, keep a short list of recent features on the homepage, and supplement it with a link to a permanent archive of all other homepage features.

USE VISUAL DESIGN TO ENHANCE, NOT DEFINE, INTERACTION DESIGN

9. Don't Over-Format Critical Content, Such as Navigation Areas

You might think that important homepage items require elaborate illustrations, boxes, and colors. However, users **often dismiss graphics as ads,** and focus on the parts of the homepage that look more likely to be useful.

10. Use Meaningful Graphics

Don't just decorate the page with stock art. Images are powerful communicators when they show items of interest to users, but will backfire if they seem frivolous or irrelevant. For example, it's almost always best to show photos of real people actually connected to the topic, rather than pictures of models.

ResumeMaker's 25 Tips—Interviewing

ResumeMaker

This article contains 25 useful tips for handling a job interview. The article comes from ResumeMaker's Web site (http://www.resumemaker.com). Notice that ResumeMaker also uses the article to market its software and other services for dealing with all aspects of the job search.

The job interviewing stage of your job search is the most critical. You can make or break your chance of being hired in the short amount of time it takes to be interviewed. Anyone can learn to interview well, however, and most mistakes can be anticipated and corrected. Learn the following top 25 interviewing techniques to give you that winning edge.

1. Bring extra copies of your résumé to the interview. Nothing shows less preparation and readiness than being asked for another copy of your résumé and not having one. Come prepared with extra copies of your résumé. You may be asked to interview with more than one person and it demonstrates professionalism and preparedness to anticipate needing extra copies.

2. Dress conservatively and professionally. You can establish your uniqueness through other ways, but what you wear to an interview can make a tremendous difference. It is better to overdress than underdress. You can, however, wear the same clothes to see different people.

3. Be aware of your body language. Try to look alert, energetic, and focused on the interviewer. Make eye contact. Nonverbally, this communicates that you are interested in the individual.

4. First/last impressions. The first and last five minutes of the interview are the most important to the interview. It is during this time that critical first and lasting impressions are made and the interviewer decides whether or not they like you. Communicate positive behaviors during the first five minutes and be sure you are remembered when you leave.

5. Fill out company applications completely—even if you have a résumé. Even though you have brought a copy of your résumé, many companies require a completed application. Your willingness to complete one, and your thoroughness in doing so, will convey a great deal about your professionalism and ability to follow through.

6. Remember that the purpose of every interview is to get an offer. You must sufficiently impress your interviewer both professionally and personally to be offered the job. At the end of the interview, make sure you know what the next step is and when the employer expects to make a decision.

7. **Understand employers' needs.** Present yourself as someone who can really add value to an organization. Show that you can fit into the work environment.

8. **Be likeable.** Be enthusiastic. People love to hire individuals who are easy to get along with and who are excited about their company. Be professional, yet demonstrate your interest and energy.

9. **Make sure you have the right skills.** Know your competition. How do you compare with your peers in education, experience, training, salary, and career progression? Mention the things you know how to do really well. They are the keys to your next job.

10. **Display ability to work hard to pursue an organization's goals.** Assume that most interviewers need to select someone who will fit into their organization well in terms of both productivity and personality. You must confirm that you are both a productive and personable individual by stressing your benefits for the employer.

11. **Market all of your strengths.** It is important to market yourself, including your technical qualifications, general skills and experiences as well as personal traits. Recruiters care about two things—credentials and personality. Can you do the job based on past performance and will you fit in with the corporate culture? Talk about your positive personality traits and give examples of how you demonstrate each one on the job.

12. **Give definitive answers and specific results.** Whenever you make a claim of your accomplishments, it will be more believable and better remembered if you cite specific examples and support for your claims. Tell the interviewer something about business situations where you actually used this skill and elaborate on the outcome. Be specific.

13. **Don't be afraid to admit mistakes.** Employers want to know what mistakes you have made and what is wrong with you. Don't be afraid to admit making mistakes in the past, but continuously stress your positive qualities as well, and how you have turned negatives into positive traits.

14. **Relate stories or examples that heighten your past experience.** Past performance is the best indicator of future performance. If you were successful at one company, odds are you can succeed at another. Be ready to sell your own features and benefits in the interview.

15. **Know everything about your potential employer before the interview.** Customize your answers as much as possible in terms of the needs of the employer. This requires that you complete research, before the interview, about the company, its customers, and the work you anticipate doing. Talk in the employer's language.

16. **Rehearse and practice interview questions before the interview.** Prior to your interview, try to actually practice the types of questions and answers you may be asked. Even if you do not anticipate all of the questions, the process of thinking them through will help you feel less stressed and more prepared during the interview itself.

17. Know how to respond to tough questions. The majority of questions that you will be asked can be anticipated most of the time. There are always, however, those exceptional ones tailored to throw you off guard and to see how you perform under pressure. Your best strategy is to be prepared, stay calm, collect your thoughts, and respond as clearly as possible.

18. Translate your strengths into job-related language of accomplishments and benefits relevant to the needs of employers. While you no doubt have specific strengths and skills related to the position, stress the benefits you are likely to provide to the employer. Whenever possible, give examples of your strengths that relate to the language and needs of the employer.

19. Identify your strengths and what you enjoy doing. Skills that you enjoy doing are the ones that are most likely to bring benefit to an employer. Prior to the interview, know what it is that you enjoy doing most, and what benefits that brings to you and your employer.

20. Know how you communicate verbally to others. Strong verbal communications skills are highly valued by most employers. They are signs of educated and competent individuals. Know how you communicate, and practice with others to determine if you are presenting yourself in the best possible light.

21. Don't arrive on time—arrive early! No matter how sympathetic your interviewer may be to the fact that there was an accident on the freeway, it is virtually impossible to overcome a negative first impression. Do whatever it takes to be on time, including allowing extra time for unexpected emergencies.

22. Treat everyone you meet as important to the interview. Make sure you are courteous to everyone you come in contact with, no matter who they are or what their position. The opinion of everyone can be important to the interview process.

23. Answer questions with complete sentences and with substance. Remember that your interviewer is trying to determine what substance you would bring to the company and the position. Avoid answering the questions asked with simple "yes" or "no" answers. Give complete answers that show what knowledge you have concerning the company and its requirements. Let the interviewer know who you are.

24. Reduce your nervousness by practicing stress-reduction techniques. There are many stress-reducing techniques used by public speakers that can certainly aid you in your interview process. Practice some of the relaxation methods as you approach your interview, such as taking slow deep breaths to calm you down. The more you can relax, the more comfortable you will feel and the more confident you will appear.

25. Be sure to ask questions. Be prepared to ask several questions relevant to the job, employer, and the organization. These questions should be designed to elicit information to help you make a decision as well as demonstrate your interest, intelligence, and enthusiasm for the job.

If you want to practice your answers to typical interview questions, you may want to locate a copy of *ResumeMaker*™ *Deluxe Edition*. The software takes you through 500 of the most commonly asked questions in a job interview, including 40 specific salary topics. You'll interact directly with a virtual interviewer in an office setting, watch professional job seekers respond to tough questions, and learn the most effective answers.

ResumeMaker also helps you put together a more effective resume and cover letter. Instead of struggling for words, simply choose from 100,000 prewritten phrases and hundreds of samples. A Career Planner™ helps you identify your ideal career and shows you the average salary range for every job.

ResumeMaker includes some of the most powerful online job-searching features available anywhere. JobFinder™ searches throughout the Internet to locate over 1.75 million available jobs in seconds and ResumeCaster™ can post your resume to every major career web site with one click where hiring companies look for potential candidates every day. For more information, visit http://www.resumemaker.com or call 800–822–3522.

Six Tips for Talking Technical When Your Audience Isn't

Janis Robinson

Speakers often find themselves in the position of presenting technical informa-tion to people who are not familiar with all the technical terms or concepts. This article gives six suggestions for tailoring a technical presentation to a nontechnical audience.

Technology is everywhere, isn't it? Well, not necessarily. As immersed as many of us are in computers, software, projectors and other technology, many peo-ple are just now delving into those topics.

As a speaker, you may have to address nontechnical audiences or people whose expertise is in topics other than your own—during company-wide meetings, training, customer-help-desk inquiries, seminars and conferences. All the basic tenets of good communication and public speaking apply in these presentation situations, but consider these additional points:

1. DETERMINE THE AUDIENCE'S TECHNOLOGY LEVEL BEFORE YOU SPEAK

Most nontechnical audiences aren't interested in becoming specialists in your area of expertise. Some attendees may be at the introductory stage. Others need to understand the point at which your technology and theirs interact. Check with the program coordinator or, even better, ask to speak to a few likely participants. Find out what they hope to do with the information after your session and customize your presentation accordingly. This is the only way to be sure you address their specific needs. Plus, by meeting a few audi-ence members beforehand, you establish a positive attitude about you and your material.

2. PUT YOUR AUDIENCE AT EASE

Some nontechnical audience members may feel tension, even fear, at the thought of hearing your presentation. Their bosses may have asked them to attend, so they think their jobs depend on understanding the material—and they may be right. This unease can have an adverse effect, creating barriers to understanding—but you can prevent this. Try to spot anyone who looks truly fearful. If you have the time and opportunity before you present, privately and tactfully ask why they are afraid, but do not publicly address them unless they volunteer. Humor also is a good tension breaker.

3. DON'T BE A TECHNO-SNOB

Eliminate the following phrases from your repertoire: "It's obvious," "It's common knowledge" and "As you all know." Such phrases sound condescending and make a nontechnical audience feel even less knowledgeable. You were asked to speak because you are the expert and someone thought you could pass along some of your expertise. Never forget that there was a time when *you* didn't understand this technology. Make your audience feel glad they came, rather than embarrassed about their unfamiliarity.

4. AVOID USING JARGON OR ACRONYMS

I've been in sessions in which presenters used jargon correctly in sentences but, when asked to clarify or explain themselves, were not able to. They understood the concepts but couldn't articulate them in nontechnical terms. As a result, their credibility was lost and unrecoverable. Whenever possible, eliminate the use of highly specialized language. For any technical terms you do use, be prepared to explain without using more jargon, and provide a glossary handout.

5. USE VERBAL ILLUSTRATIONS

For each presentation point, plan several analogies, stories or metaphors customized for the audience. Engineers and architects may proclaim your I-beam story the best they ever heard, while poets and journalists may wonder when the first break will be. The latter group may follow your meaning, but you won't spark their imagination without using their language. By thinking ahead about how you might explain the concepts to all audience segments, you eliminate the chance of going blank just when you need a special example. Even if you use perfectly good standbys ("Think of the Internet as a highway."), put life and color into them. No matter how perfect an analogy is, the audience won't be energized unless you are.

6. WHENEVER POSSIBLE SHOW, DON'T TELL

As a presenter, you know how visuals clarify your points and improve the audience's understanding. For technical presentations, visuals are vital. You can save words and prevent confusion simply by showing what you mean in a demonstration. A well-prepared and practiced performance using your choice of presentation software or multimedia tools can say volumes more than static overheads. (Although, because anything about a presentation can go awry—including the technology—have overheads ready, a flipchart on hand and your tap shoes in your suitcase.)

There are challenges to overcome when presenting technology to an audience unfamiliar with the topic. However, it's a wonderful chance to share your enthusiasm and help your audience enjoy technology as much as you do.

Eleven Commandments for Business Meeting Etiquette

Gary M. Smith

Although general business etiquette has had much attention during the past decade, little has been written about showing courtesy and respect while attending business meetings. These 11 tips for good business meeting practices are also effective tips for taking a productive approach to your classes.

Here's a knee-slapper: What did the employee say when his boss asked why he missed a recent meeting?

Answer: "Sorry, I had to get some actual work done."

What's that? I don't hear you laughing. Could be that your sense of humor has been worn down sitting through endless presentations, disorganized gripe sessions, or business meetings where key players showed up late, if at all.

Personally, I think the business world could borrow a page from the book of Emily Post, the maven of politeness and etiquette. A good business meeting is one where all the players show courtesy and respect. This approach conveys a simple message: We're all professionals here, so let's have a productive meeting.

Recently, I researched the topic of business meeting etiquette but found virtually no established rules on holding courteous meetings. So I've gathered what I've learned from my own experience into the eleven commandments listed below.

1. **R.S.V.P.** When asked via phone, email, or electronic calendar to attend a business meeting, be sure to reply if a reply is requested. Some meetings are structured and spaces secured on the basis of expected attendance.

2. **Arrive Early.** If this is not possible, arrive at the scheduled time at the latest—but never late. Do not assume that the beginning of a meeting will be delayed until all those planning to attend are present. If you arrive late, you risk missing valuable information and lose the chance to provide your input. Also, you should not expect others to fill you in during or after the meeting; everyone is busy, and those who were conscientious enough to arrive on time should not have to recap the meeting for you.

3. **Come Prepared.** Always bring something to write on as well as to write with. Meetings usually are called to convey information, and it is disruptive to ask others for paper and pen if you decide to take notes. If you know you will be presenting information, ensure that your handouts, view foils, *PowerPoint* slides, etc., are organized and ready.

4. **Do Not Interrupt.** Hold your comments to the speaker until the meeting has adjourned or until the speaker asks for comments, unless, of course, the speaker has encouraged open discourse throughout the meeting. Also, do

not interrupt other attendees. Hold your comments to others in the meeting until after the meeting is adjourned. Conversation during a meeting is disruptive to other attendees and inconsiderate of the speaker.

5. Abstain from Electronics. As the notice posted at the beginning of films in movie theaters requests, "Please silence cell phones and pagers." Activate voice mail if you have it, or forward messages to another phone.

6. Speak in Turn. When asking a question, it usually is more appropriate to raise your hand than to blurt out your question. Other attendees may have questions, and the speaker needs to acknowledge everyone.

7. Keep Your Questions Brief. When asking questions, be succinct and clear. If your question is detailed, break it into parts or several questions. But be sure to ask only one question at a time; others may have questions as well.

8. Pay Attention. Listen to the issues the speaker addresses, the questions from the attendees, and the answers provided. You do not want to waste meeting time asking a question that has already been asked.

9. Be Patient and Calm. Do not fidget, drum your fingers, tap your pen, flip through or read materials not concerning the meeting, or otherwise act in a disruptive manner.

10. Attend the Entire Meeting. Leave only when the meeting is adjourned. Leaving before the end of the meeting—unless absolutely necessary and unless you have prior permission—can be disruptive to other attendees and inconsiderate of the speaker.

11. Respond to Action Items. After the meeting, be sure to complete any tasks assigned to you as expeditiously as possible; file your meeting notes or any formalized minutes for later review or to prepare for future meetings.

Taking Your Presentation Abroad

Edmond H. Weiss

Speaking styles differ across cultures. When making an oral presentation in another country, speakers must adjust their styles and present information or solicit business in a manner that will be acceptable to the new audience. This article discusses the shifts that have to take place, including establishing the status of the speaker and creating appropriate openings, tone, and closings.

In an era of global business and world markets, the hardest thing for an accomplished professional communicator to accept might be that the "universal" rules of effective communication are not universal at all. There are scores of cultures in which, for example, ambiguity is prized over precision, in which a clearly articulated purpose is seen as pushy or immature, or in which a list of sentence fragments with bullets is perceived as condescending to the audience.

In the United States, the current emphasis is on *localization*—that is, doing everything possible to adapt to the culture of the prospective client or business partner. If a U.S. technical communicator is preparing to make a presentation overseas, even such basic communication tools as Alan Monroe's "motivated sequence" (defined in his book *Principles and Types of Speech*) or Abraham Maslow's "hierarchy of needs" (defined in his book *Motivation and Personality*) may prove to be provincial and ineffective. Indeed, any assumptions we make concerning the style of our presentation, our relationship to our audience, or appropriate ways to close a presentation may only be veiled extensions of the "ugly American" notion that everyone should speak English. Even the concept of *effectiveness* is, as I have been advised by a Korean colleague, peculiarly U.S. and male.

Those who make business and technical presentations are obliged to research the culture of the nation or community for which the presentation is intended—not just for relevant business and technical facts but also for communication practices and expectations. Presenters should learn that, in some places at least, directness is perceived as brusqueness, and personal confidence and assertiveness (the hallmarks of U.S. style) may be perceived as arrogant. Precise language may be ineffectual in some cultures, and simplicity and clarity, the goals of professional communicators throughout North America, may be viewed as ingenuousness in Europe or lack of manners in Asia.

In short, U.S. speakers who come across as talented and resourceful at home can appear ill mannered and unsophisticated elsewhere. The greatest clashes are between the West and the Far East, of course. But important cultural issues also affect presentations in Europe and South America.

When I advise my clients on international presentations—or presentations in the increasing multicultural U.S.—I urge them to study the culture to be addressed and to research seven design questions. Their answers should be based on fact (including information from informants with first-hand knowledge of the culture), rather than on standard notions of effective presentation.

What Is the Objective of the Nth Meeting?

It is characteristic of U.S. businesspeople to expect a first presentation to result, after just a few minutes, in achieving a trusting relationship and cementing a deal. In much of the world, however, business professionals expect a relationship to grow over several—or many—meetings. In some places, one does not mention business at all for the first several gatherings, and one certainly refrains from the U.S. practice called *closing*.

The prudent plan is to schedule a series of meetings/presentations, each with a particular objective:

- *Greeting*—Making yourself and your company known
- *Charming*—Establishing a pleasant atmosphere
- *Representing*—Clarifying your company's history, character, and business plans
- *Educating*—Introducing new products, technologies, ways of doing business, and opportunities for collaboration
- *Supplicating*—Emphasizing your need to win favor and approval from the audience, even though you may be undeserving
- *Selling*—U.S.-style pitching of benefits and comparative advantage

Most courses in professional communication teach techniques for educating and selling, but little else. U.S. professionals abroad, finding that they are expected to comment on the beauty of the host's surroundings, are often tongue-tied and painfully graceless.

As in almost all professional presentations, the main cause of failure in international presentations is uncertainty of purpose. And in international presentations, there is more chance to get the purpose wrong.

How Are the Presenters Related to Their Organization?

In most presentations, the character and credibility of the speaker count for much more than the attractiveness of the slides. This is especially true in international presentations, in which the title or perceived role of the speaker communicates a strong message to the audience.

While U.S. firms are quick to send a bright young man or woman to a critical meeting, there are places where the choice of a young or low-powered representative may come across as a lack of seriousness or even as an insult. In preparing for international presentations, then, it is essential to pick the right spokesperson—a leader, founder, expert, or specialist. In many international settings, youthful or inexperienced representatives—even if they are excellent communicators or fully competent—will be unable to win the confidence and support of their audience. Countries differ with respect to how much "power distance" (as defined by Geert Hofstede in the book *Cultures and Organizations: Software of the Mind*) they can abide—that is, how flat an organizational chart they can tolerate.

One of the most intransigent and frustrating problems in international presentations is the perception of women in many countries. There are places where a woman speaker cannot be taken seriously as an executive or technical expert. Indeed, there are places where women may be spoken to in a manner that would occasion litigation in North America.

What to do when you travel to a place where women are expected to play a less-than-equal role in business proceedings is an ongoing ethical dilemma in international business communication.

HOW ARE THE PRESENTERS RELATED TO THE AUDIENCE?

Business transactions are fraught with fictions. No U.S. salesperson ever says, "Buy my product because I want your money." Rather, people who sell things claim to be "helping" us by assessing our needs, or providing consumer information, or even giving us gifts. In much of the world, these U.S.-style selling ploys—these little dramas in which everyone pretends that something other than commerce is taking place—are regarded with amusement or contempt.

Although U.S. business texts stress the need to establish "relationships" with prospective customers, most U.S. sellers are far too impatient and quota-driven to pay more than lip service to this idea. But in most countries, relationships count far more in business than sales incentives, or balance sheets, or clever *PowerPoint* shows. (And, interestingly, there are also countries where relationships count for even *less* than in the U.S.)

International presenters should decide carefully what relationship is under development. Do they want to be *vendors* or *contractors*—in a strict contractual relationship? Or, alternatively, do they want to be perceived as *partners* or *collaborators*—sharing the risk and return of the venture? In some countries, it is best to be perceived as having no business relationship at all, but rather as being a *friend, political ally,* or even a *guest.*

Presentations aimed at establishing these relationships contain none of the usual objectives, benefits, plans, or budgets. Indeed, it may be completely inappropriate in some cases to offer what we think of as a formal presentation—at least not for the first several meetings.

How Should We Begin?

Some U.S. presenters can be painfully awkward when commenting on the weather, paying compliments, or making other kinds of urbane conversation. They like to get right to business—an approach they've learned in business school and communications seminars. But in most of the world, this directness will be seen as impatience and lack of civility. Those who expect to make presentations abroad—even on highly technical subjects—should learn to pronounce greetings in the host country's language, to compliment the host's meeting arrangements, to comment on the beauty of the surroundings or the change of seasons, to refer in an appropriate way to recent world events—in general, to seem poised and cosmopolitan.

Although not all cultures judge people as quickly as people in the U.S. do, first impressions are still critically important in presentations worldwide. Presenters should research the following matters:

- *Most Interesting Topics*—Favorite subjects, traditional ways of beginning gatherings and delivering addresses

- *Forbidden Topics*—Taboo subjects, such as comments concerning unfriendly neighboring countries or remarks about a person's wife

- *Etiquette and Protocol*—Titles and honorifics, the correct way to make introductions, the accepted order of speaking and deference

- *Occasions and Events*—Holidays, festivals, sporting events, celebrations (both religious and secular)

What Should Our Style Be?

The relationship between the presenters and audience should also inform the *style* or tone of the presentation. The direct and assertive style favored in the U.S. may be perceived as brusque, arrogant, or boring in other cultures. After some thoughtful research, presenters might choose one of these styles:

- *Philosophical or Religious*—Supported by logical argument, citations, proverbs, even bits of verse and appropriate literature

- *Scholarly and Technical*—Rich with information and statistical analysis, presented without gung-ho selling efforts or partisan enthusiasm

- *Humble*—Modest and self-effacing

- *Glitzy*—Filled with high-tech presentation tools and expensive communication products

There are countries where only one of these styles will carry the day: The others will estrange the audience. These alternative styles may present a problem for many U.S. presenters, but practice can lead to successful—even elegant—delivery.

WHAT SHOULD THE SUBSTANCE BE?

Professional communicators learn standard paradigms for deliberative speeches, believing them to be powerful enough for any occasion. But the countries of the world differ vastly on what they consider appropriate topics for a meeting or business presentation. U.S. businesspeople are impatient with small talk; they have no time for "philosophy" or "theory"; they prefer to cut to the chase.

Although these attitudes may account somewhat for the robustness of the U.S. economy, they also account for much misunderstanding in international communication. Researching the host culture will often lead researchers to include topics such as these:

- *General Relationships and Shared Friends*—Exploration of the links between the presenter and the audience; exploration of the "degrees of separation"; search for common origins and experiences

- *Weather and Incidentals*—Climate, change of seasons; festivals and holidays associated with the changes

- *Visions of the Future and Reflections on History*—Intelligent (not super-ficial) assessments of the historical context of the meeting; appreciation for the historical and cultural events that enabled the meeting; imagina-tive speculation on the long-range meaning of the emerging association

- *Technical Details*—Highly technical information that would ordinar-ily not be part of a business presentation

- *Money*—Not price or cost-benefit, but the *meaning* of the money in-volved; a discussion of values, including "nonmaterial" costs

- *Feelings*—The presenter's emotional responses to the situation, spoken with appropriate intonation and intensity

HOW SHOULD WE CLOSE?

The close of the presentation may vary the least across cultures. Although it may be inappropriate to end the presentation by asking for a sale (the typical procedure in the U.S.), other endings work well nearly everywhere: planning the next meeting; agreeing on what needs to be done; exchanging thanks and honoring the protocol of the situation.

U.S. businesspeople should be prepared, however, for cultures that wish to leave matters *unresolved*. Typically, this means that the hosts (or other group) wish to discuss things privately before they make any commitment. Also, people in some cultures are reluctant to express disapproval of a plan in public: They will end a meeting in a vague way and later communicate their dissatisfaction or difficulties out of public view. U.S. businesspeople are often astonished to learn that their audience—the same people who nodded and smiled politely throughout the presentation and answered "yes" to every question—actually hated the idea and wanted nothing more to do with it.

PITFALLS IN ADAPTATION

Adapting presentations to international audiences is not without risks. As in all forms of cultural adaptation, generalizations about the habits and expectations of a culture can lead to naïve or offensive stereotypes. Any statement that begins "The Chinese . . ." is likely to be a facile generalization with hundreds of relevant exceptions. There are important cultural variations within countries, within companies, even within departments. Granted, first-hand reports about a particular culture are more reliable than popular business compendiums. But even with the most reliable sources, attempts to characterize cultures can degenerate into simplistic representations that injure the cause of international communication.

Even if cultural adaptation is based on research, there is always a chance that the research is incorrect or out of date. U.S. businesspeople sometimes find it difficult to understand that a country's political and economic system can change dramatically in a few months, or, conversely, that ancient values and beliefs could be entrenched beneath a new "official" culture. In short, adaptation must be based on *sound* research, not shallow or casual impressions from a short visit, a chapter in a business text, or a brief conversation with a foreign associate.

The subtlest problem of all is that adapting to another culture can be, in itself, a form of condescension. Whenever we feel that we understand another culture well enough to satisfy its expectations and win its approval, we have, to some extent, trivialized the culture. We have made it less rich and unpredictable than our own. Reducing the culture of another country to a few easily learned rules or tips implies that the speaker's culture is the more powerful and advanced of the two.

Presenters who feel that they are being "tolerant" of other cultures also communicate an unwelcome sense of superiority. To "tolerate" a culture, after all, is to assert a kind of dominion over it. (Imagine how you would feel if you learned that Chinese or Brazilian students were being urged to "tolerate" the cultural eccentricities of the United States—as a way of getting business from it!)

Because of these complexities, a case can be made for being an effective and honorable representative of *one's own culture,* for *not* trying to adjust to another culture much beyond the simple courtesies. But, at the moment, the stronger case is for adaptation and localization. Postmodernists and businesspeople agree: The customer's way of communicating is, as always, right.

Ethical Reasoning in Technical Communication: A Practical Framework

Mark R. Wicclair and David K. Farkas

Technical writers often confront ethical problems just as lawyers and other professionals do. This article discusses three types of ethical principles—goal based, duty based, and rights based—and then describes and analyzes two cases in which a writer had to face an ethical dilemma. Notice that the third ethical case has no suggested solutions. Consider what you might do in these circumstances.

Professionals in technical communication confront ethical problems at times, just as in law, medicine, engineering, and other fields. In recent years STC has increased its effort to generate a greater awareness and understanding of the ethical dimension of the profession.[1] This article is intended to contribute to that effort.

To clarify the nature of ethical problems, we first distinguish between the ethical perspective and several other perspectives. We then discuss three types of ethical principles. Together these principles make up a conceptual framework that will help illuminate almost any ethical problem. Finally, we demonstrate the application of these principles using hypothetical case studies.

THE NATURE OF ETHICAL PROBLEMS

Typically, when faced with an ethical problem, we ask, "What should I do?" But it is important to recognize that this question can be asked from a number of perspectives. One perspective is an attempt to discover the course of action that will best promote a person's own interests. This is not, however, a question of ethics. Indeed, as most of us have discovered, there is often a conflict between ethical requirements and considerations of self-interest. Who hasn't had the experience of being tempted to do something enjoyable or profitable even while knowing it would be wrong?

A second nonethical perspective is associated with the law. When someone asks, "What should I do?" he may want to know whether a course of action is required or prohibited by law or is subject to legal sanction. There is often a connection between the perspective of law and the perspective of self-interest, for the desire to discover what the law requires is often motivated by the desire to avoid punishment and other legal sanctions. But between the perspectives of law and ethics there are several significant differences.

First, although many laws correspond to moral rules (laws against murder, rape, and kidnapping, for instance), other laws do not (such as technicalities of corporate law and various provisions of the tax code). Furthermore, almost every legal system has at one time or another included some unjust laws—such as laws in this country that institutionalized racism or laws passed in Nazi Germany. Finally, unethical actions are not necessarily illegal. For example, though it is morally wrong to lie or break a promise, only certain instances of lying (e.g., lying under oath) or promise breaking (e.g., breach of contract) are punishable under law. For all these reasons, then, it is important to recognize that law and ethics represent significantly different perspectives.

One additional perspective should be mentioned: religion. Religious doctrines, like laws, often correspond to moral rules, and for many people religion is an important motive for ethical behavior. But religion is ultimately distinct from ethics. When a Catholic, Protestant, or Jew asks, "What should I do?" he or she may want to know how to act as a good Catholic, Protestant, or Jew. A question of this type is significantly different from the corresponding ethical question: "How do I behave as a good human being?" In answering this latter question, one cannot refer to principles that would be accepted by a member of one faith and rejected by a member of another or a nonbeliever. Consequently, unlike religiously based rules of conduct, ethical principles cannot be derived from or justified by the doctrines or teachings of a particular religious faith.

THREE TYPES OF ETHICAL PRINCIPLES

To resolve ethical problems, then, we must employ ethical principles. We discuss here three types of ethical principles: goal-based, duty-based, and rights-based. Although these do not provide a simple formula for instantly resolving ethical problems, they do offer a means to reason about ethical problems in a systematic and sophisticated manner.[2]

Goal-Based Principles

Public policies, corporate decisions, and the actions of individuals all produce certain changes in the world. Directly or indirectly they affect the lives of human beings. These effects can be good or bad or a combination of both. According to goal-based principles, the rightness or wrongness of an action is a function of the goodness or badness of its consequences.

Goal-based principles vary according to the particular standard of value that is used to evaluate consequences. But the most widely known goal-based principle is probably the principle of utility. Utilitarians claim that we should assess the rightness of an action according to the degree to which it promotes the general welfare. We should, in other words, select the course of action that produces the greatest amount of aggregate good (the greatest good for the greatest number of people) or the least amount of aggregate harm. Public-policy decisions are often evaluated on the basis of this principle.

Duty-Based Principles

In the case of duty-based principles, the focus shifts from the consequences of our actions to the actions themselves. Some actions are wrong, it is claimed, just for what they are and not because of their bad consequences. Many moral judgments about sexual behavior are in part duty-based. From this perspective, if patronizing a prostitute is wrong, it is not because of harm that might come to the patron, the prostitute, or society, but simply because a moral duty is violated. Likewise, an individual might make a duty-based assertion that it is inherently wrong to lie or break a promise even if no harm would result or even if these actions would produce good consequences.

Rights-Based Principles

A right is an entitlement that creates corresponding obligations. For example, the right to free speech—a right that is valued and protected in our society—entitles people to say what they want to and imposes an obligation on others to let them speak. If what a person says would be likely to offend and upset people, there would be a goal-based reason for not permitting the speech. But from a rights-based perspective, the person is entitled to speak regardless of these negative consequences. In this respect, rights-based principles are like duty-based principles.

While many people believe strongly in the right to free speech, few would argue that this or other rights can never be overridden by considerations of likely consequences. To cite a classic example, the consequence of needless injury and death overrides anyone's right to stand up and cry "Fire!" in a crowded theater. Nevertheless, if there is a right to free speech, it is not permissible to impose restrictions on speech every time there is a goal-based reason for doing so.

APPLYING THESE ETHICAL PRINCIPLES

Together these three types of ethical principles provide technical communicators with a means of identifying and then resolving ethical problems associated with their work. We should begin by asking if a situation has an ethical dimension. To do this, we ask whether the situation involves any relevant goals, duties, or rights. Next, we should make sure that ethical considerations are not being confused with considerations of self-interest, law, or religion. If we choose to allow nonethical considerations to affect our decisions, we should at least recognize that we are doing so. Finally, we should see what course, or courses, of action the relevant ethical principles point to. Sometimes, all point unequivocally to one course of action. However, in some cases these principles will conflict, some pointing in one direction, others in another. This is termed an "ethical dilemma." When faced with an ethical dilemma, we must assign priorities to the various conflicting ethical considerations—often a difficult and demanding process.

We now present and discuss two hypothetical case studies. In our discussion we refer to several goal-, duty-, and rights-based principles. We believe that the principles we cite are uncontroversial and generally accepted. In saying this, we do not mean to suggest that there are no significant disagreements in ethics. But it is important to recognize that disputes about specific moral issues often do not emanate from disagreements about ethical principles. For example, although there is much controversy about the morality of abortion, the disagreement is not over the acceptability of the ethical principle that all persons have a right to life; the disagreement is over the nonethical question of whether fetuses are "persons."

We believe, then, that there is a broad consensus about many important ethical principles; and it is such uncontroversial principles that we cite in our discussion of the following two cases.

Case 1

Martin Yost is employed as a staff writer by Montgomery Kitchens, a highly reputable processed foods corporation. He works under Dr. Justin Zarkoff, a brilliant organic chemist who has had a series of major successes as Director of Section A of the New Products Division. Dr. Zarkoff is currently working on a formula for an improved salad dressing. The company is quite interested in this project and has requested that the lab work be completed by the end of the year.

It is time to write Section A's third-quarter progress report. However, for the first time in his career with Montgomery Kitchens, Zarkoff is having difficulty finishing a major project. Unexpectedly, the new dressing has turned out to have an inadequate shelf life.

When Yost receives Zarkoff's notes for the report, he sees that Zarkoff is claiming that the shelf-life problem was not discovered earlier because a group of cultures prepared by Section C was formulated improperly. Section C, Yost realizes, is a good target for Zarkoff, because it has a history of problems and because its most recent director, Dr. Rebecca Ross, is very new with the company and has not yet established any sort of "track record." It is no longer possible to establish whether the cultures were good or bad, but since Zarkoff's opinion will carry a great deal of weight, Dr. Ross and her subordinates will surely be held responsible.

Yost mentions Zarkoff's claim about the cultures to his very close and trustworthy friend Bob Smithson, Senior Chemist in Section A. Smithson tells Yost that he personally examined the cultures when they were brought in from Section C and that he is absolutely certain they were OK. Since he knows that Zarkoff also realizes that the cultures were OK, he strongly suspects that Zarkoff must have made some sort of miscalculation that he is now trying to cover up.

Yost tries to defuse the issue by talking to Zarkoff, suggesting to him that it is unprofessional and unwise to accuse Ross and Section C on the basis of mere speculation. Showing irritation, Zarkoff reminds Yost that his job is simply to write up the notes clearly and effectively. Yost leaves Zarkoff's office wondering whether he should write the report.

Analysis.   If Yost prepares the report, he will not suffer in any way. If he refuses, he will damage his relationship with Zarkoff, perhaps irreparably, and he may lose his job. But these are matters of self-interest rather than ethics. The relevant ethical question is this: "Would it be *morally wrong* to write the report?"

From the perspective of goal-based principles, one would want to know whether the report would give rise to any bad consequences. It is obvious that it would, for Dr. Ross and her subordinates would be wrongly blamed for the foul-up. Thus, since there do not appear to be any overriding good consequences, one would conclude that writing the report would contribute to the violation of a goal-based ethical principle that prohibits actions that produce more bad than good.

Turning next to the duty-based perspective, we recognize that there is a duty not to harm people. A goal-based principle might permit writing the report if some good would follow that would outweigh the harm to Dr. Ross and her subordinates. But duty-based principles operate differently: Producing more good than bad wouldn't justify violating the duty not to harm individuals.

This duty would make it wrong to write the report unless some overriding duty could be identified. There may be a duty to obey one's boss, but neither this nor any other duty would be strong enough to override the duty not to harm others. In fact, there is another duty that favors not writing the report: the duty not to knowingly communicate false information.

Finally, there is the rights-based perspective. There appear to be two relevant rights: the right of Dr. Ross and her subordinates not to have their reputations wrongly tarnished and the company's right to know what is actually going on in its labs. Unless there is some overriding right, it would be morally wrong, from a rights-based perspective, to prepare the report.

In this case, then, goal-, duty-, and rights-based principles all support the same conclusion: Yost should not write the report. Since none of the principles furnishes a strong argument for writing the report, Yost is not faced with an ethical dilemma. But he is faced with another type of dilemma: Since refusing to write the report will anger Zarkoff and possibly bring about his own dismissal, Yost has to decide whether to act ethically or to protect his self-interest. If he writes the report, he will have to recognize that he is violating important goal-, duty-, and rights-based ethical principles.

Case 2

Susan Donovan works for Acme Power Equipment preparing manuals that instruct consumers on the safe operation and maintenance of power tools. For the first time in her career, one of her draft manuals has been returned with extensive changes: Numerous complex cautions and safety considerations have been added. The manual now stipulates an extensive list of conditions under which the piece of equipment should not be used and includes elaborate procedures for its use and maintenance. Because the manual is now much longer and more complex, the really important safety information is lost amid the expanded list of cautions. Moreover, after looking at all the overly elaborate procedures in the manual, the average consumer is apt to ignore the manual altogether. If it is prepared in this way, Donovan is convinced, the manual will actually lead to increased numbers of accidents and injuries.

Donovan expresses her concern in a meeting with her boss, Joe Hollingwood, Manager for Technical Information Services. Hollingwood responds that the revisions reflect a new policy initiated by the Legal Department in order to reduce the number of successful accident-related claims against the company. Almost any accident that could occur now would be in direct violation of stipulations and procedures described in the manual. Hollingwood acknowledges that the new style of manual will probably cause some people not to use the manuals at all, but he points out that most people can use the equipment safely without even looking at a manual. He adds that he too is concerned about the safety of consumers, but that the company needs to protect itself against costly lawsuits. Donovan responds that easily readable manuals lead to fewer accidents and, hence, fewer lawsuits. Hollingwood replies that in the expert opinion of the Legal Department the total cost to the company would be less if the manuals provided the extra legal protection they recommend. He then instructs Donovan to use the Legal Department's revisions as a model for all subsequent manuals. Troubled by Hollingwood's response, Donovan wants to know whether it is ethically permissible to follow his instructions.

Analysis.   From the perspective of goal-based principles, writing manuals that will lead to increased injuries is ethically wrong. A possible good consequence is that a reduction in the number of successful lawsuits could result in lower prices. But neither this nor any other evident good consequence can justify injuries that might have been easily prevented.

A similar conclusion is arrived at from a duty-based perspective. Preparing these manuals would violate the important duty to prevent unnecessary and easily avoidable harm. Moreover, there appear to be no overriding duties that would justify violating this duty.

Finally, from a rights-based perspective, it is apparent that important rights such as the right to life and the right to health are at stake. It might also be claimed that people have a right to manuals that are designed to maximize their safety. Thus, ethical principles of each of the three types indicate it would be morally wrong to prepare the manuals according to Hollingwood's instructions.

Because Case 2 involves moral wrongs that are quite a bit more serious than those in Case 1, we now go on to consider an additional question, one not raised in the first case: What course of action should the technical communicator follow?

An obvious first step is to go over Hollingwood's head and speak to higher-level people in the company. This entails some risk, but might enable Donovan to reverse the new policy.

But what if this step fails? Donovan could look for a position with a more ethically responsible company. On the other hand, there is a goal-based reason for not quitting: By staying, Donovan could attempt to prepare manuals that would be safer than those that might be prepared by a less ethically sensitive successor. But very often this argument is merely a rationalization that masks the real motive of self-interest. The company, after all, is still engaging in an unethical practice, and the writer is participating in that activity.

Would leaving the company be a fully adequate step? While this would end the writer's involvement, the company's manuals would still be prepared in an unethical manner. There may indeed be an obligation to take further steps to have the practice stopped. To this end, Donovan might approach the media, or a government agency, or a consumer group. The obvious problem is that successively stronger steps usually entail greater degrees of risk and sacrifice. Taking a complaint outside the organization for which one works can jeopardize a technical communicator's entire career, since many organizations are reluctant to hire "whistle blowers."

Just how much can be reasonably asked of an individual in response to an immoral situation? To this question there is no clear answer, except to say that the greater the moral wrong, the greater the obligation to take strong—and perhaps risky—action against it.

CONCLUSION

The analyses offered here may strike the reader as very demanding. Naturally, we are all very reluctant to refuse assigned work, quit our jobs, or make complaints outside the organizations that employ us. No one wishes to be confronted by circumstances that would call for these kinds of responses, and many people simply would not respond ethically if significant risk and sacrifice were called for. This article provides a means of identifying and analyzing ethical problems, but deciding to make the appropriate ethical response to a situation is still a matter of individual conscience and will. It appears to be a condition of human existence that to live a highly ethical life usually exacts from us a certain price.[3]

Case 3

This case is presented without analysis so that readers can resolve it for themselves using the conceptual framework described in this essay.

A technical writer works for a government agency preparing instructional materials on fighting fires in industrial settings. Technical inaccuracies in these materials could lead to serious injury or death. The technical writer is primarily updating and expanding older, unreliable material that was published 30 years ago. He is trying to incorporate recently published material into the older material, but much of the recent information is highly technical and some of it is contradictory.

The technical writer has developed some familiarity with firefighting through his work, but has no special training in this field or in such related fields as chemistry. He was hired with the understanding that firefighting specialists in the agency as well as paid outside consultants would review drafts of all the materials in order to catch and correct any technical inaccuracies. He has come to realize, however, that neither the agency specialists nor the outside consultants do more than skim the drafts. Moreover, when he calls attention to special problems in the drafts, he receives replies that are hasty and sometimes evasive. In effect, whatever he writes will be printed and distributed to municipal fire departments, safety departments of industrial corporations, and other groups throughout the United States.

Is there an ethical problem here? If so, what is it, what ethical principles are involved, and what kinds of responses are called for?

NOTES

1.   Significant activities include the re-establishment of the STC Committee on Ethics, the preparation of the STC "Code for Communicators," the group of articles on ethics published in the third-quarter 1980 issue of *Technical Communication* (as well as several articles published in other places), and the continuing series of cases and reader responses that have appeared in *Intercom*.

2.   The arguments justifying ethical principles as well as ethics itself are beyond the scope of this essay but can be examined in Richard B. Brandt, *Ethical Theory: The Problems of Normative and Critical Ethics* (Englewood Cliffs, NJ: Prentice-Hall, 1959) and William K. Frankena, *Ethics*, 2nd ed. (Englewood Cliffs, NJ: Prentice-Hall, 1973). Other informative books are Fred Feldman, *Introductory Ethics* (Englewood Cliffs, NJ: Prentice-Hall, 1978) and Paul W. Taylor, *Principles of Ethics: An Introduction* (Encino, CA: Dickenson Publishing Company, 1975). All of these books are addressed to the general reader.

3.   STC might develop mechanisms designed to reduce the price that individual technical communicators have to pay for acting ethically. For their part, individuals may have an obligation to work for the development and implementation of such mechanisms.

Guidelines for Grammar, Punctuation, and Mechanics

These guidelines for grammar, punctuation, and mechanical matters, such as using numbers, will help you revise your writing according to generally accepted conventions for correctness. These guidelines cover the most frequent questions writers have when they edit their final drafts. Your instructor may tell you to read about specific topics before revising. In addition, use this Appendix to check your writing before you submit it to your instructor.

GRAMMAR

Dangling Modifiers

Dangling modifiers are verbal phrases, prepositional phrases, or dependent clauses that do not refer to a subject in the sentence in which they occur. These modifiers are most often at the beginning of a sentence, but they also may appear at the end. Correct by rewriting the sentence to include the subject of the modifier.

Incorrect: Realizing the connections between neutron stars, pulsars, and supernovas, the explanation of the birth and death of stars is complete. (The writer does not indicate *who* realized the connections.)

Correct: Realizing the connection between neutron stars, pulsars, and supernovas, many astronomers believe that the explanation of the birth and death of stars is complete.

Incorrect: To obtain a slender blade, a cylindrical flint core was chipped into long slivers with a hammerstone. (The writer does not indicate *who* worked to obtain the blade.)

Correct:	To obtain a slender blade, Cro-Magnon man chipped a cylindrical flint core into slivers with a hammerstone.
Incorrect:	Born with cerebral palsy, only minimal mobility was possible. (The writer does not indicate *who* was born with cerebral palsy.)
Correct:	Born with cerebral palsy, the child had only minimal mobility.
Incorrect:	Their sensitivity to prostaglandins remained substantially lowered hours after leaving a smoke-filled room. (The writer does not indicate *who* or *what* left the room.)
Correct:	Their sensitivity to prostaglandins remained substantially lowered hours after the nonsmokers left a smoke-filled room.

Misplaced Modifiers

Misplaced modifiers are words, phrases, or clauses that do not refer logically to the nearest word in the sentence in which they appear. Correct by rewriting the sentence to place the modifier next to the word to which it refers.

Incorrect:	The financial analysts presented statistics to their clients that showed net margins were twice the industry average. (The phrase about *net margins* does not modify *clients*.)
Correct:	The financial analysts presented their clients with statistics that showed net margins were twice the industry average.

Squinting Modifiers

Squinting modifiers are words or phrases that could logically refer to either a preceding or a following word in the sentence in which they appear. Correct by rewriting the sentence so that the modifier refers to only one word in the sentence.

Incorrect:	Physicians who use a nuclear magnetic resonance machine frequently can identify stroke damage in older patients easily. (The word *frequently* could refer to *use* or *identify*.)
Correct:	Physicians who frequently use a nuclear magnetic resonance machine can identify stroke damage in older patients easily.
Correct:	Physicians who use a nuclear magnetic resonance machine can identify stroke damage in older patients frequently and easily.

Parallel Construction

Elements that are equal in a sentence should be expressed in the same grammatical form.

Incorrect:	With professional care, bulimia can be treated and is controllable. (The words *treated* and *controllable* are not parallel.)

Correct: With professional care, bulimia can be treated and controlled.

Incorrect: The existence of two types of Neanderthal tool kits indicates that one group was engaged in scraping hides, while the other group carved wood. (The verbal phrases after the word *group* are not parallel.)

Correct: The existence of two types of Neanderthal tool kits indicates that one group scraped hides, while the other group carved wood.

Incorrect: The social worker counseled the family about preparing meals, cleaning the house, and gave advice about childcare. (The list of actions should be in parallel phrases.)

Correct: The social worker counseled the family about preparing meals, cleaning the house, and caring for children.

Elements linked by correlative conjunctions (*either . . . or, neither . . . nor, not only . . . but also*) also should be in parallel structure.

Incorrect: Such symptoms as rocking and staring vacantly are seen not only in monkeys that are deprived of their babies but also when something frightens mentally disturbed children. (The phrases following *not only* and *but also* are not parallel.)

Correct: Such symptoms as rocking and staring vacantly are seen not only in monkeys that are deprived of their babies but also in mentally disturbed children who are frightened.

Pronoun Agreement

A pronoun must agree in number, person, and gender with the noun or pronoun to which it refers.

Incorrect: Each firefighter must record their use of equipment. (The noun *firefighter* is singular, and the pronoun *their* is plural.)

Correct: Each firefighter must record his or her use of equipment.

Correct: All firefighters must record their use of equipment.

Incorrect: Everyone needs carbohydrates in their diet. (*Everyone* is singular, and *their* is plural.)

Correct: Everyone needs carbohydrates in the diet.

Correct: Everyone needs to eat carbohydrates.

Correct: Everyone needs carbohydrates in his or her diet.

Incorrect: The sales department explained their training methods to a group of college students. (*Sales department* is singular and requires an *it*. Correct by clarifying *who* did the explaining.)

Correct: The sales manager explained the department training methods to a group of college students.

Correct: The sales manager explained her training methods to a group of college students.

Incorrect: The International Microbiologists Association met at their traditional site. (*Association* is singular and requires an *it*.)

Correct: The International Microbiologists Association met at its traditional site.

Pronoun Reference

A pronoun must clearly refer to only one antecedent.

Incorrect: Biologists have shown that all living organisms depend on two kinds of molecules—amino acids and nucleotides. They are the building blocks of life. (*They* could refer to either *amino acids, nucleotides,* or both.)

Correct: Biologists have shown that all living organisms depend on two kinds of molecules—amino acids and nucleotides. Both are the building blocks of life.

Incorrect: Present plans call for a new parking deck, a new entry area, and an addition to the parking garage. It will add $810,000 to the cost. (*It* could refer to any of the new items.)

Correct: Present plans call for a new parking deck, a new entry area, and an addition to the parking garage. The parking deck will add $810,000 to the cost.

Reflexive Pronouns

A *reflexive pronoun* (ending in *-self* or *-selves*) must refer to the subject of the sentence when the subject also receives the action in the sentence. (Example: She cut herself when the camera lens cracked.) A reflexive pronoun cannot serve instead of *I* or *me* as a subject or an object in a sentence.

Incorrect: The governor presented the science award to Dr. Yasmin Rashid and myself. (*Myself* cannot serve as the indirect object.)

Correct: The governor presented the science award to Dr. Yasmin Rashid and me.

Incorrect: Mark Burnwood, Christina Hayward, and myself conducted the experiment. (*Myself* cannot function as the subject of the sentence.)

Correct: Mark Burnwood, Christina Hayward, and I conducted the experiment.

Reflexive pronouns are also used to make the antecedent more emphatic:

The patient himself asked for another medication.

Venus itself is covered by a heavy layer of carbon dioxide.

Sentence Faults

Comma Splices

A *comma splice* results when the writer incorrectly joins two independent clauses with a comma. Correct by (1) placing a semicolon between the clauses,

(2) adding a coordinating conjunction after the comma, (3) rewriting the sentence, or (4) creating two sentences.

Incorrect: The zinc coating on galvanized steel gums up a welding gun's electrode, resistance welding, therefore, is not ideal for the steel increasingly used in autos today.

Correct: The zinc coating on galvanized steel gums up a welding gun's electrode; resistance welding, therefore, is not ideal for the steel increasingly used in autos today.

Correct: The zinc coating on galvanized steel gums up a welding gun's electrode, so resistance welding, therefore, is not ideal for the steel increasingly used in autos today.

Correct: Because the zinc coating on galvanized steel gums up a welding gun's electrode, resistance welding is not ideal for the steel increasingly used in autos today.

Correct: The zinc coating on galvanized steel gums up a welding gun's electrode. Resistance welding, therefore, is not ideal for the steel increasingly used in autos today.

Fragments

A *sentence fragment* is an incomplete sentence because it lacks a subject or a verb or both. Correct by writing a full sentence or by adding the fragment to another sentence.

Incorrect: As the universe expands and the galaxies fly farther apart, the force of gravity is decreasing everywhere. According to one imaginative theory. (The phrase beginning with *According* is a fragment.)

Correct: According to one imaginative theory, as the universe expands and the galaxies fly farther apart, the force of gravity is decreasing everywhere.

Incorrect: Apes are afraid of water when they cannot see the stream bottom. Which prevents them from entering the water and crossing a stream more than 1 ft deep. (The phrase beginning with *which* is a fragment.)

Correct: Apes are afraid of water when they cannot see the stream bottom. This fear prevents them from entering the water and crossing a stream more than 1 ft deep.

Run-on Sentences

A *run-on sentence* occurs when a writer links two or more sentences together without punctuation between them. Correct by placing a semicolon between the sentences or by writing two separate sentences.

Incorrect: The galaxy is flattened by its rotating motion into the shape of a disk most of the stars in the galaxy are in this disk.

Correct: The galaxy is flattened by its rotating motion into the shape of a disk; most of the stars in the galaxy are in this disk.

Correct: The galaxy is flattened by its rotating motion into the shape of a disk. Most of the stars in the galaxy are in this disk.

Subject/Verb Agreement

The verb in a sentence must agree with its subject in person and number. Correct by rewriting.

Incorrect: The report of a joint team of Canadian and American geologists suggest that some dormant volcanos on the West Coast may be reawakening. (The subject is *report*, which requires a singular verb.)

Correct: The report of a joint team of Canadian and American geologists suggests that some dormant volcanos on the West Coast may be reawakening.

Incorrect: The high number of experiments that failed were disappointing. (The subject is *number*, which requires a singular verb.)

Correct: The high number of experiments that failed was disappointing.

Incorrect: Sixteen inches are the deepest we can drill. (The subject is a single unit and requires a singular verb.)

Correct: Sixteen inches is the deepest we can drill.

When compound subjects are joined by *and*, the verb is plural.

> Computer-security techniques and plans for evading hackers require large expenditures.

When one of the compound subjects is plural and one is singular, the verb agrees with the nearest subject.

> Neither the split casings nor the cracked layer of plywood was to blame.

When a compound subject is preceded by *each* or *every*, the verb is singular.

> Every case aide and social worker is scheduled for a training session.

When a compound subject is considered a single unit or person, the verb is singular.

> The vice president and guiding force of the company is meeting with the Nuclear Regulatory Commission.

PUNCTUATION

Apostrophe

An apostrophe shows possession or marks the omission of letters in a word or in dates.

The engineer's analysis of the city's water system shows a high pollution danger from raw sewage.

The extent of the outbreak of measles wasn't known until all area hospitals' medical records were coordinated with those of private physicians.

Chemist Charles Frakes' experiments were funded by the National Institute of Health throughout the '90s.

Do not confuse *its* (possessive) with *it's* (contraction of *it is*).

The animal pricked up its ears before feeding.

Professors Jones, Higgins, and Carlton are examining the ancient terrain at Casper Mountain. It's located on a geologic fault, and the professors hope to study its changes over time.

Colon

The colon introduces explanations or lists. An independent clause must precede the colon.

The Olympic gymnastic team used three brands of equipment: Acme, Dakota, and Shelby.

Do not place a colon directly after a verb.

Incorrect: The environmentalists reported: decreased oxygen, pollution-tolerant sludgeworms, and high bacteria levels in the lake. (The colon does not have an independent clause preceding it.)

Correct: The environmentalists reported decreased oxygen, pollution-tolerant sludgeworms, and high bacteria levels in the lake.

Comma

To Link. The comma links independent clauses joined by a coordinating conjunction (*and, but, or, for, so, yet,* and *nor*).

The supervisor stopped the production line, but the damage was done.

Rheumatic fever is controllable once it has been diagnosed, yet continued treatment is often needed for years.

To Enclose. The comma encloses parenthetical information, simple definitions, or interrupting expressions in a sentence.

Gloria Anderson, our best technician, submitted the winning suggestion.

The research team, of course, will return to the site after the monsoon season.

> The condition stems from the body's inability to produce enough hemoglobin, a component of red blood cells, to carry oxygen to all body tissues.

If the passage is essential to the sentence (restrictive), do not enclose it in commas.

> Acme shipped the toys that have red tags to the African relief center. (The phrase *that have red tags* is essential for identifying which toys were shipped.)

To Separate. The comma separates introductory phrases or clauses from the rest of the sentence and also separates items in a list.

> Accurate to within 30 seconds a year, the electronic quartz watch is also water resistant.

> Because so many patients experienced side effects, the FDA refused to approve the drug.

> To calculate the three points, add the static head, friction losses, velocity head, and minor losses. (Retain the final comma in a list for clarity.)

The comma also separates elements in addresses and dates.

> The shipment went to Hong Kong on May 29, 2004, but the transfer forms went to Marjorie Howard, District Manager, Transworld Exports, 350 Michigan Avenue, Chicago, Illinois 60616.

Dash

The dash sets off words or phrases that interrupt a sentence or that indicate sharp emphasis. The dash also encloses simple definitions. The dash is a more emphatic punctuation mark than parentheses or commas.

> The foundation has only one problem—no funds.

> The majority of those surveyed—those who were willing to fill out the questionnaire—named television parents in nuclear families as ideal mother and father images.

Exclamation Point

The exclamation point indicates strong emotion or sharp commands. It is not often used in technical writing, except for warnings or cautions.

> Insert the cable into the double outlet. Warning! Do not allow water to touch the cable.

Hyphen

The hyphen divides a word at the end of a typed line and also forms compound words. If in doubt about where to divide a word or whether to use a hyphen in a compound word, consult a dictionary.

> Scientists at the state-sponsored research laboratory are using cross-pollination to produce hybrid grains.

Modifiers of two or three words require hyphens when they precede a noun.

> Tycho Brahe, sixteenth-century astronomer, measured the night sky with giant quadrants.

> The microwave-antenna system records the differing temperatures coming from Earth and the cooler sky.

Parentheses

Parentheses enclose (1) nonessential information in a sentence or (2) letters and figures that enumerate items in a list.

> The striped lobsters (along with solid blue varieties) have been bred from rare red and blue lobsters that occur in nature.

> Early television content analyses (from 1950 to 1980) reported that (1) men appeared more often than women, (2) men were more active problem solvers than women, (3) men were older than women, and (4) men were more violent than women.

Parentheses also enclose simple definitions.

> The patient suffered from hereditary trichromatic deuteranomaly (red-green color blindness).

Question Mark

The question mark follows direct questions but not indirect questions.

> The colonel asked, "Where are the orders for troop deployment?"

> The colonel asked whether the orders for troop deployment had arrived.

Place a question mark inside quotation marks if the quotation is a question, as shown above. If the quotation is not a question but is included in a question, place the question mark outside the quotation marks.

> Why did the chemist state "The research is unnecessary"?

Quotation Marks

Quotation marks enclose (1) direct quotations and (2) titles of journal articles, book chapters, reports, songs, poems, and individual episodes of radio and television programs.

> "Planets are similar to giant petri dishes," stated astrophysicist Miranda Caliban at the Gatewood Research Institute.

> "Combat Readiness: Naval Air vs. Air Force" (journal article)

> "Science and Politics of Human Differences" (book chapter)

> "Feasibility Study of Intrastate Expansion" (report)

> "Kariba: The Lake That Made a Dent" (television episode)

Place commas and periods inside quotation marks. Place colons and semicolons outside quotation marks. See also "Italics" under "Mechanics" in this appendix.

Semicolon

The semicolon links two independent clauses without a coordinating conjunction or other punctuation.

> The defense lawyer suggested that the arsenic could have originated in the embalming fluid; Bateman dismissed this theory.

> All the cabinets are wood; however, the countertops are of artificial marble.

The semicolon also separates items in a series if the items contain internal commas.

> The most desirable experts for the federal environmental task force are Sophia Timmons, Chief Biologist, Ohio Water Commission; Clinton Buchanan, Professor of Microbiology, Central Missouri State University; and Hugo Wysoki, Director of the California Wildlife Research Institute, San Jose.

Slash

The slash separates choices and parts of dates and numbers.

> For information about the property tax deduction, call 555–6323/7697. (The reader can choose between two telephone numbers.)

> The crew began digging on 9/28/02 and increased the drill pressure 3/4 psi every hour.

MECHANICS

Acronyms and Initialisms

An *acronym* is an abbreviation formed from the first letters of the words in a name or phrase. The acronym is written in all capitals with no periods and is pronounced as a word.

> The crew gathered in the NASA conference room.

Some acronyms eventually become words in themselves. For example, *radar* was once an acronym for *radio detecting and ranging*. Now the word *radar* is a noun meaning a system for sensing the presence of objects.

Initialisms also are abbreviations formed from the first letters of the words in a name or phrase; here each letter is pronounced separately. Some initialisms are written in all capital letters; others are written in capital and lowercase letters. Some are written with periods; some are not. If you are uncertain about the conventional form for an initialism, consult a dictionary.

> The technician increased the heat by 4 Btu.

> The C.P.A. took the tax returns to the IRS office.

When using acronyms or initialisms, spell out the full term the first time it appears in the text and place the acronym or initialism in parentheses immediately after to be sure your readers understand it. In subsequent references, use the acronym or initialism alone.

> The Federal Communication Commission (FCC) has authorized several thousand licenses for low-power television (LPTV) stations.

In a few cases, the acronym or initialism may be so well known by your readers that you can use it without stating the full term first.

> The experiment took place at 10:15 A.M. with the application of DDT at specified intervals.

Plurals of acronyms and initialisms are formed by adding an *s* without an apostrophe, such as YMCAs, MBAs, COLAs, and EKGs.

Brackets

Brackets enclose words that are inserted into quotations by writers or editors. The inserted words are intended to add information to the quoted material.

> "Thank you for the splendid photographic service [to the Navy, 1980–2002] and the very comprehensive collection donated to the library," said Admiral Allen Sumner as he dedicated the new library wing.

Brackets also enclose a phrase or word inserted into a quotation to substitute for a longer, more complicated phrase or clause.

> "The surgical procedure," explained Dr. Moreno, "allows [the kidneys] to cleanse the blood and maintain a healthy chemical balance."

Capital Letters

Capital letters mark the first word of a sentence and the first word of a quotation. A full sentence after a colon and full sentences in a numbered list may also begin with capital letters if the writer wishes to emphasize them.

> Many of the photographs were close-ups.
>
> Pilots are concerned about mountain wave turbulence and want a major change: Higher altitude clearance is essential.

Capital letters also mark proper names and initials of people and objects.

> Dr. Jonathan L. Hazzard
> Ford Taurus
> USS *Fairfax*

In addition, capital letters mark nationalities, religions, tribal affiliations, and linguistic groups.

> British (nationality)
> Lutheran (religion)
> Cherokee (tribal affiliation)
> Celtic (linguistic group)

Capital letters also mark (1) place names, (2) geographical and astronomical areas, (3) organizations, (4) events, (5) historical times, (6) software, and (7) some calendar designations.

> San Diego, California (place)
> Great Plains (geographic area)
> Milky Way (astronomical area)
> League of Women Voters (organization)
> Robinson Helicopter Company (company)

World Series (event)

Paleocene Epoch (historical time)

WordPerfect (software)

Tuesday, July 12 (calendar designation)

Capital letters also mark brand names but not generic names.

Tylenol (acetaminophen)

Capital letters mark the first, last, and main words in the titles of (1) books, (2) articles, (3) reports, (4) films, (5) television and radio programs, (6) music, and (7) art objects. Do not capitalize short conjunctions, prepositions, or articles (*a, an, the*) unless they begin the title or follow a colon. If you are preparing a reference list, see Chapter 11 for the correct format.

The Panda's Thumb (book)

"Sea Turtle Secrets" (article)

Out of the Past (film)

"The Search for the Dinosaurs" (television episode)

Concerto in D Major, Opus 77, for Violin and Orchestra (music)

The Third of May (art)

Companies and institutions often use capital letters in specific circumstances not covered in grammar handbooks. If in doubt, check the company or organization style guide.

Ellipsis

An *ellipsis* (three spaced periods) indicates an omission of one or more words in a quotation. When the omission is at the end of a sentence or includes an intervening sentence, the ellipsis follows the final period.

General Donald Mills stated, "It is appalling that those remarks . . . by someone without technical, military, or intelligence credentials should be published."

At the fund-raising banquet, Chief of Staff Audrey Spaulding commented, "All of the volunteers—more than 600—made this evening very special. . . ."

Italics

Italics set off or emphasize specific words or phrases. If italic type is not available, underline the words or phrases. Italics set off titles of (1) books; (2) periodicals; (3) plays, films, and television and radio programs; (4) long musical works; (5) complete art objects; and (6) ships, planes, trains, and aircraft. The Bible and its books are not italicized.

The Cell in Development and Inheritance (book)

Chicago Tribune (newspaper)

Scientific American (journal)

Inherit the Wind (play)

Blade Runner (film)

60 Minutes (television program)

Carmen (long musical work)

The Thinker (art object)

City of New Orleans (train)

Apollo III (spacecraft)

Italics also set off words and phrases discussed as words and words and phrases in foreign languages or Latin. Scientific terms for plants and animals also are italicized.

The term *mycobacterium* refers to any one of several rod-shaped aerobic bacteria.

These fossils may be closer to the African *Australopithecus* than to *Homo*.

See also "Quotation Marks" under "Punctuation" in this Appendix.

Measurements

Measurements of physical quantities, sizes, and temperatures are expressed in figures. Use abbreviations for measurements only when also using figures and when your readers are certain to understand them. If the abbreviations are not common, identify them by spelling out the term on first use and placing the abbreviation in parentheses immediately after. Do not use periods with measurement abbreviations unless the abbreviation might be confused with a full word. For example, *inch* abbreviated needs a period to avoid confusion with the word *in*. Use a hyphen when a measurement functions as a compound adjective.

The pressure rose to 12 pounds per square inch (psi).

Workers used 14-in. pipes at the site.

The designer wanted a 4-oz decanter on the sideboard.

Ship 13 lb of feed. (Do not add *s* to make abbreviations plural.)

Ship several pounds of feed. (Do not use abbreviations if no numbers are involved.)

If the measurement involves two numbers, spell out the first or the shorter word.

The supervisor called for three 12-in. rods.

Here are some common abbreviations for measurements:

C	centigrade	kW	kilowatt
cm^3	cubic centimeter	L	liter
cm	centimeter	lb	pound
cps	cycles per second	m	meter
cu ft	cubic foot	mg	milligram
dm	decimeter	ml	milliliter
F	Fahrenheit	mm	millimeter
fl oz	fluid ounce	mph	miles per hour
fpm	feet per minute	oz	ounce
ft	foot	psf	pounds per square foot
g	gram	psi	pounds per square inch
gal	gallon	qt	quart
gpm	gallons per minute	rpm	revolutions per minute
hp	horsepower	rps	revolutions per second
in.	inch	sq	square
kg	kilogram	t	ton
km	kilometer	yd	yard

Numbers

Use figures for all numbers over ten. If a document contains many numbers, use figures for all amounts, even those under ten, so that readers do not overlook them.

> The testing procedure required 27 test tubes, 2 covers, and 16 sets of gloves.

Do not begin a sentence with a figure. Rewrite to place the number later in the sentence.

Incorrect:　42 tests were scheduled.

Correct:　We scheduled 42 tests.

Write very large numbers in figures and words.

> 2.3 billion people
> $16.5 million

Use figures for references to (1) money, (2) temperatures, (3) decimals and fractions, (4) percentages, (5) addresses, and (6) book parts.

$4230

$.75

72°F

0°C

3.67

¾

12.4%

80%

2555 N. 12th Street

Chapter 7, page 178

Use figures for times of day. Because most people are never certain about P.M. and A.M. in connection with 12 o'clock, indicate noon or midnight in parentheses.

4:15 P.M.

12:00 (noon)

12:00 (midnight)

Add an *s* without an apostrophe to make numbers plural.

4s   1980s   33s

Symbols and Equations

Symbols should appear in parenthetical statistical information but not in the narrative.

Incorrect: The *M* were 2.64 (Test 1) and 3.19 (Test 2).

Correct: The means were 2.64 (Test 1) and 3.19 (Test 2).

Incorrect: The sample population (number = 450) was selected from children in the second grade.

Correct: The sample population ($N = 450$) was selected from children in the second grade.

Incorrect: We found a high % of error.

Correct: We found a high percentage of error.

Do not try to create symbols on a typewriter or computer by combining overlapping characters. Write symbols not on the keyboard by hand in ink.

$$\chi^2 \ (N = 916) = 142.64 \quad p < 0.001$$

Place simple or short equations in the text by including them in a sentence.

The equation for the required rectangle is $y = a + 2b - x$.

To display (set off on a separate line) equations, start them on a new line after the text and double space twice above and below the equation. Displayed equations must be numbered for reference. The reference number appears in parentheses flush with the right margin. In the text, use the reference number.

$$F\,(9,740) = 2.06 \quad p < 0.05 \tag{1}$$

If the equation is too long for one line, break it before an operational sign. Place a space between the elements in an equation as if they were words.

$$3v - 5b + 66x$$
$$+2x - y = 11.5$$

Very difficult equations that require handwritten symbols should always be displayed for clarity. Highly technical symbols and long equations in documents intended for multiple readers are best placed in appendixes rather than in the main body of the document.

APPENDIX B

Frequently Confused Words

Even careful writers sometimes confuse one word with another that sounds or looks similar. To help you edit your writing, here are definitions for some easily confused words.

accept/except: Use the verb **accept** to mean receiving or approving of something or someone and also to mean regarding something as true. The preposition **except** means "other than" or "excluding" something.

- The president will **accept** the award.
- The association **accepted** the biologist's application for membership.
- Doctors **accepted** the results of the nutrition study, **except** for the soybean data.
- The patient did not speak **except** to ask for water.

advice/advise: The noun **advice** refers to recommendations or opinions. The verb **advise** means to present those recommendations or opinions. Avoid using **advise** to mean "notify" or "inform."

- The lawyer's **advice** to the client was to accept a plea bargain.
- The doctor **advised** the patient to have back surgery.

affect/effect: The verb **affect** means to influence. The noun **effect** means result; the verb **effect** means to make something happen.

- The strong winds **affected** (influenced) airport traffic.
- The new drugs have a positive **effect** (result) on angina.
- The engineer was able to **effect** (make happen) a transfer of equipment.

a lot: Do not write **a lot** as one word. The nonstandard "alot" is not acceptable.

already/all ready: Use the one word **already** to mean previous to a certain time. Use the two words **all ready** to mean completely prepared.

- The satellite was **already** breaking up when it reentered Earth's atmosphere.
- The sales packets were **all ready** for the meeting.

all right: Do not write **all right** as one word. The nonstandard "alright" is not acceptable.

amount/number: Use **amount** when the quantity is bulk and cannot be counted. Use **number** when the quantity can be separated and counted.

- A large **amount** of snow blocked truck access.
- The new telemedicine link to Mozambique offers a **number** of services.

being as/being that: Do not use **being as** or **being that** to substitute for the conjunctions "because" or "since" that introduce dependent clauses. The phrases **being as** and **being that** are nonstandard English and are unacceptable.

- **Because** the drill broke down, work stopped on all lines for two hours.
- **Since** the clerks want computer training, we have hired a consultant.

biweekly/bimonthly: These terms mean twice a week or month and also every other week or month. Because readers will not be sure what you mean, avoid using them.

- The company newsletter appeared **every other week**.

can/may/might: Use **can** to indicate the ability to do something. Use **may** to mean getting permission. **Might** indicates uncertainty or a situation contrary to fact.

- At full speed, the train **can** reach Munich by 2:00 P.M.
- The clerks **may** be able to take a longer lunch hour.
- The tour guide **might** have helped if he had been told of the problem.

continuous/continual: Use **continuous** to mean unceasing without interruption, and use **continual** to mean recurring in rapid succession.

- The **continuous** whine of the drill was annoying.
- The **continual** equipment breakdowns held up production.

e.g./i.e.: The abbreviation **e.g.** means "for example," and **i.e.** means "that is." Because readers may be confused, avoid both of these abbreviations and use the English words.

flammable/inflammable: These words both mean capable of burning. Safety experts prefer **flammable** because consumers (used to *active* and *inactive*) may think **inflammable** means something that will not burn.

flounder/founder: Use the verb **flounder** to mean moving in a clumsy manner or thrashing around, as in attempting to regain one's balance. The verb **founder** means to sink or collapse completely,

- The water buffalo **floundered** in the muddy river.
- After hitting an iceberg, the *Titanic* **foundered**.
- The computer scheme **foundered** when the bank discovered the discrepancies.

imply/infer: The speaker or writer may **imply** (suggest without stating), but the listener or reader must **infer** (draw a conclusion).

- The CEO **implied** that the new European currency would encourage company mergers.
- The employees **inferred** that more layoffs were coming.

in/into: Use **in** to mean a location inside of something. Use **into** to mean the movement from outside to inside.

- Corporate offices were **in** the industrial park.
- Dr. Hernandez ran **into** the emergency room.

its/it's: Use **its** without an apostrophe to show possession. **It's** with an apostrophe is a contraction of *it is*. The apostrophe marks the missing letter (*i*). Never place an apostrophe after the *s* in either word.

- The company opened **its** new headquarters.
- **It's** near the intersection.

lay/lie: Use **lay** to indicate placing something somewhere. The verb always requires a direct object. Use **lie** to mean to recline or to be situated somewhere.

- She **lay** the report on the supervisor's desk.
- He is **lying** on the sofa.
- Hillsdale **lies** at the junction of two rivers.

loose/lose: Use the adjective **loose** to mean "not tight" or "not restrained." Use the verb **lose** to mean "misplaced" or "deprived of something."

- The rope was too **loose** to hold the net tightly in place.
- President Barkley said he expected the project to **lose** money the first year.

oral/verbal: Spoken communication is **oral**. **Verbal** means consisting of words, either written or spoken. For clarity, use either **oral** or **written**.

- The trainer gave **oral** instructions to the group.
- He prepared **written** guidelines for newcomers.

raise/raze/rise: Use the verb **raise** to mean moving something to a higher position. **Raise** always takes a direct object. The verb **raze** also requires a direct object and means to tear down something. Use the verb **rise** to mean ascend or increase in volume or size. This verb does not require a direct object.

- **Raise** the screen higher.
- The contractor will **raze** the building tomorrow.
- The river may **rise** to flood level by Thursday.

sight/site/cite: Use **sight** when referring to a view or something that is seen. Use **site** when referring to a specific location. The verb **cite** means to refer to or mention.

- The landfill, neglected for years, was a dreadful **sight.**
- The council selected Center Plaza as the **site** for the Civil War monument.
- The speaker **cited** Thomas Jefferson as an outstanding writer.

than/then: Use **than** to introduce the second item in a comparison. Use **then** to indicate time or sequence.

- The new generator is bigger **than** our old one.
- The game starts at 4:00 P.M.; we will leave **then.**
- The clerk scanned the groceries and **then** put them in a plastic sack.

their/there/they're: Use **their** as a pronoun showing possession. Use **there** to indicate a specific location or point in time. **They're** is a contraction meaning "they are." The apostrophe marks the omitted letter (*a*).

- The supervisors held **their** annual meeting in March.
- The director refused to open a dealership **there.**
- Start **there** in the test process.
- **They're** upset over cost increases.

whose/who's: Use **whose** to signal possession. The contraction **who's** means "who is." The apostrophe marks the omitted letter (*i*).

- The engineer **whose** patent we have acquired is joining the firm.
- **Who's** handling the merger plans?
- The chemist **who's** doing the experiment will speak to the directors.

Internet Resources for Technical Communication

These Internet sites provide useful information about technical communication issues and often provide links to other relevant sites.

WRITING FOR THE WEB

About Desktop Publishing for All Platforms (http://www.desktoppub. about.com). This extensive site offers free materials, such as clip art, downloads, photos, fonts, and links to other relevant sites.

Graphic Arts Information Network (http://www.gain.org). GAIN serves the printing and graphics industries. The site features an extensive directory with links to training programs, suppliers, and printing news.

Internet-Resources.Com (http://www.internet-resources.com/writers). This site provides links to topics that concern writers, such as organization, science and medical information, grammar tips, and e-publishing.

U.S. Copyright Office (http://www.copyright.gov). This site explains copyright law, registration procedures, and includes pending legislation.

U.S. Department of Health & Human Services (http://www.usability.gov). This site covers research in Web design and usability.

Useit (http://www.useit.com). Web authority Jakob Nielsen's site is entirely devoted to Web design information. Featured are archives of past columns, interviews, updated links, and tips on avoiding the big mistakes of Web design.

Web Design & Review Magazine (http://www.graphic-design.com/Web). This Web site provides a way to exchange opinions and information on Web design. Users can post a Web site for review by other readers and receive comments and suggestions for improving the site's effectiveness. This is an excellent place for reviewing multiple sites categorized by content, such as business, education, and pastimes.

Web Developer's Journal (http://www.webdevelopers/journal.com). This site features information about Web graphics, design tools, books, and software.

W3C The World Wide Web Consortium (http://www.w3.org). This site covers varied Web guidelines, indexed A–Z, with linked topics ranging from Accessibility to XSLT.

ORAL PRESENTATIONS

On-line Technical Writing: Oral Presentations (http://www.io.com/~hcexres/tcm1603/acchtml/oral.html). The site offers guidelines on preparing and delivering an oral presentation, understanding the audience, and using visual aids. It provides links to university sites that cover oral presentations.

RICE OWL (http://www.ruf.rice.edu/~riceowl/oralpres.html). This site at Rice University provides strategies for presenters of business and technical information.

Toastmasters International (http://www.toastmasters.org). The site provides club membership information and tips on effective speaking.

JOB SEARCH

CareerBuilder (http://www.careerbuilder.com). This site searches other sites for specific jobs. Features include job-search tips, résumé writing, and personal accounts that allow users to post up to five different résumés.

Career Magazine (http://www.careermag.com). The site allows job seekers and employers to post résumés and available jobs and search by location. Also featured are columns with job tips and career advice.

CollegeGrad.Com (http://collegegrad.com). The site contains job listings and is easy to search. Users can post their résumés for free. Included are tips on cover letters, résumés, and interview questions.

Computer Jobs (http://www.computerjobs.com). The user can search this site by job type or location. Site is updated hourly.

Dice (http://www.dice.com).   This site is a job board for information technology professionals and contains an extensive Careers Resources Center.

Job Star Central (http://www.jobstar.org).   This site focuses on California and includes career advice, résumé instructions, and salary information.

Monster (http://www.monster.com).   Monster has a global network for job seekers and employers. "First-timers" to the site can build results tailored toward specific industries (e.g., financial services, health care). The Career Center features tips on résumés and interviews.

ResumeMaker.com (http://www.resumemaker.com).   The site features sample résumés for different job types, expert advice, templates for creating a résumé, and tips on interviewing and cover letters.

Salary.Com (http://www.salary.com).   This site provides typical salary information for various jobs. Users can select a job category and location and find regional salaries.

U.S. Department of Labor, Bureau of Labor Statistics (http://www.bls.gov). This site provides the latest government statistics on jobs available in specific categories. Includes a link to the *Occupational Handbook*.

PROFESSIONAL ASSOCIATIONS

American Medical Writers Association (http://www.amwa.org).   The site represents the leading professional organization for medical writers.

American Society for Information Science and Technology (http://www. asis.org).   The association focuses on developing technology and systems for the Internet.

Armed Forces Communications and Electronics Association (http://www. afcea.org).   The site represents the intelligence, electronics, and technical communication fields.

Association for Business Communication (http://www.businesscommunication. org).   The ABC stresses the theory and practice of effective business communication. Most members are educators.

Computer Professionals for Social Responsibility (http://www.cpsr.org).   The focus here is on computer ethics issues with chapter information, newsletters, conference information, and news.

The Council of Science Editors (http://www.councilscienceeditors.org).   The CSE is devoted to effective communication practices for science writers, editors, and publishers. The site includes job postings, publications, directories, and reference materials.

The IEEE Communications Society (http://www.comsoc.org).   The society promotes communication technologies. The site features a digital library, online career center, videos of previous conferences, and newsletters.

International Association of Business Communicators (http://www.iabc.com). The IABC is an association of communication professionals, including video producers, public relations specialists, editors, and training specialists.

Society for Technical Communication (http://www.stc.org).   The STC focuses on the art and science of technical communication for technical writers and educators.

CREDITS

Pages 45–46. Revised and adapted from Charles Groves, "Safety Guidelines for Workers Renovating Buildings Containing Lead-Based Paint." Reprinted with permission of the author.

Page 67. U.S. Environmental Protection Agency (2004). "Consumer Handbook for Reducing Solid Waste, The Four Basic Principles." Retrieved August 11, 2004, from http://www.epa.gov/epaoswer/non-hw/reduce/catbook/the4.htm.

Pages 69–72. "Arthritis (brochure)." Reprinted with permission of the American Academy of Orthopaedic Surgeons, Rosemont, IL.

Page 74. Hubbard, J. "Another Memorable Mission." From "Allied Force Debrief," *CODE ONE,* 14(4), October 1999. Reprinted with permission of Lockheed Martin Corporation.

Page 76. Ko, H. C., and Brown, R. R., from "Enthalpy of Formation of $2CdO.CdSO_4$." Report of Investigations 8751. (Washington, DC: U.S. Bureau of Mines, U.S. Department of the Interior, 1983).

Page 100. U.S. Department of Agriculture (2004). Homepage. Retrieved August 12, 2004, from http://www.usda.gov.

Page 101. U.S. Department of Health and Human Services (2004). Homepage. Retrieved August 4, 2004, from http://hhs.gov.

Page 148. "General Layout of Coal Cleaning Operations." From *Steam: Its Generation and Use,* 40th ed. Barberton, OH: Babcock & Wilcox Co., 1992. Reprinted with permission of the Babcock & Wilcox Company.

Page 149. "Mass Burning Schematic." From *Steam: Its Generation and Use,* 40th ed. Barberton, OH: Babcock & Wilcox Co., 1992. Reprinted with permission of the Babcock & Wilcox Company.

Page 150. "One-Leg-Stand Test," from U.S. Department of Transportation, National Highway Traffic Safety Administration, *Improved Sobriety Testing,* 1984 (Washington, DC: U.S. Government Printing Office, 1984).

Page 150. U.S. Congress, Office of Technology Assessment, *Marine Minerals: Exploring Our New Ocean Frontier,* OTA-0-342 (Washington, DC: U.S. Government Printing Office, July 1987).

Page 151. From *Spacelab J, Microgravity and Life Sciences,* NASA, Marshall Space Flight Center, n.d.

Page 152. U.S. Department of Commerce, U.S. Census Bureau (2001). "Census Regions and Divisions of the United States." Retrieved June 4, 2001, from http://www.doc.gov/census/maps.

Page 153. Photo of Si-Tex Neptune NT. Reprinted with permission of Si-Tex Marine Electronics.

Page 165. U.S. Department of Labor, Occupational Safety and Health Administration, "Preventing Fatalities." Retrieved August 16, 2004, from http://www.osha.gov/SLTC/etools/construction/index.html.

Page 166. U.S. Department of the Interior, Bureau of Land Management, "National Wild Horse and Burro Program." Retrieved August 16, 2004, from http://wildhorseandburro.blm.gov/index.php.

Pages 168–169. "Pure Touch, The Filtering Faucet System." Reprinted with permission of Moen Incorporated, North Olmstead, OH.

Page 185. U.S. Agency for International Development (2004). Homepage. Retrieved August 12, 2004, from http://www.usaid.gov.

Page 186. U.S. Department of the Interior, National Park Service (2004). Homepage. Retrieved September 5, 2004, from http://www.nps.gov.

Page 187. U.S. Department of the Interior, National Park Service (2004). "New Bedford Whaling." Retrieved August 23, 2004, from http://www.nps.gov/nebe.

Page 188. U.S. Environmental Protection Agency (2004). "Ocean Regulatory Programs." Retrieved August 9, 2004, from http://www.epa.gov/owow/oceans/regulatory/index.html.

Page 205. Department of Energy, Energy Information Administration (2004). "Wind Energy—Energy from Moving Air." Retrieved August 23, 2004, from http://www.eia.doe.gov/kids/energyfacts/sources/renewable/wind.html.

Page 206. FBI (2004). "Organized Crime." Retrieved August 4, 2004, from http://www.fbi.gov/hq/cid/orgcrime/glossary.htm.

Pages 225–226. U.S. Bureau of Reclamation, Hoover Dam (2004). "The Power Plant." Retrieved August 23, 2004, from http://www.usbr.gov/lc/hooverdam/History/workings/powerplant.html.

Page 228. StarMax 127mm EQ Web page (2001). Reprinted with permission of Orion Telescopes & Binoculars, Watsonville, CA (www.telescope.com).

Page 243. "Emergency Shutdown Procedures for an Operating Recovery Boiler." From *Steam: Its Generation and Use,* 40th ed. Barberton, OH: Babcock & Wilcox Co., 1992. Reprinted with permission of the Babcock & Wilcox Company.

Pages 257–258. "Preparing the ACS Bottles," *Maintenance and Troubleshooting Manual ACS:180.* Reprinted with permission of Ciba Corning Diagnostics Corp.

Page 260. From "Plan Your Dive, Steer Up." Reprinted with permission of the National Spa and Pool Institute.

Page 262. U.S. Environmental Protection Agency (2004). "Adopt Practices That Reduce Waste Toxicity." Retrieved August 11, 2004, from http://www.epa.gov/epaoswer/non-hw/Reduce/catbook/tip2.htm.

Pages 264–265. "Carbon Monoxide," National Safety Council Fact Sheet, from Web site www.nsc.org. Permission to reprint granted by the National Safety Council, a membership organization dedicated to protecting life and promoting health.

Page 267. "Combined Cycle PFBC." From *Steam: Its Generation and Use,* 40th ed. Barberton, OH: Babcock & Wilcox Co., 1992. Reprinted with permission of the Babcock & Wilcox Company.

Page 269. Hehs, E. "JSF119-611 Primer." From "JSF Propulsion System Testing," *CODE ONE,* 14(2), April 1999. Reprinted with permission of Lockheed Martin Corporation.

Page 304. "Appendix D:ACS:180 Specifications," *Reference Manual ACS:180.* Reprinted with permission of Ciba Corning Diagnostics Corp.

Pages 375–379. Based on information from "Employment Characteristics of Families in 2003." *Bureau of Statistics Web site.* Retrieved August 5, 2004, from http://www.bls.gov/news.release/famee.nr0.htm.

Pages 386–392. Revised and adapted from Novacheck, M. J., "Standby Power Improvement at Water Treatment Plant." Reprinted with permission of the author.

Pages 475–479. From Allen, Lori, and Voss, Dan, *Ethics in Technical Communication* (pp. 73–78). © 1997 by John Wiley & Sons, Inc. Reprinted by permission of John Wiley & Sons, Inc.

Pages 480–482. Bagin, C. B., and Van Doren, J., "How to Avoid Costly Proofreading Errors," from *Simply Stated* 65 (April 1986). Reprinted with permission of Document Design Center, American Institutes for Research.

Index